PHYSICS:

an ebb and flow of ideas

Einleitung.

Dass die Elektrodynamik Maxwells — wie dieselbe gegenwärtig aufgefasst zu werden pflegt — in ihrer Anwendung auf bewegte Körper zu Asymmetrien führt, welche den Phänomenen nicht anzuhaften scheinen, ist bekannt. Man denke z. B. an die elektrodynamische Wechselwirkung zwischen einem Magneten und einem Leiter. Das beobachtbare Phänomen hängt hier nur ab von der Relativbewegung von Leiter und Magnet, während nach der üblichen Auffassung die beiden Fälle, dass der eine oder der andere dieser Körper der bewegte sei, streng voneinander zu trennen sind. Bewegt sich nämlich der Magnet und ruht der Leiter, so entsteht in der Umgebung des Magneten ein elektrisches Feld von gewissem Energiewerte, welches an den Orten, wo sich Teile des Leiters befinden, einen Strom erzeugt. Ruht aber der Magnet und bewegt sich der Leiter, so entsteht in der Umgebung des Magneten kein elektrisches Feld, dagegen im Leiter eine elektromotorische Kraft, welcher an sich keine Energie entspricht, die aber — Gleichheit der Relativbewegung bei den beiden ins Auge gefassten Fällen vorausgesetzt — zu elektrischen Strömen von derselben Grösse und demselben Verlaufe Veranlassung gibt, wie im ersten Falle die elektrischen Kräfte.

Beispiele ähnlicher Art, sowie die misslungenen Versuche, eine Bewegung der Erde relativ zum „Lichtmedium" zu konstatieren, führen zu der Vermutung, dass dem Begriffe der absoluten Ruhe nicht nur in der Mechanik, sondern auch in der Elektrodynamik keine Eigenschaften der Erscheinungen entsprechen, sondern dass vielmehr für alle Koordinatensysteme, für welche die mechanischen Gleichungen gelten, auch die gleichen elektrodynamischen und optischen Gesetze gelten, wie dies für die Grössen erster Ordnung bereits erwiesen ist. Wir wollen diese Vermutung (deren Inhalt im Folgenden „Prinzip der Relativität" genannt werden wird) zur Voraussetzung erheben und ausserdem die mit ihm nur scheinbar unverträgliche Voraussetzung einführen, dass sich das Licht im leeren Raume stets mit einer bestimmten, vom Bewegungszustande des emittierenden Körpers unabhängigen Geschwindigkeit V fortpflanze. Diese beiden Voraussetzungen genügen,

The opening page in Einstein's handwriting of his article, *Zur Elecktrodynamik bewegter Körper (On the Electrodynamics of Moving Bodies)*. The article is the first in a series of papers developing the theory of special relativity and originally appeared in 1905. This holograph copy was made in 1943. (By permission of the estate of Albert Einstein.)

PHYSICS:

an ebb and flow of ideas

Stuart J. Inglis

Chabot College

JOHN WILEY & SONS, INC.

NEW YORK | LONDON | SYDNEY | TORONTO

Library of Congress Catalogue Card Number: 70–101973

SBN 471 42734 9

Printed in the United States of America

10 9 8 7 6 5 4 3 2 1

This book is dedicated to

PROFESSOR G. F. W. MULDERS

whose inspiration early in my student days

awakened by interest in astronomy and physics

Preface

The purpose of this book is to introduce the fundamental concepts of physics to the liberal arts college student. During ten years of teaching a one-semester physics course for the nonscience major, I came to realize that this student usually finds physics more interesting if he is shown some of its development. The contributions of the great men in physics take on additional meaning when these contributions are considered as part of the development of thought.

A detailed study of the development of ideas in physics would consider the original need for each major idea, discuss the concept itself, and finally examine how the concept affected the whole structure of physics. This study would, however, be a major undertaking for any student, even the physics major (not to mention the teacher). Therefore for this book certain selections have been made. Some of the major concepts have been presented in depth, others have been quite neglected. The applications of physics have been for the most part omitted. There are no discussions of lenses, curved mirrors, resistors, capacitors, motors, generators, pumps, electrical gadgets, levers and pulleys, and the like. Coverage has less breadth so that selected areas can be studied in greater depth.

Some of the examples selected from history stress the theoretical approach to physics, for example, the "thought experiments" of Galileo and Einstein. Other examples are the actual experiments of men like Coulomb, Faraday, and Rutherford. The work of James Joule has been considered in some detail because it relates so many different aspects of physics. In addition, Joule's work has been used to illustrate the development of an idea by an individual.

Throughout the book I have tried to indicate to the student that major theories are not fashioned of thin air. Each major concept was developed because of a need to explain new observations. For example, if the student understands that failures of classical physics led to the theory of special relativity and to quantum mechanics, he will be more likely to appreciate these newer ideas. Some discoveries, however, come as "fortunate accidents," such as Galvani's work with twitching frog legs. Such events are convincing evidence that accidents lead to progress in science only when they happen to men who are curious, who wonder, who when something strange happens ask why, and who then actively seek to learn the cause.

In a course utilizing the history of physics, occasions inevitably arise in which either the history or the physics must suffer for the sake of the presentation. Since the main purpose of this book is to teach physics rather than history, the presentation of history is necessarily sometimes slighted but is, hopefully, never in error. Many of the men who contributed to the study of physics are either not discussed or mentioned only briefly in order that others can be studied in detail.

In a history of science course the scientist himself is of great importance. The tradition he inherits establishes his starting point; the times in which he lives influence the development of his thought. But in a short course for non-science majors biographical material, although no less interesting, is of secondary importance. Consequently, where biographical material is given, it consists of brief sketches. The instructor may, however, introduce additional biographical material to great advantage.

No attempt has been made to reproduce the contributions of any one physicist in their entirety. The mathematics of Galileo, that of Newton, and that of Maxwell are not given. Algebra is used instead, introduced gradually in the hope that, by exposure, the student will recall more of this subject as the course proceeds. The mathematics in the last chapters is therefore more complex than that in the beginning.

When aspects of the philosophy of physics lend themselves to the discussion, they are introduced. But physics is many things; it is made up of men, ideas, and experiments, and it has been buffeted by the times and tides of our ever-changing society. We must realize too that physics has, at times, traveled down some dead-end roads, lest we think that the course of science is always direct.

I have made a conscious effort to demonstrate that the "laws of physics" are really nothing more than the best possible descriptions of Nature as the physicist sees her. The work of Plato and Aristotle is not unrelated to the work of today; some ideas do bridge lapses of time, even if most have been altered in the process. Time and mankind have wrought great changes in man's conception of the world. As our means of observation become more and more precise, our descriptions must become more refined.

If, during the reading of this book, the liberal arts student can learn enough physics to become aware of how mankind has struggled to describe the physical world, if he can understand some of the important turning points in the history of physics, if he can learn where today's physicist stands, this teacher and this author will have achieved one of his goals.

Stuart J. Inglis
Chabot College

Hayward, California
November 1969

Acknowledgments

During the writing and preparation of this text I have benefited from the aid and advice of many people. It is now my privilege to thank them. The initial historical background for this book was acquired during a year of study at the University of California in Berkeley under a National Science Foundation Science Faculty Fellowship. It was my pleasure during that year to study intensively the history of science under the superb guidance of Professors Thomas Kuhn (now at Princeton University) and Roger Hahn. My interest in this approach to science, encouraged by them, has greatly altered my teaching methods, and if I assess the interest of my students correctly, this change has been for the better.

A number of people have read the manuscript and made many helpful suggestions. I accept full responsibility, however, should any errors still persist. Among those who made important contributions, I wish to call particular attention to Professor Roger Hahn who read most of the manuscript, with special concern for the history of science; to Professor Arnold Strassenberg of the State University of New York at Stony Brook who spent a great deal of time reviewing and discussing this work with me; to Professor Robert Sells of the State University of New York at Geneseo who read the manuscript in its earlier stages; and to Professors Jae Ballif and William Dibble, both of Brigham Young University, who read the manuscript in its final stages.

Several of the photographs in this book also appear in *An Approach to Physical Science* by the PSNS Product Staff, John Wiley and Sons, 1969. These photographs were procured by the John Wiley staff at my request while I was serving as editor of the PSNS Project.

I must also thank editors Russell Fraser and Donald Deneck of John Wiley and Sons who made the publication of this book possible. To Otto Barrett of Contra Costa College I am especially indebted for the discussions of physics we have had over the years.

But invariably an author owes his greatest thanks to his family who both help and bear with him. My daughters Jennifer and Adrienne have been very patient with a father who has spent too much time in his study. My son Jeff has shown me the same patience and has helped me directly with my work. Isa, my wife, has not only typed and read the manuscript but bolstered my spirits when they sagged. I am truly grateful for the encouragement and consideration of my family.

S. J. I.

Contents

Chapter Seven

Heat, Work, and Energy   171

Chapter Eight

Optics   197

Chapter Nine

Electromagnetic Waves   229

Chapter Ten

Four Failures of Classical Physics   257

Chapter Eleven

Einstein and Special Relativity   289

Chapter Twelve

The Atom and Quantum Mechanics   311

Chapter Thirteen

Radioactivity and the Nucleus   349

Chapter Fourteen

The Nucleus and Particles   377

PHYSICS:

an ebb and flow of ideas

Chapter One

Aristotle and Greek Physics

Art Reference Bureau

T o describe the physical world about us, to explain its varied elements, its actions, its motions, constitutes a study that has occupied and fascinated man since prehistoric times. The earliest explanations of physical phenomena told of mythical gods who played major roles in creating and maintaining the world. These myths, elaborated upon and added to by the men who told and retold them from generation to generation, reflected man's continuing need for guidance and support. His emotional reactions to his environment largely shaped these myths. The world as early man knew it was vastly different from the world we know today, even though the rivers, the oceans, the mountains, the sun and moon, the planets and stars are all essentially the same. The change has come in man himself. As he changes, he must describe his world differently.

PRE-SOCRATIC IDEAS

The change from a mythical explanation of the world to a rational explanation was not accomplished with one stroke of a wand, nor by one man, nor in one century. But in the history of Western man there was a turning point, a glimmer of objectivity, a break from the old, a beginning of the new. That turning point is ascribed to Thales (624–548 B.C.), who lived in the city of Miletus on the eastern shore of the Aegean Sea (1). During his travels he had learned the art of surveying from the Egyptians who needed to resurvey property lines each year after the Nile had overflowed and deposited silt on the land. The lines and triangles of the practical Egyptian surveyor became, for Thales, a study for its own sake, the study of geometry. He saw the isosceles triangle not merely as a convenience to surveying but as a configuration of unique and interesting properties. For example, he realized that in an isoceles triangle the angles opposite the equal sides are themselves always equal. He also realized that the diameter of a circle bisects it; that when two lines cross, the vertical angles are equal; and that all triangles inscribed in a semicircle are right triangles as long as one of the sides is a diameter. He learned a practical art and from it evolved an interesting and more fruitful study. Because he looked at triangles more objectively than did the Egyptian surveyors, he could introduce the study of geometry.

Thales also tried to be objective in his view of the physical world. He tried to step outside the mythical world order of which man was emotionally and subjectively a part. He, of course, was not as objective as we pretend to be today, but he is given credit for initiating the break from the dominance of mythology.

To Thales the world seemed to be composed of the most important substance along the arid Ionian coast—water. Water was the only substance he knew to exist as a solid, liquid, and gas; it was water that fell as rain, made the grass grow, and offered relief on hot days. Water in its varied states and with its life-giving properties became Thales' unifying element, that is, it became his *Prime Substance* out of which all other things are made. Thales

3

proposed that water pervaded all places, made up all things, and brought some order out of chaos. In starting mankind's long search for order, Thales was apparently the first to see the world objectively. From this new idea, this new concept, man could go on to develop other descriptions of the world.

Thales was regarded by his contemporaries as one of the seven wise men. Indeed, he founded a school of thought (the Ionian School) which included philosophers of his own generation as well as a number of men in succeeding centuries (2). These men grasped Thales' concept of detaching man from his physical environment and studying that environment because it was interesting and worthy of study. Many followed Thales' example in selecting a Prime Substance as a unifying element out of which all else was made.

Among the substances suggested were *apeiron*, an indefinite, undetermined, unexperienced substance out of which all things were made and into which all things turned at their destruction. The use of such a non-limited principle to explain the universe indicated the desire for a grander agent than water.

Air was suggested as the Prime Substance, for air seemed to embody the very essence or breath of life. *Fire* was another suggested Prime Substance; it was believed that the essential feature of the world and of life was change, and nothing seemed to move and change its shape as much as fire.

In contrast to the changing nature of fire, it was suggested that the changes perceived by the senses are only apparent. It was felt that the universe was full of matter and without void, that it was eternal with neither change nor motion. No material body could be created out of nothing; no object could be destroyed into nothingness. In this philosophy the very essence of the world was constancy. This concept of constancy seemed to be the first grasp at what we now call the conservation principles. Although man has come to realize that motion and change in the universe are inevitable, we have also come to recognize that certain features of the physical world are constant and not subject to change.

Some of these early philosophers advanced beyond this concept of a Prime Substance. Empedocles (*c.* 500–430 B.C.) suggested that the universe was composed not of just one substance but of four: *earth, water, air,* and *fire*. These four elements were presumed to be subject to the two universal forces: love, which combined the four elements in various ways to form all the substances in the world; and strife, which separated or tore the substances back into the basic elements. These two forces were not matter, yet they operated on and somehow changed matter. The introduction of this intangible concept, force, was a fundamental step forward.

THE ATOMISTS

Before discussing the physics of Aristotle, the culmination of Greek physics, we must consider another school of thought. The concept of an all-pervading Prime Substance came from the Ionian coast, and there too the idea of atoms apparently originated. Leucippus appears to have formulated the idea, but little is known of him except that he flourished in the middle of the fifth century B.C. and was very likely born in Miletus. His student Democritus (*c.*470–*c.* 400 B.C.) of Abdera (along the northern coast of the Aegean Sea) was the philosopher who made this atomistic theory famous.

In the ancient atomistic theory, the world was composed not of an all-pervading Prime Substance or even of four elements; it was made up of an infinite void in which an infinity of atoms were immersed. All matter was supposedly composed of atoms whose shapes determined their nature. The number of shapes involved was not infinite, but so large as to be incomprehensible. (That these early philosophers distinguished between a number large enough to be incomprehensible, but still a definite number, and *infinity*, larger than any assigned number, indicates a high degree of insight into the theory of numbers.)

The atoms of Democritus were absolutely simple and indestructible; furthermore, they supposedly moved incessantly, being propelled by collisions. The origin of the atoms and the initial cause of their motions was not considered; both were simply taken for granted. The idea that they moved, however, required that they be immersed in a void, for if the atoms were packed together so closely that all free space between them was eliminated, they would not be free to move.

Substances differed because atoms of different shape assumed various orders and positions. A clear differentiation of these terms was given by Aristotle who, in discussing the atomists, said,

these differences, they say, are three—shape and order and position. For they say the real is differentiated only by "rhythm" and "inter-contact" and "turning;" and of these rhythm is shape, inter-contact is order, and turning is position; for A differs from N in shape, AN from NA in order, ⊏ from H in position. The question of movement—whence or how it is to belong to things—these thinkers, like the others, lazily neglected (3).

These originators of the atomic theory perpetuated one other important concept, the conservation of matter. Because it was taken for granted that the atoms were indestructible, matter must be conserved. It could neither be created out of nothing nor completely destroyed during any action.

SOCRATES

Contemporary with Democritus lived one of the most profound thinkers of all time, Socrates (470–399 B.C.). It appears that Socrates listened to Anaxagoras (488–428 B.C.) who journeyed from Ionia to Athens and brought with him much of Ionian philosophy. In particular, Anaxagoras proposed that one Mind, rather than a Prime Substance, was the unifying principle. It was this one Mind (he called it *Nus*) which in the beginning turned chaos into order. Anaxagoras' claims at first impressed Socrates, but then Socrates learned that Anaxagoras could not go beyond this single statement that Mind ordered the universe. Because he could not say how it was ordered, or even what constituted order, Socrates lost interest. In fact, Socrates came to see the utter futility of such wild speculations about the world. None of the Ionian philosophers had really made any sense to Socrates. No philosopher had clearly defined his terms, no philosopher had made statements that could be irrefutably defended, and they all disagreed with one another. Socrates

FIGURE 1-1

Socrates (470–399 B.C.). British Museum.

set science on a new track by demanding that philosophers formulate clear definitions and classifications, that they argue without flaw, and that they free themselves of prejudices and superstitions.

But why talk of things celestial, said Socrates, when we cannot even answer the problems of human frailties? Let us talk of things human and try to answer those questions most important to human affairs.

What is godly, what is ungodly; what is beautiful, what is ugly; what is just, what is unjust; what is prudence, what is madness; what is courage, what is cowardice; what is a state, what is a statesman; what is government, what is a governor . . . (4).

In fact, Socrates is remembered today more for his concern for human affairs than for his contributions to science.

PLATO

Certainly the most famous of Socrates' students was Plato (427–347 B.C.), who was 28 years old when his master was condemned to death by the authorities (5). Plato's principal contributions were in the fields of philosophy and social science, but he also made a lasting mark in the field of science. His importance to scientists stems partly from the fact that his most famous

FIGURE 1-2
Plato (427–347 B.C.). Ny Carlsberg Glyptatek, Copenhagen.

student, Aristotle, developed a scientific system that lasted for nearly 2000 years. As will be seen later, Plato influenced the thinking of those in the seventeenth century who were most intent on overthrowing the Aristotelian system.

Plato was impressed by Pythagoras (born *c*. 582 B.C.) and his followers for their study of geometry. He admired their perfect circle, their perfect isosceles triangle, their perfectly parallel lines, in fact, the exactness of all their geometric figures about which they and others had built an entire system of logical thought and description. And yet no man can draw a perfect circle or a perfect isosceles triangle; however, man's inability to draw the perfect, the ideal geometrical figure, does not limit his ability to build with axioms and propositions an irrefutable, ordered system. Why then should not the rest of man's endeavors establish perfect systems? What is the Perfect, the Ideal?

As there was an Ideal circle that no one could draw, so there was an Ideal horse that no one could ride because it existed only in the mind. There was a perfect, or an Ideal chair in which no one could sit. For Plato, the Ideals constituted Reality as against the mundane reality.

Plato searched for the Ideal. His most eloquent statement of it was his allegory of the cave (6). He described a scene in which men, in a cave, spent their lifetimes in chains, seeing only the shadows of figures cast on the wall before them by a fire that blazes behind them. To them these shadows were reality. But one prisoner was freed, and gradually he saw what the shadows really were. Then, when led up and out of the cave into the sunlight, he was dazzled and became, finally, enlightened.

Those in the cave had a practice of searching for some order in the shadows so that they might be able to predict which shadow would appear next on the wall. Prizes were offered to the one who predicted the most successfully. (Plato, here, anticipated the very essence of science, the ability to predict.) But now that the freed prisoner was enlightened, he saw little value in predictions made from just watching and remembering the order of the shadows as they passed. This empirical approach was not enough anymore. The enlightened one now saw the *reason* for the shadows; he had seen and understood the Ideal.

In Plato's world, and in our world, the sun, stars, moon, and planets all rise and set. But the planets move erratically against the background of stars, and such erratic motion must be only a shadow of Reality. The motions that cause these shadows must be pure, must be celestial motions: they must, proclaimed Plato, be circular (the perfect geometric figure) and proceed at a constant rate. Plato's "decree" of planetary motion, issued to mathematicians to "save the appearances" of the planets by determining their Real motion, influenced not only Aristotle but all who searched for the best description of planetary motion even as late as the seventeenth century.

ARISTOTLE

Of all the ancient Greeks the one who most influenced the scientific world was Aristotle (384–322 B.C.), for twenty years a student and colleague of Plato. We know of his ideas because notes of his lectures were kept and transcribed; it is possible that some of his very own lecture notes were also transcribed and have come down to us. His lectures covered a wide range of subjects: logic, physics, metaphysics, the soul, biology, ethics, politics, literature. In each field he was influential, in some more than others. In the field of physics his system overwhelmed all others. The contribution of the Ionian philosophers, great as it was, was rightly criticized by Socrates, for no further progress could have been made through sheer speculation and incomplete philosophies. Aristotle's philosophy was complete; he based his conclusions upon observations (7).

FIGURE 1-3
Aristotle (384–322 B.C.). Alinari Art Reference Bureau, New York.

If Thales was the first man to look at the world objectively, Aristotle was the first to make keen and discriminating observations. His observations in the field of biology were unsurpassed in some respects until quite recent times. They even received the highest praise from the founder of modern biology, Charles Darwin. Aristotle's observations in physics were less complete but still thorough enough to enable him to establish a complete, logically intradependent, and adequate (for the times) description of the world.

In studying Aristotle, we must realize that his description of the world does not satisfy us today, for we see the world differently. We have refined instruments to make precise measurements; Aristotle had no instruments, nor even the concept of making a measurement. Today we have a long tradition of scientific research and many standard reference books. Aristotle's age had no tradition of keen and discriminating observation, for Aristotle established that tradition; his students wrote the first standard texts in physics. Aristotelian physics was adequate for the world for many centuries. As a science its qualities still command our respect, for we can ask no more of any science than that it be adequate and complete, and that its parts be logically intradependent. In fact, these very qualities ensured its longevity and made it difficult to replace.

Aristotle, like Plato, searched for the Ideal, but Aristotle was less impressed by geometry and more impressed by the perfection everything should achieve if allowed to do so. Such a philosophy, based on the ultimate goal or end, is called a teleological philosophy (teleology is the study of final causes). Aristotle wove his physics upon the loom of ultimate perfection. To achieve his goal, he had to divide the universe into two parts and to formulate different laws for each. The first, the astronomical world, was by definition perfect; the second, the sublunar world (the regions beneath the moon) sought perfection.

The Sublunar World and Its Motions

All matter in the sublunar world of Aristotle was composed of the four elements of Empedocles: fire, air, water, and earth. These four elements in turn established the two basic contrary directions of up and down. Fire and air went up; water and earth went down. The causes of these motions were the same; fire and air were more perfect up; water and earth were more perfect down. Thus, a rock falls because it is striving for perfection; fire rises because it, too, is striving for perfection. Lightness is as significant as weight (Figure 1-4). Since Aristotle endowed each element with a "knowledge" of what for it was perfection, and a striving to achieve that perfection, we say that Aristotle's physics is *animistic* in nature. His elements have an animallike quality.

Motion in the sublunar world, as conceived by Aristotle, was much more encompassing than motion as we know it today; it included all changes. He conceived of four distinct categories of motion: local motion, creation and destruction, alteration, and increase and decrease. Local motion was what we today consider to be motion—that is, movement of an object from one place to another. Creation and destruction (Aristotle called them generation and corruption) constituted the birth and death of things. But creation and destruction were never complete; that is, Aristotle's sublunar world was governed by the principle of conservation of matter. Alteration was the motion of an object that changed within itself, as when a leaf turns yellow in the autumn. Increase and decrease were the motions of an object, such as a baby or a tree, as it grows.

Every motion had a natural direction and a purpose. The natural direction was toward something better, toward perfection; the purpose was to get there, to fulfill potentiality. In falling, a piece of earth fulfills its potentiality as much as it can, that is, by falling as far as it can. An acorn has the potential to become a big oak tree and will move in that direction to fulfill its potential, to become more beautiful, more nearly perfect.

Of the four kinds of motions, only local motion has any relevance for physics today and is therefore of interest to us here. For Aristotle, local motion consisted of three kinds: straight, circular, and mixed. Straight motion had contraries, up and down, and constituted the *natural motion* of the sublunar world. Up was contrary to down; fire and air moved up, water and earth moved down. When earth was forced up, it was moving in its contrary direction, away from perfection, and this was unnatural or *violent motion*. Circular motion had no contrary (a circle has no beginning or end), hence motion in a circle could not have a goal.

Aristotle's concept of locomotion (local motion) was the result of his observations of the world about him. He never observed (nor have we for that matter) an object on the Earth start to move without a force acting on

FIGURE 1-4

Aristotle maintained that fire and air rise for the same reason that earth and water fall. As the satchel, striving for perfection, tends to fall, so does the balloon rise. (Reprinted with permission from *The Red Balloon* by A. Lamorisse, Finedita S. A. Glaris, Switzerland.)

it. And it seemed to Aristotle that all moving objects required a force to keep them moving. All moving things, Aristotle observed, come to rest soon after the force that causes the motion stops acting. The state of rest in

Aristotelian physics was the natural state of matter, and motion of sublunar material was a temporary state resulting from a force. Aristotle observed that an oxcart moved on a level road only when the oxen pulled it; it would not move by itself, and when the oxen stopped pulling the oxcart, it returned to its natural state of rest.

The motion of all things that are moved by something else [violent motion] must proceed in one of four ways . . . pulling, pushing, carrying, and twirling . . . [but] of these four, carrying and twirling are reducible to pulling or pushing (8).

Aristotle, therefore, defined a force as a push or a pull. This was a more objective concept of a force than that of Empedocles. It was not as animistic in quality, and it had a more specific function. Force, as Aristotle conceived it, caused motion.

Aristotle felt that if an object moves "there must always be a certain amount of distance that has been traversed and a certain amount of time that has been occupied" (9). Hence his concept of velocity is close, but not equivalent, to our present concept of *average velocity:* the distance traversed divided by the time required to traverse that distance. Aristotle, however, did not conceive of a *ratio* of two different entities, such as that of distance to time.

Having defined force and velocity, Aristotle related the two. For a given object, if the force were doubled the velocity would double. Presumably he felt that two oxen could pull a cart twice as fast as one. Using present-day symbols, we would write the proportion

$$v \propto F$$

to indicate that velocity v is directly proportional to force F.

Furthermore, Aristotle recognized that objects in locomotion moved through a medium.

Now the medium causes a difference [in the velocity] because it impedes the moving thing . . . in proportion to the density [resistance offered] of the hindering body.

. . . if air is twice as thin [as water], the body will traverse [the same distance in water] in twice the time it does [in air] (10).

That is, as the resistance to motion decreased, the velocity increased. Today we would write

$$\mathbf{v} \propto \frac{1}{R}$$

to indicate that the velocity is inversely proportional to the resistance.

If both of these proportions are combined into one, we obtain what might be called *Aristotle's Law of motion:*

$$v \propto \frac{F}{R}$$

For a given object in a given medium, if the force applied is doubled, tripled, or quadrupled, its velocity will be doubled, tripled, or quadrupled. If the resistance of the medium through which it travels is changed so that it is doubled, tripled, or quadrupled, then for a given force applied to a given object, the velocity becomes one-half, one-third, or one-fourth.

This law of motion had to apply to bodies moving with their natural motion as well as to those moving with violent motion.

It is apparent, I mean, that fire, in whatever quantity . . . moves upward, and earth downward; and, if the quantity is increased, the movement is the same, though swifter. (11).

The heavier the earthy body, the faster it moved down; the larger the fire, the faster it moved up. This greater speed had to be related to the cause of motion.

Now, that which produces upward and downward movement is that which produces weight and lightness, and that which is moved is that which is potentially heavy or light, and the movement of each body to its own place is motion towards its own form (12).

The body moved toward perfection. In other words, "the bodies are thought to have a spring of change within themselves" (13). And the greater the amount of material, the greater was this spring or desire to achieve perfection. As the weight or lightness of a body was increased, the spring, this inner desire causing it to move, increased. This increase was in direct proportion to the weight or lightness: "for instance, if one weight is twice another, it will take half as long over a given movement" (14). Since this inner desire was the cause of motion, it was as Aristotle himself said, "the force responsible for the downward motion of the heavy body" (15).

But to say that Aristotle's law of motion is a quantitative approach to the study of motion because it can be expressed as a proportion distorts the meaning of the word quantitative. Aristotle had absolutely no way of measuring speed; in fact, the thought never occurred to him. Some objects traveled faster than others—this was as close as Aristotle came to quantifying speed. Even using the phrase "twice as fast" does not serve to quantify speed, for we are not told *how* fast since the speed to which the comparison is being made is never specified.

That heavier objects do, indeed, fall faster than lighter ones can be seen by dropping a handful of loose dirt. Try it. If the dirt falls too fast to determine which pieces reach the ground first, drop a handful of dirt into water where it is easy to see that the heavier chunks fall faster.

This very simple demonstration was Aristotle's evidence that the rate of fall depends on the weight of the object falling. If the speed was too great to observe carefully, the only means, within the framework of Aristotelian physics, of slowing that motion was to increase the resistance. Aristotle always maintained that observation should support theory; he could say that his theory of falling bodies was confirmed by observation.

There is one exception to the law of motion expressed as the proportion

$$v \propto \frac{F}{R}$$

and Aristotle recognized it himself: the resistance may be so great that the application of a force does not cause the object to move at all. In other words, if the resistance R is greater than the force F, the velocity is zero, not some quantity between zero and one, as we might expect from the proportion. Under these conditions, even if a force is applied,

it might well be that it will cause no motion at all; for it does not follow that, if a given motive power causes a certain amount of motion, half that power will cause motion either of any particular amount or in any length of time; otherwise one man might move a ship . . . (16).

Aristotle used his law of motion to prove one of the most important concepts in his description of the universe. In contradistinction to the atomists, Aristotle was convinced that a void (essentially a perfect vacuum) could not exist. A void, he argued, would offer zero resistance to an object passing through it. But, as Aristotle pointed out, if the resistance to motion were to decrease, the velocity would increase; if the resistance were to decrease to zero, the velocity would become infinitely great. However, an infinite velocity, Aristotle maintained, was impossible, because an object would travel from one place to another in no time at all, that is, it would have to be in two places at the same time, and that is impossible. Consequently, a void is impossible. This argument proved to Aristotle's satisfaction that the universe cannot be composed of atoms immersed in a void.

Hence, Aristotle based a good part of his description of the world on the observation that objects do indeed move through a medium and that medium impedes motion. His law of motion was one of his few attempts to incorporate mathematics into his physics—and his law was only a proportion that did not always apply. Furthermore, where applicable, it could not be verified with quantitative observations, that is, by measurements.

Projectile motion, that is, motion with no apparent force actively propelling the object, was not so easily explained by Aristotle's law. This law demanded that a force, internal for natural motion and external for violent motion, be present to cause motion. Without this force the object would come to a state of rest.

Where is the force that pushes a stone (or discus) after it has been hurled into the air? What force makes a ball continue rolling on a level floor after the projector has released it? What keeps it rolling? These fundamental questions perplexed all those who tried to answer them until the seventeenth century, and even today they perplex many who have not studied physics.

Aristotle knew that his law did not solve the problem of projectile motion; he proposed two possible solutions and accepted one of them, but with reservations.

In point of fact things that are thrown move through that which gave them their impulse . . . either by reason of mutual replacement, as some maintain, or because the air that has been pushed pushes them with a movement quicker than the natural locomotion of the projectile . . . (17).

These, then, were the two possible solutions. (1) Air pushed aside by the projectile goes around behind it and pushes it forward, a process that came to be known as *antiperistasis*. (2) While pushing the projectile, the projector also pushes some air and gives that air some kind of motive power which it in turn imparts to the projectile to keep it moving.

In the very last pages of his *Physics,* Aristotle admitted the difficulty posed by the problem of projectile motion.

... it will be well to discuss a difficulty that arises in connection with locomotion. If everything that is in motion with the exception of things that move themselves is moved by something else, how is it that some things, e.g., things thrown, continue to be in motion when their movent [cause of motion] is no longer in contact with them? (18)

The ensuing discussion amplified the solution by which air was given some kind of motive force by the projector and kept the projectile going. Of this solution Aristotle said,

therefore, while we must accept this explanation to the extent of saying that the original movent gives the power of being a movent either to air or to water or to something else of the kind, naturally adopted for imparting and undergoing motion ... (19).

Seldom was Aristotle so vague; seldom did he express such doubts and seldom did he accept explanations with reservations. In fact, it was his lack of vagueness, his certainty, his explanations without reservations that made his description of the sublunar world so acceptable to those who lived in the centuries that followed.

The Celestial World and Its Motions

The *astronomical world* of Aristotle was a thing apart from the sublunar world. It could not be composed of any of the four elements, which moved either straight up or straight down. The motions in the heaven were circular by Plato's decree. Circular motion was perfect and eternal, for such motion had no beginning or end. Thus the heavens had to be composed of a fifth element capable of perfect circular motion. This element, *aether*—the quintessence—was by definition perfection, since only a perfect substance could undergo circular motion. Because aether was already perfection, it had no potential; it did not undergo any of the imperfect motions (e.g., generation and corruption or alteration). Aether did not strive.

To account for the motion of the stars, the sun, the moon, and the planets, Aristotle adapted and expanded a system originated by Plato's mathematician, Eudoxos. The Aristotelian system was composed of fifty-five spheres concentric with the center of the Earth, each of which moved at a constant speed. Yet these spheres were so related and moved in such a manner that the planets, each attached to one sphere, moved as the planets in the skies. To Plato these spheres and the uniform circular motion of the planets were Reality; the complex motion of the planets in the sky was only a shadow of that Reality.

To Aristotle the outer sphere was the sphere of the stars and the cause of all the motion in the universe; it was itself the Prime Mover. It moved the aether beneath it, which in turn rotated the sphere of Saturn, which in turn moved the sphere of Jupiter, and so on down to the sphere of the moon. The sphere of the moon, which was at the bottom of the astronomical world agitated the uppermost regions of the sublunar world. Since fire was the lightest of the four sublunar elements, fire occupied the upper regions of the sublunar sphere and was thus agitated by the sphere of the moon. Fire in turn agitated the air, which then agitated the water and the earth. Hence, the four elements in the sublunar world were prevented from forming four concentric shells by the rotation of the lunar sphere.

The Earth was spherical and at the center of All. No other worlds were possible because if earth existed anywhere beyond the sphere of stars, that earth would have the inner desire to come to the center of this world. Aristotle, therefore, could not detach himself from a concept so easily arrived at: this world, with mankind, was the very center, the most important thing in the universe. And man, the ruling form of life on the Earth, was even more important. Such a system has two names, *geocentric* or Earth-centered and *anthropocentric* or man-centered. From the study of the history of science, it is easy to believe that the geocentric concept is secondary to the anthropocentric.

Not only was the Earth at the center of everything, but it was also immovable. The natural motion of earth was straight down, but because it was already at the center, it could move no closer. Were the Earth to rotate, the material of which it is composed would necessarily move in circles. But circular motion is unnatural for earth, so it could not persist without a force acting on it. However, reasoned Aristotle, there is no outside force acting on the Earth causing it to rotate, and since the Earth is eternal, it cannot possibly be rotating. Therefore, the Earth is neither rotating nor falling toward the center, so it must be immovable. The logic of Aristotle is of this nature; if his premises are accepted, there is no escaping his conclusions.

Aristotle found the conclusion of a stationary Earth to be supported by observation. By his law of motion any movement of the Earth had to be the consequence of a force, and so too were the movements of all objects on the Earth. Objects at rest on the Earth had to be moving with it, requiring a force to be applied to them. By this reasoning an object thrown up or falling from a tower had at that moment no observable force operating on it and, free of the Earth, should be left behind by it. However, a falling object falls not to the west but straight down. Therefore, Aristotle concluded, the Earth was at rest.

Aristotle's physics is logically intradependent; his law of motion was used over and over again as a premise by which to arrive at logical conclusions. His physics was complete; it discussed and explained all that was observed in the world as long as one neither used instruments nor made measurements. But his physics was not without loopholes. In the ensuing centuries, certain isolated statements were attacked, albeit unsuccessfully.

Many who challenged Aristotelian physics on certain points, with valid observations to back up their claims, were entirely correct. But to challenge Aristotelian physics on a few points was to challenge the entire system, because it was interwoven with threads of logic. Until the seventeenth century no one could offer a better system to replace it. Then, in 1686, Sir Isaac Newton's famous *Principia* was published, offering a system that more adequately described the world for the men of the time.

References

1. Milton K. Munitz, *Theories of the Universe,* The Free Press, Glencoe, Ill., 1957, pp. 21 ff.

2. F. M. Cornford, *Before and After Socrates,* Cambridge University Press, London, England (paperback), 1960

3. Aristotle, *Metaphysics,* Book I, Chapter 4, $985^b/15/$, Harvard University Press, Cambridge, Mass., 1933.

4. Xenophon, *Memorabilia,* as quoted in *History of Science,* Vol. I, by George Sarton, Harvard University Press, Cambridge, Mass., 1952, p. 261.

5. "The Apology of Socrates," in Hugh Tredennick, trans., *The Last Days of Socrates,* Penguin Books, Baltimore, Md., 1954.

6. Plato, *The Republic,* Book VII/514/, Oxford University Press, New York (paperback), 1945, p. 227.

7. A. E. Taylor, *Aristotle,* Dover Publications, New York (paperback), 1955.

8. Aristotle, *Physics,* Book VII, Chapter 2, $243^a/15/$, Harvard University Press, Cambridge, Mass.

9. *Ibid.,* Book VII, Chapter 5, $249^b/28/$.

10. *Ibid.,* Book IV, Chapter 8, $215^a/28/$.

11. Aristotle, *On the Heavens,* Book IV, Chapter 4, $311^a/20/$, Harvard University Press, Cambridge, Mass., 1953.

12. *Ibid.,* Book IV, Chapter 3, $310^a/32/$.

13. *Ibid.,* Book IV, $310^b/25/$.

14. *Ibid.,* Book I, Chapter 6, $274^a/1/$.

15. *Ibid.,* Book IV, Chapter 6, $313^b/17/$.

16. Aristotle, *Physics, op. cit.,* Book VI, Chapter 5, $250^a/15/$.

17. *Ibid.,* Book IV, Chapter 8, $215^a/14/$.

18. *Ibid.,* Book VIII, Chapter 10, $266^b/26/$.

19. *Ibid.,* $267^a/3/$.

Chapter Two

Physics Through
the Renaissance

New York Public Library Picture Collection

A ristotelian physics became securely established after Aristotle's death in 322 B.C. In fact, proposals made in the third century B.C. that were contrary to his works were considered heretical. Nevertheless, such proposals were made, and criticisms of certain aspects of Aristotle's physics continued intermittently throughout the succeeding centuries. Most of these criticisms pertained to one of three aspects of his work: (1) violent motion, and in particular projectile motion; (2) natural local motion; and (3) celestial motion. Each of these concepts will be considered as a separate strand of thought leading from the third century B.C. to the seventeenth century. However, only the major contributors to these strands of thought will be considered.

VIOLENT MOTION

Projectile motion evoked the most controversy and no wonder, for Aristotle himself was uncertain on this point. The difficulty stemmed from Aristotle's conclusion that objects in motion require a force to maintain that motion. Hence, a ball rolling on a level floor and a stone thrown in the air require a force to keep them moving. The controversy centered on what supplied that mysterious force.

John Philoponus

In the early sixth century A.D., John Philoponus of Alexandria, sometimes called John the Grammarian, wrote many commentaries on Aristotle's work, and openly criticized him on projectile motion.

For in the case of antiperistasis there are two possibilities: (1) the air that has been pushed forward by the projected arrow or stone moves back to the rear and takes the place of the arrow or stone [since, according to Aristotle, a void cannot exist], and being thus behind it pushes it on, the process continuing until the impetus of the missile is exhausted, or (2) it is not the air pushed ahead but the air from the sides that takes the place of the missile. . . .

Let us suppose that antiperistasis takes place according to the first method indicated above, namely, that the air pushed forward by the arrow gets to the rear of the arrow and thus pushes it from behind. On that assumption, one would be hard put to it to say what it is (since there seems to be no counter force) that causes the air, once it has been pushed forward, to move back, that is along the sides of the arrow, and, after it reaches the rear of the arrow, to turn around once more and push the arrow forward. . . . Furthermore, how can this air, in so turning about, avoid being scattered into space, but instead impinge precisely on the notched end of the arrow and again push the arrow on and adhere to it? Such a view is quite incredible and borders rather on the fantastic (1).

Philoponus' criticism is extremely clear and rational, and he apparently realized that to change the direction of the air's motion requires a force. But Aristotle, too, rejected antiperistasis, so what did Philoponus have to say of the second argument?

Now there is the second argument which holds that the air which is pushed in the first instance [i.e., when the arrow is first discharged] receives an impetus to motion, and moves with a more rapid motion than the natural [downward] motion of the missile, thus pushing the missile on while remaining always in contact with it until the motive force originally impressed on this portion of air is dissipated. This explanation, though apparently more plausible, is really no different from the first explanation by anti-peristasis. . . . In the first place we must address the following question to those who hold the views indicated: "When one projects a stone by force, is it by pushing the air behind the stone that one compels the latter to move in a direction contrary to its natural direction? Or does the thrower impart a motive force to the stone, too?" Now if he does not impart any such force to the stone, but moves the stone merely by pushing the air . . . of what advantage is it for the stone to be in contact with the hand . . . ? But the fact is that even if you place the arrow or stone upon a line or point quite devoid of thickness and set in motion all the air behind the projectile with all possible force, the projectile will not be moved the distance of a single cubit.*

From these considerations and from many others we may see how impossible it is for forced motion to be caused in the way indicated. *Rather it is necessary to assume that some incorporel motive force is imparted by the projector to the projectile,*† and that the air set in motion contributes either nothing at all or else very little to this motion of the projectile (1).

In trying to answer the question "What keeps a projectile moving?" Philoponus made a new suggestion: the projector imparts a *motive force* directly to the projectile. However, this motive force was not really a part of the projectile. Philoponus went on to explain that each projectile slowed down not only because of the resistance of the air, but also because the motive force was *replaced* by the natural inclination of the projectile to fall downward.

Certainly this explanation was better than the Aristotelian theory that a motive force is imparted to the air, which in turn imparts it to the projectile. At the very end, however, Philoponus hedged by indicating that the air might help out a little. Aristotle's authority influenced even those who criticized him. But this is how science progresses. As our way of looking at nature gradually changes, a new explanation may be offered, but it is never a sudden and complete break with the old way of explaining a particular aspect of nature. Both the old and the new explanations, after all, seek to describe the same phenomenon.

---

* A cubit is a unit of length that was never standardized, simply because the need for standardized units never arose until the nineteenth century.
† Italics are those of the translator.

Simplicius

Some explanations attempt to combine the new perspective with the old. Simplicius (d. A.D. 549), an inhabitant of Cilicia* and a contemporary of Philoponus, offered such an explanation of projectile motion. He suggested, in the true spirit of compromise, that the air and projectile take turns pushing each other!

Aristotelian physics was part of the Greek heritage translated into the Arabic language and incorporated in the Arabic intellectual world that flourished from the eighth to the thirteenth century. With the fall of the Roman Empire, the Western world lost contact with all but a very small part of the Greek heritage. Increased communication with the Arabic world in the tenth and eleventh centuries, however, revealed to Western scholars the extent to which the Greek writings were being studied by the Arabs. The revelation sparked a reawakening in the Western world, and a flurry of translation of the Greek writings from the Arabic into the Latin continued from the eleventh through the thirteenth century. Major centers of translation were set up in Toledo, Spain, and in Sicily, the two locations in which the Europeans and Arabs were living peaceably together. There the Arabs, Christians, and Jews continued to translate scientific and philosophic materials even while the Crusaders waged their campaigns farther to the east. It was through these translations in the twelfth and thirteenth centuries that European scholars became acquainted with the vast writings of the ancient Greeks. By the fourteenth century most of the translations had been completed; indeed, an effort had been made to translate directly from Greek into Latin. Scholars in Europe then had the enormous but rewarding task of studying and assimilating all this newly acquired knowledge.

Jean Buridan

Jean Buridan (*c*. 1300–*c*. 1360), a recognized scholar and twice rector of the University of Paris, was one of a small group of Europeans who first studied the problem of projectile motion and proposed what we now call the *impetus theory*. Buridan referred directly to Aristotle's statements about projectile motion but did not accept his solution. Instead he described the movements of three familiar objects and claimed that in none of these instances could the air, pushed aside by the projectile, go around behind it and push it.

Buridian's first example was the smith's mill, or grindstone. A smith's mill will continue to turn once the hand that started it turning is removed, yet air is not pushed aside nor is there a behind against which any air can push to keep it going.

The second example was the flight of a lance or spear that has been sharpened on both ends. The air pushed aside by the lance would be unable to impinge against a sharpened posterior.

* The south central region of what is now Turkey.

His third example was that of a ship set in motion, perhaps by men on shore pulling it; after the ship is released, it continues to move ahead. The men on deck feel no wind pushing them; in fact, they feel a wind from the front resisting their motion.

Having successfully disproved the idea of antiperistasis, Buridan proceeded to ask why, if Aristotle could have the projector give the air the power to keep the projectile moving, the projector could not impart that same motive power to the projectile. Buridan answered that it was indeed possible for the projectile to be given motive power directly, and he called this motive power *impetus*. He went so far as to suggest that the impetus of a projectile is proportional to both its velocity and weight. He imagined that the impetus was a part of the body, *not* something outside it. Impetus was "a thing of permanent nature" as long as the object was moving. Hence it did not die of its own nature. Supporting evidence was given by the celestial motions, for

when He created the world, [He] moved each of the celestial orbs as He pleased, and in moving them He impressed in them impetuses which moved them without His having to move them any more. . . . And these impetuses which He impressed in the celestial bodies were not decreased nor corrupted afterwards. . . . Nor was there resistance which would be corruptive or repressive of that impetus. But this I do not say assertively but so that I might seek from the theological masters what they might teach me in these matters as to how these things take place . . . (2).

Buridan, who apparently did not want to ruffle the theologians, developed the most advanced explanation of projectile motion that was to be proposed until the seventeenth century. Although the impetus theory was a more advanced concept that Philoponus' motive force, to the modern reader it may actually seem more vague. Both served the same purpose, however, to keep the projectile moving. In fact, we could redefine impetus as that which keeps a projectile moving.

FALLING BODIES

The second strand or line of thought in Aristotelian physics that received some criticism before the seventeenth century was the problem of natural motion, or falling bodies. It will be recalled that Aristotle (see p. 13) asserted that objects of different weight fall the same distance in different intervals of time, the heavier falling faster.

The principal difficulty in studying falling objects is that their velocity changes—it increases. Even *with* our modern concept of measurement, it is not an easy task to measure velocity directly and hence to measure any differences in that velocity in order to see *how* it changes. Furthermore, the velocity of a falling object is fairly rapid so that we cannot, by just looking, learn much about it. We have seen that to slow the motion down Aristotle resorted to a method quite natural for him and entirely consistent with his physics; he used water instead of air as the medium. In water, objects fall much more slowly, and he saw heavier objects falling faster than light objects, as they do in any resisting medium.

Aristotle recognized the fact that the speed of an object can change; he referred to *uniform motion* in which the speed is constant, and to *nonuniform motion* in which the speed either increases or decreases. But he went no further than this. It remained for one of his intellectual heirs to begin the study of changing speed.

Strato the Physicist

In the third century B.C., Strato the Physicist (fl. c. 300 B.C.), born in the city of Lampsicus on the southern shore of the Dardanelles, contributed some valuable insights to the description of falling bodies. We know of Strato's opinions largely through the commentaries of Simplicius, for as is usual when one man (in this instance Aristotle) is granted the role of supreme authority, most of the writings of others seem to disappear, even though these writings may be significant in their own right.

Simplicius pursued the discussion of acceleration in a manner that credits his sense of need for more observation.

It is universally asserted as self-evident that bodies moving naturally to their natural places undergo acceleration. . . . But few adduce any proof of the fact itself. . . . It may therefore not be out of place to set forth the indications [of acceleration] given by Strato, the Physicist. For in his treatise On Motion,* after asserting that a body so moving completes the last stage of its trajectory in the shortest time, he adds: ". . . For if one observes water pouring down from a roof and falling from a considerable height, the flow at the top is seen to be continuous, but the water at the bottom falls to the ground in discontinuous parts. *This would never happen unless the water traversed each successive space more swiftly.*"† By "this" Strato means the breaking up of the continuity of the object as it approaches the ground.

Strato also adduces another argument, as follows: "If one drops a stone or any other weight from a height of about an inch, the impact made on the ground will not be perceptible, but if one drops the object from a height of a hundred feet or more, the impact on the ground will be a powerful one. Now there is no other cause for this powerful impact [than the greater velocity]. . . . It is merely a case of acceleration" (3).

Dropping the stone from various heights to observe the differences in impact was a crude means of estimating velocity, but these differences in impact were impressive and could be used effectively in debates. Observing impact was, in fact, the Greeks' only reasonably objective method of estimating velocity. It is fairly obvious, then, that their studies of acceleration were, of necessity, limited.

Yet Strato went so far as to relate the breaking up of a stream of water to the increased velocity of the water as it falls. We have all observed a falling

* Not extant.
† Italics are mine.

stream of water, but how many of us would draw the same conclusions that Strato did in the third century B.C.?

Strato used space as a reference for defining acceleration: an accelerating object moved so that it traversed equal spaces in succeedingly smaller intervals of time. If an object fell a distance of two feet, it took less time to travel through the second foot of fall than to travel through the first foot.

John Philoponus

Another keen observer was John Philoponus who took Aristotle to task on natural motion as well as violent motion. In commenting on Aristotle's opinion of falling bodies, Philoponus said:

But this is completely erroneous, and our view may be corroborated by actual observation more effectively than by any sort of verbal argument. *For if you let fall from the same height two weights of which one is many times as heavy as the other, you will see that the ratio of the times required for the motion does not depend on the ratio of the weights, but that the difference in time is a very small one.** And so, if the difference in the weights is not considerable, that is, if one is, let us say, double the other, there will be no difference, or else an imperceptible difference, in time, though the difference in weight is by no means negligible . . . (4).

This simple but extremely important experiment was apparently performed as early as the sixth century A.D. And even though such an experiment was conclusive proof that Aristotle was wrong about the speed with which bodies fall, Aristotelian physics continued to be accepted with very little question.

The Mertonians

In fourteenth-century England at Merton College, Oxford University (Figure 2-1), a group of scholars were the first to give a hard and fast definition to the term acceleration. This early accomplishment speaks highly of the ability of these men, but the fact that the definition was not applied to falling objects until the seventeenth century indicates continued difficulty in studying velocities and falling objects. In the 1330's many scholars at Merton College were engaged in these studies, making it difficult to give any one man credit for a particular contribution. So the group as a whole is referred to as the Mertonians; the principal members were Thomas Bradwardine, William Heytesbury, Richard Swineshead, and John Dumbleton.

This group was the first to make a clear distinction between kinematics and dynamics. *Kinematics* is the study of motion itself; *dynamics* is the study of the cause of motion. Aristotle was concerned almost entirely with dynamics but made no clear distinction between the two fields of study.

Furthermore, the Mertonians considered speed or velocity quantitatively,

---

* Italics are those of the translator.

FIGURE 2-1

The buildings of the Mob Quadrangle of Merton College, Oxford University, are among the oldest structures in Oxford. The Chapel is in the background. (Courtesy of Merton College, Oxford.)

a new perspective that permitted them to conceive of *instantaneous velocity* and to give a clear-cut definition of uniformly accelerated motion which nearly matches our definition today. Through the writings of Heytesbury, the Mertonians defined *uniform motion: that motion is called uniform in which an equal distance is continuously traversed in an equal part of time.*

With this basic definition the Mertonians were then able to define motion of a more complex nature, first, the instantaneous velocity of an object that moves with nonuniform motion or velocity. Its velocity at any instant is measured by the distance it *would* cover if it *were* to maintain the velocity of that instant for an extended period of time. We see immediately that the Mertonians defined velocity in terms of the distance traversed, and indeed they could think of it only in these terms.

Extending their logic one more step, the Mertonians realized that of all the ways in which velocity could increase, one was of special interest. They called it *uniformly accelerated motion.*

With regard to acceleration and deceleration of local motion, however, it is to be noted that there are two ways in which a motion may be accelerated or decelerated: namely, uniformly, or nonuniformly. *For any motion whatever is uniformly accelerated if, in each of any equal parts of time whatsoever, it acquires an equal increment of velocity.** And such a motion is uniformly decelerated if, in each of any equal parts of time, it loses an equal increment of velocity. But a motion is nonuniformly accelerated or decelerated, when it acquires or loses a greater increment of velocity in one part of the time than in another equal part (5).

The use of the term "increment" demands that velocity be considered quantitatively; an increment is a certain specified quantity of velocity. If an object is traveling with uniformly accelerated motion, its velocity, for example, might increase by an increment of 5 feet per second each second, to use the terminology of today. The Mertonians did not consider velocity as a ratio of distance to time, but rather as a distance covered in a certain interval of time.

The difficulty of measuring velocity becomes apparent when we consider the Mertonians' *average speed theorem.* If we must rely on distance traversed to measure velocities, how much distance does an object traveling with uniform acceleration cover? This is not an easy problem to solve, but if we accept their definition of uniform acceleration (and basically we do), then in a given interval of time the distance covered by an object traveling with *uniform acceleration* will be equal to the distance covered by an object traveling with a *uniform velocity* equal to the *average instantaneous velocity* of that accelerated motion. For example (and to use today's terminology again), if a car accelerates uniformly from 20 to 40 miles per hour in a certain interval of time, it will cover the same distance as another car traveling for the same interval of time at a uniform velocity of 30 miles per hour.

Hence, considering uniform motion quantitatively permitted the Mertonians to define uniformly accelerated motion and even to describe the distance traversed by a body traveling with uniformly accelerated motion.

Men have developed other methods of studying acceleration, not only to improve their understanding of it but also to predict the distance covered by objects moving with nonuniformly accelerated motion. One such method was a graphical representation of velocity and time; its origin stems directly from the work of the Mertonians which spread to the continent and their directly influenced an Italian in Bologna, Giovanni di Casali, and a Frenchman in Paris Nicole Oresme. Oresme, however, expanded the graphical method and applied it with the most effect.

Nicole Oresme

Nicole Oresme (*c.* 1323–1382) was a student at the University of Paris in 1348 and became a master in 1356. Consequently, there seems to be little doubt that he heard lectures given by Jean Buridan who died about 1360.

---

* Italics are mine.

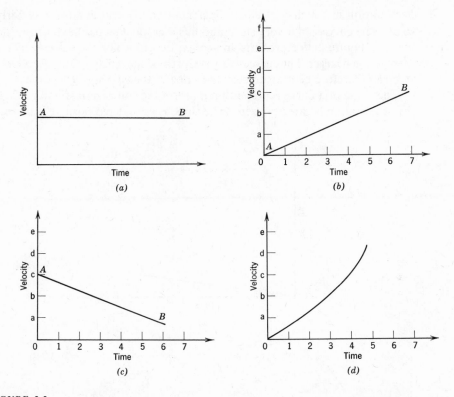

FIGURE 2-2

Velocity-time graphs of objects moving with *(a)* uniform motion—the velocity remains constant; *(b)* uniformly accelerated motion—the velocity increases in proportion to the time; *(c)* uniformly decelerated motion—the velocity decreases in proportion to the time; *(d)* nonuniformly accelerated motion.

Following the example set by the geographers in plotting locations on the Earth's surface by latitude and longitude, Oresme plotted velocity against time. If we think of a horizontal line as representing time and a vertical line as representing velocity, we can describe any type of motion we care to. Without making a measurement, let us illustrate four types of motion and use modern terminology. The graph in Figure 2-2a represents uniform motion; the line *AB* is called the *curve* and indicates a constant velocity. The curve in Figure 2-2b represents motion with uniform acceleration; in equal intervals of time (1, 2, 3, 4, . . .) the velocity increases by equal increments (a, b, c, d, . . .). The graph in Figure 2-2c represents motion with uniform deceleration. That in Figure 2-2d represents motion in which the velocity increments are not equal, that is, with succeeding equal intervals of time the velocity increments increase. This is one type of nonuniform acceleration.

The term *slope* has come to describe the steepness of the curve. Let us compare the graphs in Figure 2-3. The curve in Figure 2-3a represents an acceleration in which the velocity increment added to the motion is fairly small. The curve is not very steep and has a gradual slope. In contrast, the curve in Figure 2-3*b* represents an acceleration in which the velocity increment is much larger. The curve is very steep, the slope much greater. Acceleration can therefore be measured by the slope of the velocity-time curve.

The slope of a curve representing uniform motion is zero (Figure 2-2*a*); the acceleration is therefore zero. The slope is considered positive if the ac-

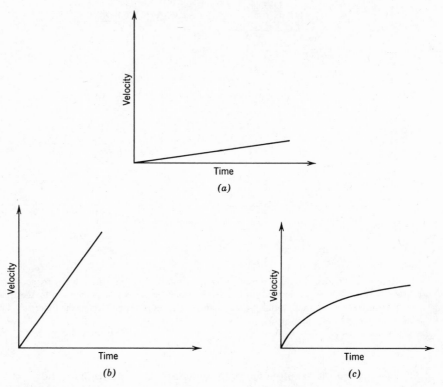

FIGURE 2-3

The acceleration of the object in *(a)* is less than that of the object in *(b)*. The object in *(c)* travels with a decreasing acceleration even if the velocity continues to increase.

celeration is positive, that is, if the speed increases. The slope is negative if the speed decreases (Figure 2-2*c*). Correspondingly, such motion is called not only deceleration but also negative acceleration.

The slope of the curve in Figure 2-3*c* changes continuously. The slope is decreasing with time, that is, the acceleration is decreasing. Even if the velocity continues to increase, the acceleration decreases. Only when the acceleration becomes zero does the velocity remain constant.

The fact that Oresme wrote of the condition in which the velocity continues to increase even while the acceleration decreases indicates the value of the graphical method; the graph allowed him to "see" this type of motion.

The graphical method also helped Oresme to recognize the meaning of the area under the velocity-time curve, the shaded areas in Figures 2-4a and 2-4b. If an·object moves with uniform motion, it traverses equal distances in equal time intervals. If the time interval is doubled, the distance traversed will be doubled and so will the area under the velocity-time curve (Figure 2-4a): area $ABCD$ equals twice area $AEFD$.

On the other hand, if two objects travel for the same time interval, one at a velocity twice the other, the faster one will traverse twice the distance traveled by the slower one (Figure 2-4b). Again the area under the velocity-time curve bears the same ratio: area $ABCD$ equals twice area $ABNM$.

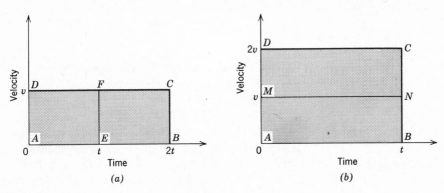

(a)

(b)

FIGURE 2-4

The area under the velocity-time curve represents the distance traversed.

Therefore, it occurred to Oresme that the area under the velocity-time curve actually represents the distance traversed. Using the algebra that was not available to Oresme, we can now say that

$$\text{area of a rectangle} = \text{height} \times \text{base}$$

and for uniform motion

$$\text{space traversed} = \text{velocity} \times \text{time}$$

$$s = v \times t$$

If the motion is not uniform as in Figure 2-5, the average speed theorem of the Mertonians is easily described. The velocity at A is zero and the velocity at the end of the time interval is v. By the average speed theorem, the distance traversed by an object traveling with a uniformly accelerated motion equals the distance traversed by an object traveling with a uniform motion equal to the average of the two extreme velocities of the accelerated motion.

The two extreme velocities of the accelerated motion are zero and v; their average is $\frac{1}{2}v$. Consequently, the distance traversed by an object moving with a uniform motion equal to $\frac{1}{2}v$ is represented by the area enclosed in rectangle $ABNM$ in Figure 2-5.

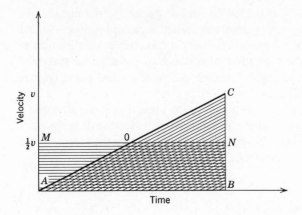

FIGURE 2-5

Area *ABC* equals area *ABNM*. The distance traversed by the object represented by curve *AC* equals the distance traversed by the object represented by the curve *MN*.

But the triangle *ABC* should represent the distance traversed by the object moving with uniformly accelerated motion, and thus the areas enclosed by triangle *ABC* should equal the area of rectangle *ABNM*. And, indeed, those areas are equal, for the triangle *ABC* and the rectangle *ABNM* differ from each other by the small triangles *AOM* and *CON*, and these two triangles are equal in area; in fact, they are congruent. Consequently, the area of triangle *ABC* equals the area of rectangle *ABNM*, and the distances traversed are indeed represented by the area under the velocity-time curves.

If we let s represent the space traversed, t the time, and $\bar{v}$ (called v-bar) the average velocity of a uniformly accelerated object, our earlier equation $s = vt$ for uniform motion (v = constant) becomes $s = \bar{v}t$. In fact, the area under *any* velocity-time curve bounded by a given interval of time can be equaled (and therefore replaced) by the area of a rectangle whose height is the average velocity, and whose length is that same interval of time. To predict the distance covered by any object whose motion is complex, we need only draw a velocity-time curve and measure the area under the curve.

PLANETARY MOTIONS

The third and final strand of thought to be traced from the time of the Greeks through the sixteenth century is planetary motion. It will be recalled that Aristotle used a system originated by Eudoxos (see page 15) in which the planets were placed on the inside of spheres so located that their combined motions would account for the apparent erratic motion of the planets. But the Aristotelian system left much to be desired since it could not account for the planetary motions with much accuracy, nor could it account for the fact that the brightness of the planets varied.

Observed Motions

As seen from the Earth, there are two general types of apparent planetary motion, one for the planets Mercury and Venus, and the other for Mars, Jupiter, and Saturn—the five planets perceptible to the naked eye and hence the only ones known before the eighteenth century.

Venus, representing the first type, moves first to the east and then to the west of the sun, but never more than about 48 degrees from the sun. When it is east of the sun, it is seen in the western sky just after sunset; when it is west of the sun, it is seen in the eastern sky just before sunrise. During these motions its brightness varies considerably.

Mars represents the second type of motion very well. It generally travels slowly eastward in the sky in relation to the background stars, but when it is high in the sky at midnight—and at its brightest—it reverses its motion and appears to travel westward in relation to the background stars (Figure 2-6). This motion westward is called *retrograde motion,* for it is contrary to the general direction of travel of the celestial objects, the sun, moon, and planets, which move in relation to the fixed stars. It was this retrograde motion accompanied by the brightening of the planet that was so difficult for early astronomers to explain.

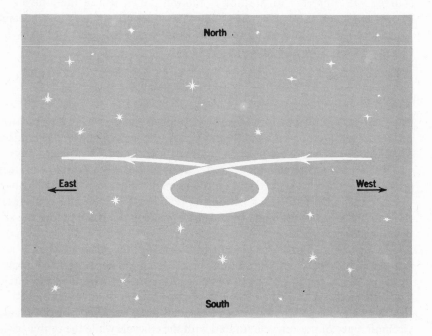

FIGURE 2-6

A schematic drawing of the motion of Mars including retrograde motion.

Aristarchus of Samos

In the century after Aristotle, a student of Strato made a very bold suggestion about the structure and motions of the universe. Aristarchus of Samos (*c.* 310–230 B.C.), from one of the islands off the Ionian coast just north of Miletus, proposed that the Earth was only a planet and revolved, along with the five known planets, about the sun.

The book of Aristarchus has been lost, and we know of his heliocentric (sun-centered) idea through the writings of a late contemporary of his Archimedes (*c.* 287–212 B.C.) who, in his brilliant work, *The Sand-Reckoner*, wrote:

Aristarchus of Samos brought out a book consisting of some hypotheses, in which the premises lead to the result that the universe is many times greater than that now so called. His hypotheses are that the fixed stars and the sun remain unmoved, that the earth revolves about the sun in the circumference of a circle, the sun lying in the middle of the orbit, and that the sphere of fixed stars, situated about the same center as the sun, is so great that the circle in which he supposes the earth to revolve bears such a proportion to the distance of the fixed stars as the center of the sphere bears to its surface (6).

Aristarchus was clearly centuries ahead of his time. It naturally follows that he was immediately criticized for proposing not only a heliocentric universe but a universe so large that the orbit of the Earth appeared as a mere point or speck. According to Plutarch (*c.* A.D. 46–120), Aristarchus was charged with impiety "for putting in motion the Hearth of the Universe." The Aristarchan universe could not survive in an atmosphere that became dependent on Aristotelian physics, for the two systems were mutually incompatible; one was heliocentric, the other geocentric.

Claudius Ptolemy

The final system by which the Greeks explained the planetary motions was proposed by Claudius Ptolemy (fl. A.D. 150), a Greco-Roman of Alexandria. The Ptolemaic system was at once highly dependent on Aristotelian physics and at the same time at odds with it. But it is the only example of Greek science that relied to any extent on measurement, and it indeed relied heavily upon measurements of the positions of planets. It is therefore considered the only truly and thoroughly quantitative work of mankind before the sixteenth century.

To be sure, the Earth was placed at the center of the system (or nearly so), but the planets were not placed on spheres. Each planet was placed upon a little circle, the center of which was placed upon another and larger circle. The little circle was called an *epicycle* (Figure 2-7*a*), and its center moved on the circumference of the bigger circle called a *deferent*. The planet on the circumference of the epicycle rotated with the epicycle while the center of the epicycle revolved to the east about the sun on the circumference of the deferent. When any planet was well inside the circumference of its deferent,

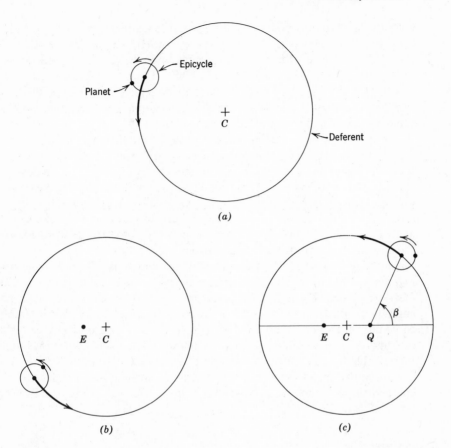

FIGURE 2-7

(a) Each planet, in the Ptolemaic system, rotates with its epicycle while the epicycle revolves about the Earth. *(b)* The Earth is located at the eccentric *E*. *(c)* The motion of the epicycle is uniform about the equant *Q*. The angle *β* increases uniformly with time.

and moving faster than the center of its deferent moved along the epicycle, that planet, as seen from the Earth, appeared to be moving in a direction opposite that in which the center of the epicycle moved; the planet appeared to move to the west or to move with retrograde motion. While in retrograde motion it was closer to the Earth and should therefore appear brighter (a conclusion that concurred with observation).

The epicycle-deferent system did not adequately account for all the planetary motions, so Ptolemy placed the Earth in a location other than at the center of the deferent. This position was called the *eccentric* (see *E* in Figure 2-7*b*). The eccentric, when combined with the epicycles, did a much better job of describing the motions of the planets.

But Ptolemy's combined eccentric and epicycle system required further refinements and improvements, for it still did not account for the more precise observations and measurements of the positions of the planets that were available to him. Fairly accurate measurements of planetary positions had been made 300 years before Ptolemy, by Hipparchus of Nicaea* (fl. 146–127 B.C.). Indeed, Ptolemy drew heavily upon the ideas and observations of Hipparchus and had great praise for him. Hence the Ptolemaic system could rightfully be called the Hipparchan-Ptolemaic system, but Ptolemy carried out extensive calculations with such accuracy that his name alone is generally associated with this complex system.

The provision entirely original with Ptolemy and the one refinement contrary to Plato's decree of uniform motion in a circle was the addition of a new center for the Planet's motion along the deferent. About this point, called the *equant* (see Q in Figure 2-7c), the center of each epicycle appeared to move with uniform speed, that is, the angle β (Greek beta) increased at a uniform rate. If the Earth was located at the eccentric E and the center of the deferent at C, the motion of the planets would certainly not correspond to Plato's Ideal of circular motion at constant speed.

This time, however, the theories of Plato and Aristotle had to be disregarded. Ptolemy wrote such a majestic work (the Arabs called it *The Majestic,* or the *Almagest* as we refer to it today), supported by such accurately measured positions and such extensive calculations, that it simply could not be rejected. It did, after all, predict the positions of the planets, the sun, and the moon with far greater accuracy than any system previous to it. Indeed, the Ptolemaic system lasted until the seventeenth century with only minor improvements (additional epicycles were included) and with only one competitor, the Copernican system, a heliocentric system proposed in the sixteenth century.

Nicolas Copernicus

A worthy competitor indeed, Nicolas Copernicus (1473–1543) initiated the movement that finally overthrew not only the Ptolemaic system but also the whole of Aristotelian physics. This great innovation was to be called the Copernican Revolution.

Copernicus was born in the city of Torun, Poland, and entered the University of Cracow in 1491. Like every promising student of his time, he journeyed to Italy in 1496 and there studied at the universities of Bologna, Padua, and Ferrara, where he came in contact with the center of intellectual life of his day. Upon his return to Poland in 1506, he began to develop his astronomical theory. In 1543, the year of his death, his book, *The Revolutions of Heavenly Spheres,* was published.

The Copernican theory differed from the Ptolemaic in two principal respects; it was a heliocentric system, and it did away with the equant. Copernicus was not, by any means, the first to conceive of a heliocentric system. In his writings he made reference to Greek authors, including Aristarchus, who propounded systems in which the Earth moves.

---

* Nicaea is on the eastern shore of the Sea of Marmara, or a little southeast of the present city of Istanbul.

However, these were not the only systems to serve as patterns for his plan; the Ptolemaic system was also used by Copernicus, whose universe had its epicycles, its deferents, and its eccentrics for each planet. But the planets moved around the sun, which simplified the mathematics somewhat by eliminating one motion for each planet. The Ptolemaic system ascribed all changes in the positions of the visible planets in relation to the background of fixed stars exclusively to the motions of the planets. To account for the changes of position actually produced by the real motion of the Earth about the sun, an additional motion for each planet had to be postulated. In the Copernican system the Earth's revolution about the sun removed the necessity of postulating this additional motion. Nevertheless, the Copernican system was not materially simpler and only slightly more accurate than that of Ptolemy. It did, however, present certain mathematical advantages.

Copernicus eliminated the equant that Ptolemy had devised, for the pure and simple reason that Plato's decree was violated by a circular motion that was not uniform about the center of the circle, but about some point near the center. This innovation, when considered with his strong reliance on Aristotle for his belief in the perfection of the spheres, and for other matters, indicates how closely Copernicus was bound to the Greek tradition (the Aristotelian system), even though the revolution he unwittingly started was to overthrow it.

Try as he might, however, Copernicus could not prove his heliocentric system; he could not prove that the Earth revolves about the sun. In fact, the Aristotelians of his day and later argued that our senses tell us the Earth is at rest and the heavens move. But that argument was questioned even 200 years before Copernicus by Oresme:

I make the supposition that local motion can be sensibly perceived only in so far as one may perceive one body to be differently disposed with respect to another. In support of this [I give the following illustration]: If a person is in one ship called *a* which is moved very carefully [smoothly]—either rapidly or slowly—and this person sees nothing except another ship called *b*, which is moved in every respect in the same manner as *a* in which he is situated, I say that it will seem to this person that neither ship is moving. And if *a* is at rest and *b* is moved, it will appear and seem to him that *b* is moved. [On the other hand], if *a* is moved and *b* is at rest, it will appear to him as before that *a* is at rest and *b* is moved. . . . Similarly, if a person were in the heavens and it were posited that they were moved with a diurnal [daily] movement, and [furthermore] that this man who is transported with the heavens could see the earth clearly and distinctly and its mountains, valleys, rivers, towns and chateaux, it would seem to him that the earth was moved with a diurnal movement, just as it seems to us who are on the earth that the heavens move. Similarly, if the earth and not the heavens were moved with a diurnal movement, it would seem to us that the earth was at rest and the heavens were moved. This can be imagined easily by anyone with good intelligence (7).

Oresme argued further that the air can be carried with the moving Earth, just as the air in an enclosed ship moves with the ship. So, too, an arrow shot straight up comes back down to the spot from which it was shot; it is not left behind to fall to the west. In short, Oresme showed conclusively that one cannot demonstrate by any experience whatever that the heavens are moved with daily movement, because, regardless of whether it has been posited that the heavens and not the earth are so moved or that the earth and not the heavens is moved, if an observer is in the heavens and sees the earth clearly, it (the earth) would seem to be moved; and if the observer were on the earth, the heavens would seem to be moved. (8)

Oresme was very clear in this statement of the concept of relative motion. In everyday life the Earth, or the ground under our feet, is tacitly used as a *frame of reference*. So we say that a car is traveling down the road at a speed of 50 miles per hour. If you are the driver of that car, you too are traveling at the same speed, using the same frame of reference, the Earth. But you are at rest with respect to your car. If an oncoming car is traveling, with respect to the Earth, at a speed of 55 miles per hour, then with respect to that oncoming car you are traveling at a speed of 105 miles per hour. You as the driver of your car have at least three velocities, depending on which frame of reference is used. Your speed is zero with respect to your car, 50 miles per hour with respect to the Earth, and 105 miles per hour with respect to the oncoming car.

Well aware of Oresme's argument, Copernicus was forced to rely for support of his theory, not on observation, not on simplicity, not on accuracy, but on good taste, "admirable symmetry," and the "clear bond of harmony in the motion and magnitude of the spheres."

References

1. M. R. Cohen and I. E. Drabkin, *A Source Book in Greek Science*, Harvard University Press, Cambridge, Mass., 1958, pp. 221 f.
2. Marshall Clagett, *The Science of Mechanics in the Middle Ages*, University of Wisconsin Press, Madison, 1959, p. 536.
3. Cohen and Drabkin, *op. cit.*, pp. 211 f.
4. *Ibid.*, p. 220.
5. Clagett, *op. cit.*, pp. 236 f.
6. Cohen and Drabkin, *op. cit.*, p. 108.
7. Clagett, *op. cit.*, p. 601
8. *Ibid.*, p. 606.

Questions

1. Draw both a distance-time and a velocity-time graph for uniform motion and for uniformly accelerated motion. Compare the two sets of graphs for each type of motion by considering the following: (*a*) initial velocity, (*b*) final velocity, (*c*) average velocity, (*d*) acceleration, and (*e*) distance traversed.

2. If an airplane travels at a constant velocity of 600 mph how far will it travel in 4 hours?

3. An automobile travels for 30 minutes at 40 mph and then 30 minutes at 50 mph.

(*a*) Draw a velocity-time graph for this motion, assuming that the time required for the acceleration is negligible.

(*b*) What is the total distance traveled?

(*c*) What is the average velocity?

4. Compare the *motive force* of John Philoponus with the *impetus theory* of Jean Buridan.

5. Read in *The Copernican Revolution,* by T. Kuhn, a Modern Library paperback, pp. 171–177, the sections "Copernican Astronomy—The Planets" and "The Harmony of the Copernican System," and comment on the mathematical and astronomical value of the Copernican system.

Chapter Three

Galileo and the Scientific Revolution

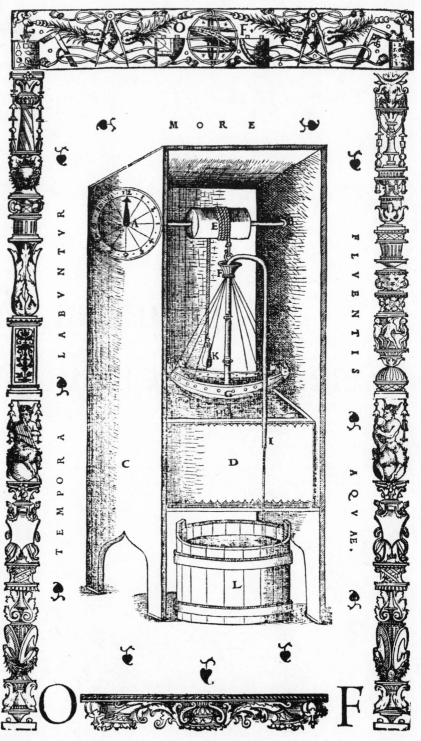

MORE

TEMPORA LABVNTVR

FLVENTIS AQVAE.

O F

Copernicus' highly complex and mathematically rigorous book did not produce much of a general reaction in the decades immediately following publication. Copernicus had written in the preface, "Mathematics is for the mathematicians; and among them, if I am not mistaken, my labors will be seen to contribute something. . . ." The complexity and difficulty of his book meant that it would at first be read only by those who could understand it, the astronomers trained in mathematics. It would not be read by the general public, that is, not unless someone really took his heliocentric theory seriously and "translated" the difficult text into common everyday language. For the next fifty years, however, almost no one considered his version of the heliocentric theory worthy of much comment.

The world of the second half of the sixteenth century was not at all inclined to accept the Copernican theory. The writings of Aristotle, fully adopted by the universities and incorporated into the theology of the Catholic church by Saint Thomas Aquinas (1225–1274), had become a major part of the teachings of both. To change a belief embedded in two such fundamental institutions is not easy. (It must be recalled that for centuries the Church and the State were essentially one.)

Most astronomers realized the mathematical advantages of Copernicus' system over that of Ptolemy; new tables of planetary motions were prepared in 1551, only eight years after Copernicus' book was published. With the new mathematics and some new observations, these tables were more accurate than any available at the time. Hence, by adopting these tables, astronomers came to rely on Copernican mathematics, but nearly all of them rejected or were silent on the heliocentric concept. They simply ignored the physical consequences of the Copernican theory and used its mathematical conveniences.

Such arbitrariness was not new to science. Ptolemy himself had never pretended that his deferents and epicycles were physically real; he was simply answering Plato and trying to "save the appearances." If one aim of science is to predict, and if certain mathematical techniques permit more accurate predictions, the techniques are used and the physical consequences ignored, at least for a time. Sooner or later, however, the physical consequences usually come to play a major role. It is at this point that the geocentric theory ultimately ran into serious trouble.

In an attempt to adopt the mathematical simplification introduced by Copernicus and yet maintain the Earth at the center of the universe, Tycho Brahe (1546–1601), a Danish astronomer with exceptional skills of observation and with an established reputation, devised a compromise in 1583. Brahe insisted that all observations left only one conclusion: the Earth is at the center. But why could not the moon and sun revolve about the Earth with the *planets revolving about the sun* (Figure 3-1)? Because Brahe's system retained the Earth and therefore man as the center of the universe, it was generally accepted by the astronomers who objected to the Copernican idea. At its best, the Tychonic system was a compromise. Brahe, like Simplicius before him (see p. 23), tried to integrate old ideas with contemporary innovations.

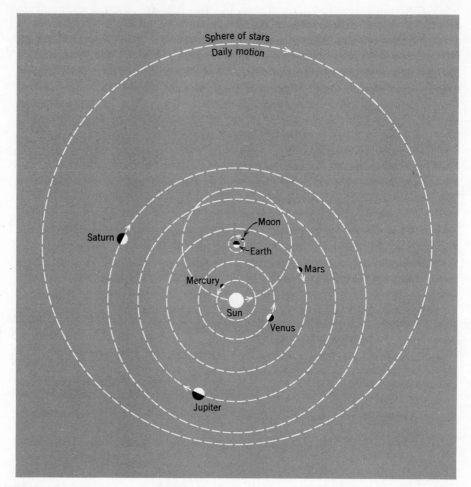

FIGURE 3-1
  The planetary system of Tycho Brahe.

PLANETARY MOTION

Among the men who ultimately became champions of Copernicanism was Galileo Galilei (1564–1642). Born in the city of Pisa, Galileo (Italians customarily refer to their famous and beloved men in history by the familiar first name) was raised in the strict Scholastic tradition, the tradition that had been passed on to the Europeans by the Arabs but which the Europeans modified to suit their own needs. This tradition became the complete authority.*

---

* A man of this tradition is called a Scholasticist or Schoolman; a man who in particular follows Aristotle is referred to as a Peripatetic because Aristotle used to walk with his students under the olive trees of his Lyceum.

FIGURE 3-2
  Galileo Galilei (1564–1642).

Telescopic Observations

Even in his student days at the University of Pisa, Galileo became known as one who rebelled against the Aristotelian authority. In 1597 he wrote to Johannes Kepler (the man who proposed the planetary theory that finally replaced the Copernican theory) that he, Galileo, had "become a convert to the opinions of Copernicus many years ago." Not until 1609, however, did he become an active and overt supporter of that heliocentric system, for in this year he first turned his telescope to the heavens and saw the celestial objects as no one before him had seen them.

He saw mountains on the moon and so knew that it was not the perfect sphere of Aristotle. He saw multitudes of previously unseen stars in the Milky Way. He saw dark spots on the sun which indicated to him that the sun rotated with a period of about 27 days and carried the spots with it. The sun was not a perfect sphere, nor were all its motions geocentric. He discovered the first four satellites of Jupiter, indicating that the Earth is not the center of all celestial motion as Aristotle had claimed. He saw that as Venus changed in brightness, it also changed phase like the moon (Figure 3-3).

To explain why Venus sometimes appeared in crescent phase, sometimes in half, and sometimes in full phase, Galileo quite logically resorted to a heliocentric system. Venus, like all planets, is seen only by reflected sunlight. Each phase indicates something of the relative positions of the sun, the

FIGURE 3-3

A composite of five photographs of Venus showing not only the changes of phase, but also the change in its apparent diameter. The planet is farthest from the Earth at full phase. (Lowell Observatory photograph.)

Earth, and Venus. When Venus is in full phase, we on the Earth face the same side of that planet as does the sun; therefore the sun must be somewhere between Venus and the Earth (Figure 3-4a). During crescent phase we see mostly the dark face of Venus, so it must be somewhere between the sun and the Earth (Figure 3-4b).

According to the Ptolemaic theory, the planet Venus would always be seen in crescent phase, for it would always be somewhere between the sun and the Earth, never beyond the sun (Figure 3-5). Observations of the phases of Venus, however, did not eliminate the Tychonic system as a possible explanation of planetary motion (see Figure 3–1).

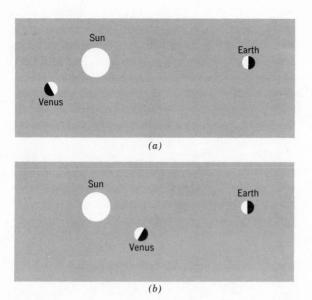

FIGURE 3-4

The relative positions of the sun, Venus, and the Earth when Venus is seen in *(a)* full phase; *(b)* crescent phase.

All these observations strongly indicated that neither Aristotle's nor Ptolemy's description of the astronomical world was adequate. To Galileo the theories of Copernicus were unquestionably correct, although none of his observations offered *proof* of the heliocentric theory and he knew it.

Ocean Tides

In his later years Galileo discovered evidence that he considered proof of the Earth's motion and indeed, when it is treated with present-day physics, it is proof. Galileo made a rather thorough analysis of the tides and concluded that they ebb and flow because the Earth rotates and revolves at the same time. Since the Earth rotates (on its axis) and revolves (about the sun) in the same direction (Figure 3-6), the portions of the Earth on the side opposite the sun (the nighttime side) are traveling faster with respect to the sun than the portions on the side facing the sun. Consequently, each portion of the Earth moves with a velocity that varies periodically every 24 hours. To Galileo the tides were comparable to water sloshing in a large basin that is being moved at a periodically varying velocity.

Galileo flatly rejected the moon as the cause of the ocean's tides, simply because the moon is so far away. He felt that the moon could not possibly influence the water in the oceans since there is no physical contact. The moon could not exert an occult force. Forty years after Galileo's death, Isaac Newton did indeed explain how the moon (and the sun) cause the tides. Today, a complete analysis of the tides is based on the works of both Newton and Galileo, since our study of them includes the "sloshing" effect

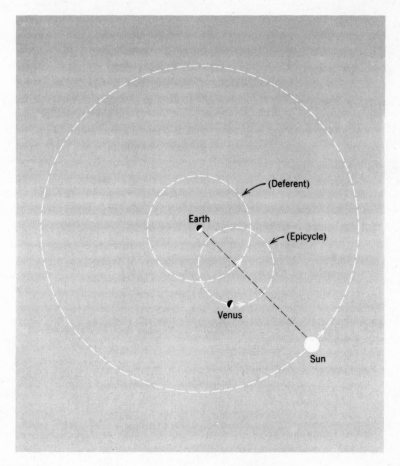

FIGURE 3-5
The motion of Venus as described by Ptolemy would not permit Venus to appear at full phase, for the planet would never appear on the far side of the sun.

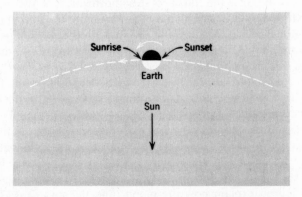

FIGURE 3-6
The waters of the oceans travel, relative to the sun, faster at midnight than at noon.

which Galileo hoped would prove that the Earth moves by both rotating and revolving.

The Dialogues

Although Galileo published a number of articles and letters, his main contribution to physics was contained in his two major books, *The Dialogue Concerning the Two Chief World Systems* (1632), and *The Dialogues Concerning the Two New Sciences* (1638). The first of these two books climaxed years of trouble for Galileo, and for writing it he was tried and convicted by the Inquisition. He was seventy years old at the time and was sentenced to house arrest for the remainder of his life!* During those nine years of house arrest, he wrote the second and more important work, *The Two New Sciences*.

Galileo was well aware of the dangers he courted in publishing ideas in defense of the Copernican theory, although he did have the permission of the Pope to publish the first book and the manuscript was read by the Catholic censors. Galileo wrote both of his books as dialogues. By choosing this form he could present both sides of a debate and thus be less likely to antagonize the authorities; at the same time he showed his great respect for Plato, who wrote in dialogue form. Socrates plays a leading role in Plato's dialogues.

Three participants assume roles in Galileo's dialogues. *Salviato* speaks as a student of the author and voices most of Galileo's own arguments. *Sagredo* is a learned nobleman, a free thinker whose mind is open to new ideas. He sometimes speaks for the practical Galileo, the experimenter. *Simplicio* is, as might be guessed, the Aristotelian. The name is double significant. The historical Simplicius (see p. 23) was indeed an Aristotelian of the sixth century A.D.; more important, however, is the attitude expressed by Galileo's Simplicio reflecting the point of view most of the Scholastics of Galileo's day.

Early in the *Dialogues* Galileo tried to establish what he considered to be Aristotle's true approach to knowledge, and in condemning the Peripatetics who followed blindly and without question, he painted a vivid picture of the Aristotelians with whom he fought so bitterly.

SALV . . . . But to give Simplicio more than satisfaction, and to reclaim him if possible from his error, I declare that we do have in our age new events and observations such that if Aristotle were now alive, I have no doubt he would change his opinion. This is easily inferred from his own manner of philosophizing, for when he writes of considering the heavens inalterable, etc., because no new thing is seen to be generated there or any old one dissolved, he seems implicitly to let us understand that if he had seen any such event he would have reversed his opinion, and properly preferred the sensible experience to natural reason . . . (1).

Does he not also declare that what sensible experience shows ought to be preferred

---

* Modern research substantiates Galileo's belief that his troubles with the Church were, in large part, a result of a personal grudge held against him by a number of men of influence in the Church. For a thorough analysis see *The Crime of Galileo* by Giorgio de Santillana, The University of Chicago Press (paperback) Chicago, 1955. See also *Galileo, Man of Science*, ed. Ernan McMullin, Basic Books, Inc., New York 1967.

over any argument* even one that seems to be extremely well founded? And does he not say this positively and without a bit of hesitation?

SIMP. . . . He does.

SALV. . . . Then of the two propositions, both of them Aristotelian doctrines, the second—which says it is necessary to prefer the senses over arguments—is a more solid and definite doctrine than the other, which holds the heavens to be inalterable. Therefore it is better Aristotelian philosophy to say, "Heaven is alterable because my senses tell me so," than to say, "Heaven is inalterable because Aristotle was so persuaded by reasoning." Add to this that we possess a better basis for reasoning about celestial things than Aristotle did. He admitted such perceptions to be very difficult for him by reason of the distance from his senses, and conceded that one whose senses could better represent them would be able to philosophize about them with more certainty. Now we, thanks to the telescope, have brought the heavens thirty or forty times closer to us than they were to Aristotle, so that we can discern many things in them that he could not see. . . .

SAGR. . . . I can put myself in Simplicio's place and see that he is deeply moved by the overwhelming force of these conclusive arguments. But seeing on the other hand the great authority that Aristotle has gained universally; considering the number of famous interpreters who have toiled to explain his meanings; and observing that the other sciences, so useful and necessary to mankind, base a large part of their value and reputation upon Aristotle's credit; Simplicio is confused and perplexed, and I seem to hear him say, "Who would there be to settle our controversies if Aristotle were to be deposed? What other author should we follow in the schools, the academies, the universities? What philosopher has written the whole of natural philosophy, so well arranged, without omitting a single conclusion? Ought we to desert that structure under which so many travelers have recuperated? Should we destroy that haven, that Prytaneum where so many scholars have taken refuge so comfortably; where, without exposing themselves to the inclemencies of the air, they can acquire a complete knowledge of the universe by merely turning over a few pages? Should that fort be leveled where one may abide in safety against all enemy assaults?" . . . (2).

SALV. . . . It is the followers of Aristotle who have crowned him with authority, not he who has usurped or appropriated it to himself. And since it is handier to conceal oneself under the cloak of another than to show one's face in open court, they dare not in their timidity get a single step away from him, and rather than put any alterations into the heavens of Aristotle, they want to deny out of hand those that they see in nature's heaven (3).

We learn immediately, in statements that probably did not endear him to some of his contemporaries, that Galileo scorned not Aristotle but the Peripatetics who followed him blindly. Indeed, Galileo paid homage to Aristotle's strong desire to square theory with sensible observation.

---

*Aristotle, *On the Generation of Animals*, Book 3, line 760^b/30. "The facts, however, have not been sufficiently grasped; if ever they are, then credit must be given rather to observation than to theories, and to theories only if what they affirm agrees with the observed facts."

FREELY FALLING BODIES

Galileo's great contribution to physics, however, lies not in his unsuccessful attempt to prove that the Earth moves, but in his analysis of motion and in particular that of falling bodies. Galileo, like every man before him and since, had to rely heavily on the past. He did not choose Aristotle's law of motion, however, but followed the Mertonians; his reliance on them is seen very clearly in his definitions of motion.

In dealing with steady or uniform motion, we need a single definition which I give as follows:

Definition. By steady or uniform motion, I mean one in which the distances traversed by the moving particle during any equal intervals of time, are themselves equal.

Caution. We must add to the old definition (which defined steady motion simply as one in which equal distances are traversed in equal times) the word "any," meaning by this, all equal intervals of time; for it may happen that the moving body will traverse equal distances during some equal intervals of time and yet the distances traversed during some small portion of these time-intervals may not be equal, even though the time-intervals be equal (4).

To make clear exactly how Galileo was extending the definition given by the Mertonians (see p. 26f) by the addition of the word "any" and to appreciate this contribution, we use the graphical method. Figure 3-7a is a velocity-time graph of a particle in perfectly uniform motion (assuming a perfectly straight line, of course). Figure 3-7b is a graph of nonuniform motion in which the average velocity is equal to that of Figure 3-7a, and thus the dis-

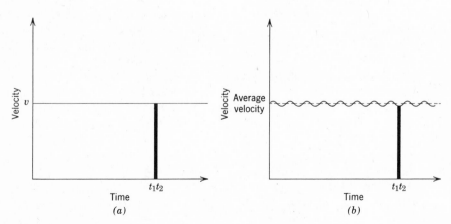

FIGURE 3-7

The object whose motion is represented in *(b)* will, after long time intervals, have traveled nearly the same distance as that in *(a)*. But for very short time intervals *(b)* may travel a distance different from that traveled by *(a)*.

tances covered (represented by the area under the curve) in a large time interval will be essentially the same for both particles. If very small time intervals are chosen, however, the shaded areas and thus the distances traversed may not be equal. Galileo, therefore, stressed the point that very small time intervals, those too small for him to measure, are important.

If Galileo relied on the past, he also made alterations in the traditional approach. It may be recalled that Aristotle found it necessary to slow the motion of falling bodies so that he could study them better. To do this, he increased the resistance of the medium through which the bodies fell. He found that in water or oil heavy objects fall faster than light objects. Galileo, however, was more impressed with Plato's concept of the Ideal than with Aristotle's concept of perfection. What would be Plato's Ideal of a falling object? Galileo chose to define the Ideal fall as that fall in which the object would be free of the resistance of the air. Today we call this *free fall* and recognize that satellites and space probes do indeed fall in this Ideal way. But vacuum pumps had not been invented by Galileo's time, so how did he study the motion of freely falling objects?

Galileo was the first to realize that the resistance of air to the passage of a body falling through it depends on the velocity of the body. Consequently, reasoned Galileo, if the velocity of an object was reduced, the resistance would also be reduced. Perhaps the velocity of a falling body could be reduced enough to eliminate for all practical purposes the effect of air resistance.

Falling Bodies and the Pendulum

In order to reduce the speed of falling objects and at the same time to compare the motions of objects of different weights, Galileo resorted to a technique that stemmed from a discovery about pendulum motion made during his first year as a student at the University of Pisa. In that year, 1581, when he was but seventeen years old, Galileo observed that the *period* (time for one complete back and forth oscillation) with which a pendulum swings depends not on the weight of the pendulum bob but on the length of the pendulum string. If the length of the string is increased by a factor of four, the period is only doubled; if the length is increased by a factor of nine, the period is but tripled. That is, the period is proportional to the square root of the length of the pendulum. Today we write this relation as

$$T \propto \sqrt{L}$$

Hence, if pendulums of the same length are set in motion, they oscillate with the same period, even though they have bobs of different weight. Furthermore, if the strings are long and if they are pulled aside only slightly (i.e., displaced slightly from the vertical position), their speed is very slow and the resistance of air negligible.

SALV. . . . The experiment made to ascertain whether two bodies, differing greatly in weight will fall from a given height with the same speed offers some difficulty; because,

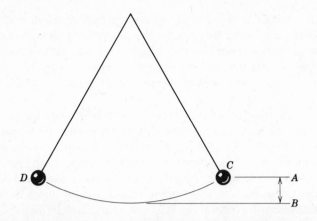

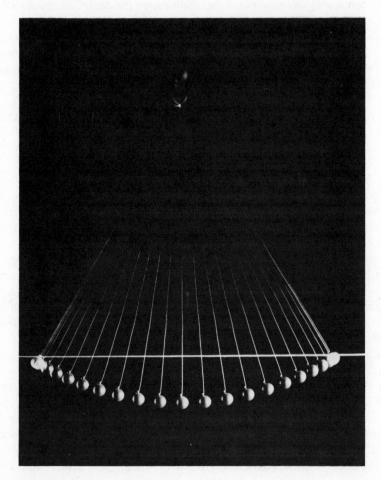

FIGURE 3-8
The swinging pendulum bob actually makes repeated falls through the distance *AB*. (Fundamental photographs.)

if the height is considerable, the retarding effect of the medium, which must be penetrated and thrust aside by the falling body, will be greater in the case of the small momentum of the very light body than in the case of the great force* of the heavy body; so that, in a long distance, the light body will be left behind; if the height be small, one may well doubt whether there is any difference; and if there be a difference it will be inappreciable.

It occurred to me therefore to repeat many times the fall through a small height in such a way that I might accumulate all those small intervals of time that elapse between the arrival of the heavy and light bodies respectively at their common terminus, so that this sum makes an interval of time which is not only observable, but easily observable. . . . Accordingly I took two balls, one of lead and one of cork, the former more than a hundred times heavier than the latter, and suspended them by means of two equal fine threads, each four or five cubits† long. Pulling each ball aside from the perpendicular, I let them go at the same instant, and they, falling along the circumferences of circles having these equal strings for semidiameters, passed beyond the perpendicular and returned along the same path. This free vibration repeated a hundred times showed clearly that the heavy body maintains so nearly the period of the light body that neither in a hundred swings or even in a thousand will the former anticipate the latter by as much as a single moment so perfectly do they keep step (5).

Each pendulum falls repeatedly and successively through a distance AB (Figure 3-8). If there were the slightest difference in the rate of fall of the two bodies, that is, if the heavier body did indeed fall faster than the lighter one, the small difference in time would not be apparent after one or even several swings of the pendulums; but after a hundred swings this supposed difference in the time of fall would accumulate so that the pendulums would obviously no longer swing together. As Galileo observed, the pendulums always maintained the same period (as long as the strings were of equal length).

We might ask whether Galileo was justified in comparing the falling pendulum bob to an object falling free of any constraints. The answer is yes. In fact, it was Galileo himself who first proved that the velocity attained by the falling bob of a pendulum is the same as the velocity attained by a body falling freely through the same vertical height; that is, the velocity attained by each depends not on the path but on the vertical height only.

If a pendulum is released from a given height, say AB (Figure 3-9), it will swing down to C where the string is vertical and then swing up to D, at very nearly the same height as A. It does not quite make the same height again, as Galileo pointed out, because of the resistance of the air and of the string. But he was searching for the Ideal motion, that motion free from resistance. Hence, the Ideal velocity acquired in falling from D to C is just enough to carry it back up to A again. If an obstacle is placed in the path of the string at M, the bob will swing along a different path to E and still achieve the same

---

* Note the use of the terms momentum and force to refer to essentially the same aspect of a moving body, namely, its ability to alter other bodies upon collision.

† Galileo's cubit is roughly $1\frac{1}{2}$ to 2 feet long.

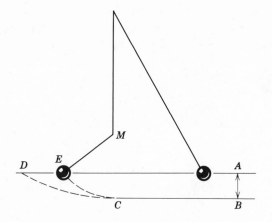

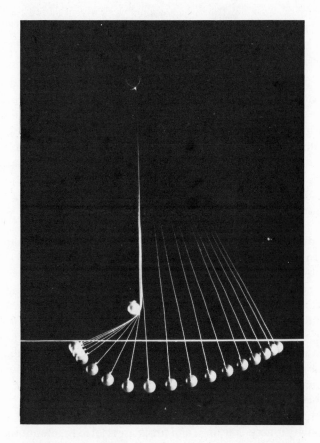

FIGURE 3-9

In falling from *D* to *C* the pendulum bob acquires the same velocity as it does in falling from *E* to *C*. (Fundamental photographs.)

height. If the bob is released from *E*, it will attain enough velocity at *C* to swing all the way up to point *A* again. Consequently, the velocity attained in falling from *D* to *C* equals the velocity attained in falling from *E* to *C*; the path through which the body falls does not alter the velocity; the velocity depends only on the height through which the body has fallen.

By means of a simple pendulum, then, Galileo discovered that objects of different weight fall with the same acceleration. Nowhere in his writings or in any of the records of his day is there any suggestion that he took weights to the top of the leaning tower of Pisa to display before a large crowd the truth of his belief. In fact, not only did Galileo probably know of John Philoponus' sixth-century experiment, but Simon Stevin (1548–1620), a Flemish scientist, reported the following experiment in 1586 (when Galileo was twenty-two years old):

The experiment against Aristotle is this. . . . Let us take . . . two leaden balls, one ten times greater in weight than the other, which allow to fall together from the height of thirty feet upon a board or something from which a sound is clearly given out, and it shall appear that the lightest does not take ten times longer to fall than the heavier, but that they fall so equally upon the board that both noises appear as a single sensation of sound (6).

Galileo had no need to redo this experiment in the same fashion that it had been done before; he improved the technique by using a pendulum. Indeed, Galileo was searching for the Ideal motion; this he could approximate with the short swing of his long pendulums. He knew he could not neglect the effect of the air in dropping a lead and a cork ball from the top of the leaning tower of Pisa. A fall from a height of 180 feet would certainly not be Ideal.

Distance, Time, Velocity, and Acceleration

We have not yet discussed the particular type of acceleration that falling objects assume. In Galileo's time, some philosophers thought that the velocity was proportional to the *time* the object had fallen, that is, $v \propto t$. If, for example, an object acquired a certain velocity in falling for two seconds, it would acquire a velocity twice as great in falling for four seconds, three times as great in falling for six seconds, and so on. Others claimed that the velocity was proportional to the *distance* through which the object had fallen, that is, $v \propto s$. If it acquired, for example, a certain velocity in falling two feet, it would acquire twice the velocity after falling four feet, three times the velocity after falling six feet, and so on.

To establish whether a body falling freely from rest acquires a velocity in proportion to the time it falls or the distance fallen, Galileo first considered how distance *s*, velocity *v*, acceleration *a*, and time *t* are related to one another by studying the distance-time and velocity-time graphs developed in the fourteenth century. Before discussing this aspect of Galileo's work, we shall review these graphs and extend the ideas of motion by employing algebra.

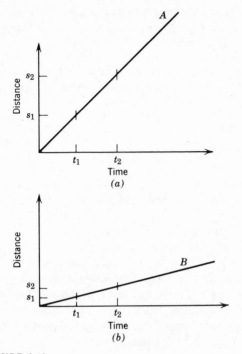

FIGURE 3-10

Distance-time graphs for two objects with different velocities.

Galileo did not know algebra, which had been invented by the Arabs but did not come into general use in Europe until the late seventeenth and early eighteenth centuries. Until that time the Europeans used the geometry of the Greeks. The algebraic expressions are indeed simpler, however, and the relationships easier to see than those of geometry.

We recall that the distance-time graph of an object moving with uniform motion (constant velocity) is a straight line (see Figure 3-10). The position of each object at time t_1 is s_1; at time t_2 both are at s_2. The time intervals selected in both Figures 3-10a and 3-10b are represented by $t_2 - t_1$ and are equal. The distances traversed, $s_2 - s_1$, are not equal, however. Object A, whose motion is represented in the graph in Figure 3-10a, will have moved farther during the time interval $t_2 - t_1$ than will object B whose motion is represented in Figure 3-10b. Clearly, object A is moving faster than object B, but each is moving with a constant velocity.

Since the velocity of each object is constant, the distance covered by each is proportional to the time:

$$s_2 - s_1 \propto t_2 - t_1$$

Any proportionality, either direct or inverse, can be expressed as an algebraic equation by the inclusion of a constant of proportionality. For example, the circumference C of a circle is directly proportional to the circle's diameter D:

$$C \propto D$$

If the diameter is doubled, so is the circumference; if the diameter is tripled, so is the circumference, and so on. This proportionality is changed into an equation by the inclusion of a constant coefficient π (pi):

$$C = \pi D$$

In the same manner, the proportion

$$s_2 - s_1 \propto t_2 - t_1$$

can be changed into an equation by the inclusion of a constant coefficient:

$$s_2 - s_1 = K(t_2 - t_1)$$

The meaning of the constant remains to be determined.

For any one given time interval, K tells us how much distance has been traversed. The coefficient K represents the slope of the straight line in Figure 3-10; the greater the slope, the larger K is. But in uniform motion, for any one given time interval the velocity v also tells us how much distance has been traversed. Thus, for motion with constant v, coefficient K is seen to be identical to the velocity of the object and the equation can be written

$$s_2 - s_1 = v(t_2 - t_1)$$

It is common practice to write the time interval not as $t_2 - t_1$ but as Δt, where Δ, the Greek letter delta, generally means *change in*, such as the change in time. The distance traversed can similarly be written as Δs. The equation now takes the form

$$\Delta s = v \, \Delta t$$

By dividing both sides of this equation by Δt, we obtain

$$\frac{\Delta s}{\Delta t} = v$$

or

$$v = \frac{\Delta s}{\Delta t}$$

This equation can define velocity even when the velocity is not constant. If the intervals of time are large, the velocity is the average velocity $\bar{v}$. For example, if on a trip you drive 200 miles in 5 hours, your average velocity will be 40 miles per hour. Chances are, however, that you will not proceed at a steady 40 miles per hour but drive faster for a certain time interval, more slowly for another, and you will probably stop for a brief rest somewhere along the way.

If the time intervals in the above equation are extremely small, so small that they approach being only an instant of time, the velocity defined by the equation is the instantaneous velocity. The velocity read on the speedometer of a car is nearly the instantaneous velocity.

The previous arguments developing the relationships of distance, time, and velocity were all based on the Mertonian definition of uniform motion. Presumably, then, these same arguments also work for uniformly accelerated motion. Galileo followed the example of the Mertonians and defined "a motion as uniformly . . . accelerated when, during any equal intervals of time whatever, equal increments of speed are given to it."

In the velocity-time graphs of Figure 3-11 are straight lines, each of which represents the motion of an object accelerating uniformly, that is, moving with constant acceleration. For a given interval of time, $\Delta t = t_2 - t_1$, the velocity of object A, whose motion is represented in Figure 3-11a, increases more rapidly than the velocity of object B (in Figure 3-11b). Since the velocity of each is proportional to the time, we may write

$$v_2 - v_1 \propto t_2 - t_1$$
$$v_2 - v_1 = K'(t_2 - t_1)$$

where the coefficient K' is different from the K used in the distance-time relationship. Again we must determine the physical meaning of the coefficient K' which represents the slope of the velocity-time graph. From an examination of the two curves, we see that each object acquires an equal increment

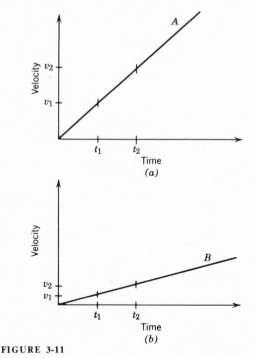

FIGURE 3-11

Velocity-time graphs for two objects with different accelerations.

of velocity in equal time intervals, which is our definition of uniform acceleration. Object A, however, acquires a greater increment of velocity than object B since the slope of the v-t line of A is greater than that of B. From these considerations we conclude that the coefficient K' is identical to the acceleration a, and we can write

$$v_2 - v_1 = a(t_2 - t_1)$$

This equation may also be written using the delta notation, where Δv represents an increment of velocity (in the sense that the Mertonians used it) or a change in velocity:

$$\Delta v = a \Delta t$$

Solving this equation for the acceleration a, we obtain a definition of acceleration

$$a = \frac{\Delta v}{\Delta t}$$

Acceleration is defined as the rate of change of velocity with respect to time.

Using this review of the distance-time and velocity-time graphs, and applying algebra, we can now more easily describe the logic that led Galileo to speculate about the acceleration of freely falling bodies. Let us assume a particular type of acceleration, for example $v \propto t$, derive a relationship for the distance and the time, and then rely on Galileo's experimental work to see whether our assumption is correct. If it is not correct, we can assume another type of acceleration, for example $v \propto s$, and derive another expression relating distance and time; we could then test that relationship against the observations of experiments. Such a process is often the way science progresses, and Galileo was the first to make it clear that mathematical statements should be supported by experimental results.

Let us assume that freely falling objects accelerate uniformly, so that their velocity is proportional to the time of fall, that is,

$$v_2 - v_1 = a \Delta t$$

We can represent the motion of such an object by the velocity-time graph shown in Figure 3-12. The distance traveled during the time interval $\Delta t = t_2 - t_1$ is represented in the velocity-time graph as the area under the curve. From the graph it is evident that we have assumed that the object started from rest, that is,

$$v_1 = 0$$

and therefore

$$v_2 = a \Delta t$$

The distance traveled during this time is represented by the shaded triangle ODE under the curve. This triangle is a right triangle with an area equal to one half the product of the base and the height, or

$$\text{area of triangle } ODE = \tfrac{1}{2} v_2 \Delta t$$

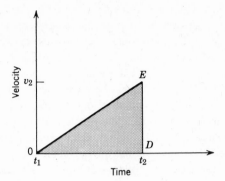

FIGURE 3-12

The area under the velocity-time curve represents the distance traveled.

Since the area of this triangle equals the distance traveled s during the interval of time Δt, and furthermore, since the velocity v_2 is equal to $a \, \Delta t$, we can substitute equals for equals and write

$$s = \tfrac{1}{2}(a \, \Delta t) \, \Delta t$$

or

$$s = \tfrac{1}{2} a \, (\Delta t)^2$$

This equation is the relationship between the distance traveled and the interval of time. Galileo derived the same relationship but used geometry rather than algebra. He suggested that objects do fall in such a way that their velocity is proportional to the time, and he proposed testing his suggestion by measuring the distance traveled by balls rolling down an inclined plane to see whether the distance traveled is proportional to the time squared, as in the above equation.

Falling Bodies and the Inclined Plane

To demonstrate that rolling balls down inclined planes is a justifiable way of studying the acceleration of freely falling objects, Galileo asked his readers to compare the motion of a ball rolling down the incline with a pendulum bob on the end of a string. If the string is long and the bob is pulled aside only slightly, the bob moves in a path that is not appreciably different from that followed by a ball rolling down an inclined plane. Anyway, he had demonstrated that the shape of the path is not important; the speed of the bob, or of the ball, or of a freely falling object depends only on the vertical height through which each falls.

Galileo did not recognize that since the ball rolls and the pendulum bob does not, there is a slight difference. In his experiment, however, this difference is smaller than the inaccuracies of his measurements. He reported this experiment in some detail.

A piece of wooden moulding or scantling, about 12 cubits long, half a cubit wide, and three finger-breadths thick, was taken; on its edge was cut a channel a little more than one finger in breadth; having made this groove very straight, smooth, and polished, and having lined with with parchment, also as smooth and polished as possible, we rolled along it a hard, smooth, and very round bronze ball. Having placed this board in a sloping position, by lifting one end some one or two cubits above the other, we rolled the ball . . . along the channel, noting, in a manner presently to be described, the time required to make the descent. We repeated this experiment more than once in order to measure the time with an accuracy such that the deviation between two observations never exceeded one-tenth of a pulse-beat. Having performed this operation and having assured ourselves of its reliability, we now rolled the ball only one-quarter the length of the channel; and having measured the time of its descent, we found it precisely one-half of the former. Next we tried other distances, comparing the time for the whole length with that for the half, or with that for two-thirds, or three-fourths, or indeed for any fraction; in such experiments, repeated a full hundred times, we always found that the spaces traversed were to each other as the squares of the times, and this was true for all inclinations of the plane, i.e., of the channel, along which we rolled the ball. We also observed that the times of descent, for various inclinations of the plane, bore to one another precisely that ratio which . . . the Author [Galileo] had predicted and demonstrated for them.

For the measurement of time, we employed a large vessel of water placed in an elevated position; to the bottom of this vessel was soldered a pipe of small diameter giving a thin jet of water, which we collected in a small glass during the time of each descent, whether for the whole length of the channel or for a part of its length; the water thus collected was weighed, after each descent, on a very accurate balance; the differences and ratios of these weights gave us the differences and ratios of the times and this with such accuracy that although the operation was repeated many, many times, there was no appreciable discrepancy in the results (7).

By rolling a ball down an inclined plane, Galileo was therefore able to determine that the distance traversed by a falling body was proportional to the square of the time it had fallen:

$$s \propto (\Delta t)^2$$

In demonstrating that this relationship holds true for balls rolling down planes of any inclination, Galileo proved that the velocity of a falling object increases in direct proportion to the time the object has fallen, that is,

$$v \propto \Delta t$$

Some Examples of Motion Problems

Galileo showed that for freely falling bodies the rates of fall are identical, that is, the acceleration is constant. This particular acceleration, the acceleration produced by gravity, is therefore given a special symbol g. For freely falling bodies

$$\Delta v = g \, \Delta t$$

The value of the acceleration produced by gravity has been determined experimentally to be equal to nearly 32 feet per second per second (or 32 feet per second²) the increment of velocity added each second is 32 feet per second. If instead of falling from rest the object falls with some initial velocity v_0 (rather than v_1), the velocity v after any interval of time Δt is given by the equation

$$v - v_0 = g\,\Delta t$$

or

$$v = v_0 + g\,\Delta t$$

To demonstrate these algebraic relationships, it will help to work a few examples. In these examples we use the foot as a unit of distance; velocity will then be measured in feet per second and acceleration in feet per second squared. Suppose that a stone is dropped from rest ($v_0 = 0$) from the Leaning Tower of Pisa and you are there to time its fall. You might find that the stone takes 3.4 seconds to fall. How fast is the stone traveling when it hits the ground? (Do not forget that the acceleration produced by gravity is 32 feet per second².)

$$v = v_0 + g\,\Delta t$$
$$v = 0 + (32\ \text{ft/sec}^2)(3.4\ \text{sec})$$
$$v = 109\ \text{ft/sec}$$

To find how far the stone will fall during this time, we need the relationship (see problem 3-9)

$$s = v_0\,\Delta t + \tfrac{1}{2}a\,(\Delta t)^2$$

This equation is more general than equation $s = \tfrac{1}{2}a(\Delta t)^2$ because it permits the calculation of the distance traveled by an object which has an initial velocity v_0. But the stone in our example fell from rest, $v_0 = 0$, and was released from a height s. Thus,

$$s = v_0\,\Delta t + \tfrac{1}{2}a(\Delta t)^2$$
$$s = 0 + \tfrac{1}{2}(32\ \text{ft/sec}^2)(3.4\ \text{sec})^2$$
$$s = (16\ \text{ft/sec}^2)(11.5\ \text{sec}^2)$$
$$s = 185\ \text{ft}$$

The tower is 180 feet tall, so the person who dropped the stone must have held it 5 feet above the top of the tower.

What was the average velocity $\bar{v}$ of the stone as it fell?

$$\bar{v} = \frac{s}{\Delta t}$$

$$\bar{v} = \frac{185\ \text{ft}}{3.4\ \text{sec}}$$

$$\bar{v} = 54\ \text{ft/sec}$$

The average velocity, it will be noted, is very nearly

$$\tfrac{1}{2}(v_1 + v_2) = \tfrac{1}{2}(0 + 109)$$

Another mathematical expression relates the final velocity to the initial velocity, the acceleration, and the distance traveled:

$$v^2 = v_0^2 + 2as$$

This equation is useful when the final velocity of an accelerated object is sought and the time of travel is not known. For example, what is the velocity of an object dropped from rest from a height of 100 feet?

$$v^2 = v_0^2 + 2as$$
$$v^2 = 0 + 2(32 \text{ ft/sec}^2)(100 \text{ ft})$$
$$v^2 = 6400 \text{ ft}^2/\text{sec}^2$$
$$v = 80 \text{ ft/sec}$$

It should be noted that 88 feet per second is the same as 60 miles per hour.

These problems are done much more easily with algebra than with the geometry Galileo used; and, of course, Galileo did not have a stopwatch to time the fall. To study the motion of a falling object, Galileo slowed down that motion in order to reduce the resistance; Aristotle had increased the resistance in order to slow the motion down. This single but extremely significant change in approach permitted Galileo, using equipment that could have been available to Aristotle, to describe the motion of freely falling bodies. So it is not necessarily the equipment at the scientist's disposal, but the approach he takes to the problem, that leads to the best solution.

PROJECTILE MOTION

Galileo also made significant gains in studying the age-old problem of projectile motion. Through all these centuries the question—What keeps a projectile in "violent motion?" had remained unanswered. Although Galileo did not solve the problem, he placed it in the proper perspective so that it could be solved by others.

What Is Going To Stop It?

It will be recalled that, according to Aristotle's law of motion, to keep any object moving, for example, a ball rolling on a horizontal surface, a force is required. According to this law, a stone dropped from the top of the mast of a moving ship would cease its forward motion the instant it was released, for there would be no force propelling it forward. It would therefore fall on the deck to the rear of the foot of the mast if the ship were moving forward. The distance from the point of impact to the foot of the mast would equal the distance the ship had moved forward while the stone was falling.

A similar argument was used by the Peripatetics to prove that the Earth did not move, for a stone dropped from the top of a tower hits the ground at the foot of the tower; if indeed the Earth were moving, Aristotelian physics predicted that the stone would land a short distance away from the base, a distance dependent on the speed of the Earth.

To refute this argument, Galileo resorted not to experiment but to armchair logic; he presented a "thought experiment," a technique he often used.

SALV. . . . Now tell me: suppose you have a plane surface as smooth as a mirror and made of some hard material like steel. This is not parallel to the horizon, but somewhat inclined, and upon it you have placed a ball which is perfectly spherical and of some hard and heavy material like bronze. What do you believe this will do when released? . . .

SIMP. . . . I am sure that it would spontaneously roll down. . . .

SALV. . . . Now how long would the ball continue to roll, and how fast? Remember that I said a perfectly round ball and a highly polished surface, in order to remove all external and accidental impediments. Similarly I want you to take away any impediment of the air caused by its resistance to separation, and all other accidental obstacles, if there are any.

SIMP. I completely understood you, and to your question I reply that the ball would continue to move indefinitely, as far as the slope of the surface extended, and with a continually accelerated motion. For such is the nature of heavy bodies . . . and the greater the slope, the greater would be the velocity.

SALV. But if one wanted the ball to move upward on this same surface, do you think it would go?

SIMP. Not spontaneously, no; but drawn or thrown forcibly, it would.

SALV. And if it were thrust along with some impetus impressed forcibly upon it, what would its motion be, and how great?

SIMP. The motion would constantly slow down and be retarded, being contrary to nature, and would be of longer or shorter duration according to the greater or lesser impulse and the lesser or greater slope upward.

SALV. Very well. . . . Now tell me what would happen to the same movable body placed upon a surface with no slope upward or downward.

SIMP. Here I must think a moment about my reply. There being no downward slope, there can be no natural tendency toward motion; and there being no upward slope, there can be no resistance to being moved, so there would be an indifference between propensity and resistance to motion. Therefore it seems to me that it ought naturally to remain stable. . . .

SALV. I believe it would do so if one set the ball down firmly. But what would happen if it were given an impetus in any direction?

SIMP. It must follow that it would move in that direction.

SALV. But with what sort of movement? One continually accelerated, as on the downward plane, or increasingly retarded as on the upward one?

SIMP. *I cannot see any cause for acceleration or deceleration, there being no slope upward or downward.**

---

* The italics are mine.

SALV. Exactly so. But if there is no cause for the ball's retardation, there ought to be still less for its coming to rest; so how far would have the ball continue to move?

SIMP. As far as the extension of the surface continued without rising or falling.

SALV. Then if such a space were unbounded, the motion on it would likewise be boundless? That is, perpetual?

SIMP. It seems so to me. . . . (8)

Galileo had Simplicio say, "I cannot see any cause for acceleration or deceleration. . . .," and by this statement Galileo placed the projectile problem into its proper perspective; Galileo did not ask, "What keeps the projectile going?" but rather, *"What is going to stop it?"* Again, by taking an entirely new approach to an old problem, Galileo came upon a solution that is far more satisfactory. He claimed that a body will continue to move with uniform velocity as long as that body is not acted on by any external agent which would cause it to accelerate or decelerate, and as long as no resistances or accidental impediments act on it. In effect, he postulated that an external agent is required to alter the motion of a body.

Galileo then asked how long a ball would continue to roll on a smooth surface. He concluded that in Ideal motion it would continue to roll with a uniform velocity as long as that surface goes neither up an inclined plane away from the center of the Earth nor down an inclined plane toward the Earth's center.

SALV. Then in order for a surface to be neither downward nor upward, all its parts must be equally distant from the center. Are there any such surfaces in the world?

SIMP. Plenty of them; such would be the surface of our terrestrial globe if it were smooth, and not rough and mountainous as it is. But there is that of the water, when it is placid and tranquil.

SALV. Then a ship, when it moves over a calm sea, is one of these movables which courses over a surface that is tilted neither up nor down, and if all external and accidental obstacles were removed, it would thus be disposed to move incessantly and uniformly from an impulse once received?

SIMP. It seems that it ought to be (9).

Faithful to Plato's decree that all celestial motion must be circular and of constant rate, and indeed applying that decree to motion on the Earth, Galileo predicted that a ball that meets neither resistances nor obstacles will continue to move with uniform motion only if it is allowed to proceed in a circle about the center of the Earth. Galileo could not free his imagination from the confines of the Earth; his impetus or impressed force or momentum —whatever one chooses to call it—is not only circular in nature but also geocentric.

The Moving Ship and the Falling Stone

Returning to the problem of the stone dropped from the top of a mast, Salviati continued his discussion with Simplicio.

SALV. Now as to that stone which is on top of the mast; does it not move, carried by the ship, both of them going along the circumference of the circle about its center? And consequently is there not in it an ineradicable motion, all external impediments being removed? And is not this motion as fast as that of the ship?

SIMP. All this is true, but what next?

SALV. Go on and draw the final consequence by yourself, if by yourself you have known all the premises.

SIMP. By the final conclusion you mean that the stone, moving with an indelibly impressed motion, is not going to leave the ship, but will follow it, and finally will fall at the same place where it fell when the ship remained motionless. And I, too, say that this would follow if there were no external impediments to disturb the motion of the stone after it was set free. But there are two such impediments; one is the inability of the movable body to split the air with its own impetus alone, once it has lost the force from the oars which it shared as part of the ship while it was on the mast; the other is the new motion of falling downward, which must impede its other, forward, motion.

SALV. As for the impediment of the air, I do not deny that to you. . . . As for the other, the supervening motion downward, in the first place it is obvious that these two motions (I mean the circular around the center and the straight motion toward the center) are not contraries, nor are they destructive of one another, nor incompatible. As to the moving body, it has no resistance whatever to such a motion. . . . Hence the cause of motion is not a single one which must be weakened by the new action, but there exist two distinct causes. Of these, heaviness attends only to the drawing of the movable body toward the center, and impressed force only to its being led around the center, so no occasion remains for any impediment (9).

Galileo again broke with tradition and introduced the extremely important concept that the projectile has two motions which it accomplishes simultaneously: the downward motion caused by gravity and the horizontal motion resulting from the original projection. These two motions proceed independently of one another. This concept is essential to any detailed study of motion, and its introduction signified a great step forward (Figure 3-13).

The stone initially held at the top of the mast of a ship moving with uniform motion has the same uniform motion of the ship. After release, the stone acquires a second motion; it retains the initial forward motion and in addition assumes a downward motion toward the deck. Although the stone has two motions during its fall, the ship takes part in only one of these motions. Both the ship and the stone continue moving forward with the same (horizontal) velocity. The sailor on top of the mast, moving forward with the ship, sees the stone fall "straight down" and strike the deck at the foot of the mast. A man ashore watching the same stone falling does not take part in the horizontal motion of the ship; he sees the stone move forward at a constant velocity while it falls with accelerated motion. Galileo went on to prove that the man ashore sees the stone fall along a curved path.

The two men see different things. One claims that the stone falls along a straight line, the other that it falls along a curved path. Who is right or can

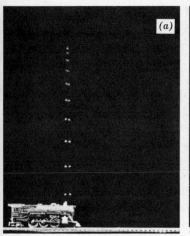
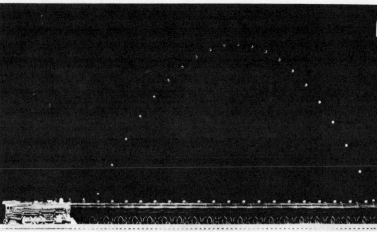

FIGURE 3-13

(a) A ball projected vertically upward from the smokestack of a train at rest falls vertically back into the smokestack. *(b)* A ball projected vertically from a moving train maintains the horizontal velocity it had at the moment of release. It again falls into the smokestack (or nearly so) which has moved horizontally during the flight. (From *The Birth of a New Physics* by I. Bernard Cohen. Copyright 1960 by Educational Services, Inc. Science Study Series. Reprinted with permission of Doubleday & Company, Inc.)

they both be? The answer is that both claims are correct, for each man views the falling stone from a different vantage point. One moves forward with the stone and sees it fall straight down; the other remains at rest on the shore and sees the stone move forward as it falls.

To return to the arguments of Salviati, a discussion ensued to the effect that the Peripatetics (in spite of Jean Buridan) detested the impressed force and claimed that the air carried the stone. Salviati entertained a counterclaim and briefly returned to his pendulum experiments.

SALV. . . . There must be something conserved in the stone. . . . If two strings of equal length were suspended from that rafter, with a lead ball attached to the end of one and a cotton ball to the other, and then if both were drawn an equal distance from the perpendicular and set free . . . which . . . do you believe would continue to move the longer before stopping vertically?

SIMP. The lead ball would go back and forth a great many times; the cotton ball, two or three at the most.

SALV. So that whatever the cause of that impetus and mobility, it is conserved longer in the heavy material than in the light (10).

Galileo here admitted that he did not know the cause of this impetus or impressed force, but he was quite clear that it is conserved.

Time Required for Projectile Fall

The dialogue continued with a discussion showing how the ship-stone example may be applied to the problem of a stone falling from a tower. Galileo thus demonstrated that the "vertical" descent of the stone from the tower does not prove the immobility of the Earth. Then Sagredo thought of an additional question about the stone falling from the ship's mast.

SAGR. Excuse me, Salviati, but before going on to the others let me bring up a certain difficulty that has been going round in my head while you were so patiently going into such detail with Simplicio on this ship experiment.

SALV. What we are here for is to discuss things, and it is good for everyone to raise his objections as they occur to him, for this is the road to knowledge. So speak up.

SAGR. If it is true that the impetus of the ship's motion remains indelibly impressed on the stone after it has separated from the mast, and that furthermore this motion occasions no hindrance or slowing in the straight-downward motion which is natural to the stone, then an effect of remarkable nature must take place.

Let the ship be motionless and the fall of the stone from the mast take two pulse beats. Then cause the ship to move, and drop the same stone from the same place; from what has been said, it will still take two pulse beats to arrive at the deck. In these two pulse beats the ship will have gone, say, twenty yards, so that the actual motion of the stone will have been a diagonal line [actually a parabola as Galileo himself proved and discussed in the *Two New Sciences*] much longer than the first straight and perpendicular one, which was merely the length of the mast; nevertheless, it will have traversed this distance in the same time. Now, assuming the ship to be speeded up still more, so that the stone in falling must follow a diagonal line [parabolic path] very much longer still than the other, eventually the velocity of the ship may be increased by any amount, while the falling rock will describe always longer and longer diagonals, and still pass over them in the same two pulse beats. Similarly, if a perfectly level cannon on a tower were fired parallel to the horizon, it would not matter whether a small charge or a great one was put in, so that the ball would fall a thousand yards away, or four thousand, or six thousand, or ten thousand, or more; all these shots would require equal times, and each time would be equal to that which the ball would have taken in going from the mouth of the cannon to the ground if it were allowed to fall straight down without any other impulse. . . .

SALV. This reflection is very beautiful by reason of its novelty, and if the effect is true it is most remarkable. And I have no doubt as to its correctness (11).

Nor do we have any doubt about its correctness today, as long as the cannon ball does not travel so great a distance that the curvature of the Earth must be considered. The horizontal speed of the cannon ball depends on the size of the charge that impels it, but the vertical speed depends *only* on the acceleration produced by gravity; as long as two cannon balls are fired from the same vertical height above the surface of the Earth, each falls

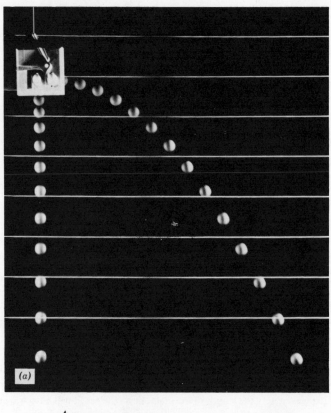

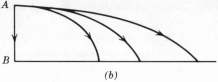

FIGURE 3-14

(a) A ball projected horizontally will accelerate downward at the same rate as one that is dropped and falls vertically. (From *The Birth of a New Physics* by I. Bernard Cohen. Copyright 1960 by Educational Services, Inc. Science Study Series. Reprinted by permission of Doubleday & Company, Inc.) *(b)* The greater the horizontal velocity, the farther a projectile travels before striking the ground.

in the same amount of time. The *horizontal* velocity remains constant since nothing can change it (neglecting air resistance, of course). The *vertical* velocity downward increases at the rate of the acceleration produced by gravity. The two motions are independent of each other and proceed simultaneously; the resulting path, called a parabola, is shown in Figure 3-14.

The Falling Leaf

We must remember, however, that the actual motion of falling objects, including that of projectiles, is affected by air resistance. Galileo realized this fact and furthermore, as mentioned earlier, he knew that resistance increases with the speed of the object, for as the body falls through the medium, it

is by nature continuously accelerated so that it meets with more and more resistance in the medium and hence a dimunition in its rate of gain of speed until finally the speed reaches such a point and the resistance of the medium becomes so great that, balancing each other, they prevent any further acceleration and reduce the motion of the body to one which is uniform and which will thereafter maintain a constant value (12).

This example clearly shows that Galileo understood motion better than anyone before him. He said that there are two agents or causes or forces acting on a falling body. One is the natural tendency for it to fall toward the center of the Earth ("by nature continuously accelerated"); the other is the resistance of the air through which it falls. This resistance increases as the speed of the falling object increases. If an object falls from rest, the resistance of the air is initially zero (Figure 3-15). As its speed increases, the air resistance increases. Finally, the falling body reaches a speed at which the resist-

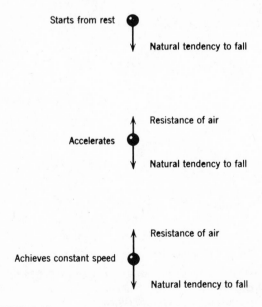

FIGURE 3-15

An object starting from rest will encounter no resistance from the air until it starts to fall. As it falls faster and faster that resistance increases.

ance of the air equals the natural tendency for the object to accelerate downward. When this speed is reached, the object ceases accelerating and maintains that speed; it then falls with uniform motion. Galileo must have watched leaves fall in autumn.

In this example of a body falling through an air medium, the motion continues along a straight line, whereas in the example of the ball rolling along a surface everywhere equidistant from the center of the Earth, the motion is circular. Galileo seems never to have brought these two examples together, although each expresses his feeling that, in contradistinction to Aristotle's law of motion, moving objects upon which no net forces are exerted continue at a constant speed.

References

1. Galileo, *Dialogue Concerning the Two Chief World Systems,* University of California Press, Berkeley (paperback), 1962, pp. 50 ff.
2. *Ibid.,* pp. 55 f.
3. *Ibid.,* p. 111.
4. Galileo, *Dialogues Concerning Two New Sciences,* McGraw-Hill Book Company, New York (paperback), 1963, p. 148.
5. *Ibid.,* p. 80 f.
6. Simon Stevin, as quoted in *Main Currents of Scientific Thought,* by Stephen Mason, Abelard-Schuman, New York, 1956, p. 119.
7. Galileo, *Two New Sciences, op. cit.,* pp. 171 f.
8. Galileo, *Two Chief World Systems, op. cit.,* pp. 145 ff.
9. *Ibid.,* pp. 148 f.
10. *Ibid.,* pp. 151 f.
11. *Ibid.,* pp. 154 f.
12. Galileo, *Two New Sciences, op. cit.,* p. 71.

Questions

1. Define (*a*) uniform motion and (*b*) uniformly accelerated motion.

2. Compare the world systems of Ptolemy, Copernicus, and Tycho Brahe.

3. If Galileo's observations of the phases of Venus cannot be explained by the Ptolemaic theory, are those observations necessarily a *proof* of the Copernican theory? Could those observations be explained by the Tychonic system?

4. How justified are scientists in using mathematics to describe an event and to predict the outcome of a particular experiment, even if they cannot relate their mathematics to a physical model?

5. Read pp. 9–14 of the first day of Galileo's *Dialogue Concerning the Two Chief World Systems,* University of California Press paperback, and compare the arguments of Simplicio with those of Salviati.

6. Read from p. 144, line 4, to the top of p. 158 in the *Dialogue Concerning the Two Chief World Systems,* University of California Press paperback, and then answer the following questions.

(*a*) Was Galileo, who admittedly never dropped a stone from the top of the mast of a moving ship, justified in using this "thought experiment" to prove his point?

(*b*) Did Galileo successfully refute Aristotle's arguments on projectile motion? Why?

7. Read the arguments advanced by Francesco Sizi against Galileo's telescopic observations, as presented on p. 13 of *The Crime of Galileo,* by Giorgio de Santillana, University of Chicago paperback, and comment on the justification of Galileo's attitude toward the Peripatetics of his day.

8. To extend your understanding of these dialogues, read the foreword by Albert Einstein and the translator Stillman Drake's preface to *Dialogue Concerning the Two Chief World Systems,* University of California Press paperback.

Problems

1. How far will an object travel if it moves with an average velocity of 40 mph for 3 hours?

2. What will be the average velocity of an airplane if it travels 150 miles in 30 minutes? Express the answer in miles per hour.

3. An object travels from rest for 6 sec with an acceleration of 8 ft/sec². What is its final velocity?

4. Find the acceleration of an object for each of the following situations, assuming the acceleration to be constant for each.

(*a*) Its velocity changes from 15 ft/sec to 25 ft/sec in 4 sec.

(*b*) Its velocity changes from 36 ft/sec to 14 ft/sec in 3 sec.

5. An object traveling with a velocity of 24 ft/sec accelerates at a rate of 5 ft/sec² for 10 sec. What is the final velocity?

6. An object falls from rest for 4 sec. How far will it fall?

7. How long will it take an object to fall from rest through a distance of 100 ft?

8. (*a*) Of two pendulums, the first with a string 12 inches long and the second with a string 48 inches long, which has the longer period of oscillation?

(*b*) If the period of oscillation of the first pendulum is 1.1 sec, what is the period of oscillation of the second one?

9. Using the velocity-time graph, geometry, and algebra, demonstrate that if the initial velocity v_0, is not zero, then

$$s = v_0 \, \Delta t + \tfrac{1}{2} a \, (\Delta t)^2$$

10. From the following measurements made of a ball rolling from rest down an inclined plane, demonstrate that $v \propto t$.

s, ft	t, sec
0.5	0.11
1.0	0.50
1.5	0.61
2.0	0.71
2.5	0.79
3.0	0.86
3.5	0.94
4.0	1.00

Chapter Four

Newton and the Laws of Motion

A DUTCH PLEASURE WAGON.

G alileo's work was the last stage of the transition between Aristotle and what is now called Newtonian physics. Galileo utilized the works of the medieval philosophers, the Mertonians, Jean Buridan, Nicole Oresme, and others, and he demonstrated the great value of experiment. Although Aristotle had been a keen observer, Galileo showed that simple observations of nature's world are not enough. The scientist must make nature perform certain actions so that not only can these actions be closely observed, but measurements of them can also be made. The scientist must do more than just observe; he must experiment.

Hence Galileo is often referred to as the father of experimental science, but as with many generalizations of this sort, by giving him this title we neglect others who were also very influential in establishing the experimental nature of science. Of great importance was the work of Francis Bacon* (1561–1620), a contemporary of Galileo. Although Bacon was an "armchair scientist," he was certainly very influential.

Nevertheless, it was Galileo who set the stage for the discoveries of Isaac Newton (1642–1727), born coincidentally in the year of Galileo's death. Newton not only vastly extended the concept of experimental science, but also established a theoretical mathematical basis for physics that allowed scientists for the next two hundred years to attain ever greater heights. It was not until the beginning of the twentieth century that scientists realized a more adequate mathematical basis was needed. By its very nature, the Newtonian system is more profound than those that preceded it. Since the physics of Newton is more important to us today than the physics of either Aristotle or Galileo, we study it in much greater detail.

To devise such a complete system of thought requires first that new terms be defined and old terms redefined to fit the new system. Newton began his *Mathematical Principles of Natural Philosophy* (1686), more often called simply *The Principia,* with a section of definitions and followed them with his axioms or laws of motion (1). To make these definitions more understandable, they are paraphrased in modern English and stated in terms that are currently applicable.

Definition I. *The quantity of matter depends on its density and volume.*

Definition II. *The quantity of motion, momentum, depends on both the quantity of matter and the velocity.*

Definition III. *The inertia of matter is its natural resistance to a change in motion, whether it be from a state of rest to a state of motion, or from one state of uniform motion in a straight line to another.*

Definition IV. *An impressed force is an action exerted on a body that may change its motion either from a state of rest to a state of motion, or from one state of uniform motion in a straight line to another.*

---

*In his works *Advancement of Learning* (1605) and *Novum Organum* (1620), Bacon established the necessity of (1) collecting facts, (2) forming hypotheses that link a group of these facts together, and (3) finally testing these hypotheses.

FIGURE 4-1
    Sir Isaac Newton (1642–1727).

Definition V. A centripetal force is the force that causes the motion of an object to deviate from straight-line motion.

The analysis of these definitions and their relationships to Newton's three laws of motion and his law of gravity will constitute our study of Newtonian mechanics, that is, our study of the causes of the changes in motion.

FUNDAMENTAL QUANTITIES

In considering the properties of matter, Galileo attempted to establish man as an objective observer of nature. With true Platonistic spirit, he differentiated properties that are independent of man the observer—the primary properties (e.g., weight, volume, shape)—from properties that are dependent on man the observer—the secondary properties (e.g., color, taste, texture, and odor). Galileo emphasized that man should become a detached observer of nature; he must strive not to influence or alter what he chooses to study.

The properties that most interest the modern scientist are those that can be measured, the quantitative properties rather than the qualitative ones. Measurements yield numbers, numbers fit into equations, and theories have been vastly more successful when they describe an aspect of nature with suitable equations.

Establishing the quantitative aspect of any field of study demands the adoption of certain fundamental quantities, quantities that are not defined in terms of any others used in the study. All other quantities must be derived from them or depend on them. Fundamental quantities are to physics what axioms are to geometry; they constitute the premises upon which a logical structure is created.

In selecting the most appropriate fundamental quantities, a number of factors must be considered. First, their number must be held to a minimum, and yet the quantities selected must be able to form the basis of a logical system of derived quantities. The system of derived quantities must in turn supply the field of study with an unambiguous and logical framework against which all measurements can be made.

Second, standard units of measure must be agreed upon and adopted for the numerical determination of the fundamental quantities. These units should, for convenience sake, be reproducible in any laboratory in the world where studies that involve the use of these standards are carried out.

Third, the system of measurement must be adaptable to the size of objects studied. In physics, the range varies from the very small, the atomic nucleus, to the very large, the galaxies.

The number of fundamental quantities adopted in physics has been held to four: time, length, mass, and electric charge. The units of electric charge will be discussed in Chapter 6, because the others are important for the study of mechanics, they will be considered now. The establishment of units of measurement is a completely arbitrary affair, but for accurate work the system of units must be well standardized.

Time

The unit of measure for time is the *second*, which was previously defined as 1/86,400th part of a mean solar day. The length of time from one midnight to the next is one day, but that interval of time is not constant throughout the year. The day varies in length because a combination of several of the motions of the Earth determine it. Astronomers have therefore established a mean solar day, which is the average length of a day throughout the year. The vibrations of the cesium atom, however, have proved to be more accurate than the rotation of the Earth, so it is now being used to standardize the second of time.

Length

The unit of length is well standardized but not universally applied. Most people in the world, and the entire scientific community, use the metric system in which the basic unit of length is the *meter*. Originally, the meter was defined as 1/10,000,000th part of the distance from the north pole to the equator, but later measurements of the Earth have revealed that the standard meter presently accepted is not such a neat fraction of the Earth's quarter-circumference. In recent years a more convenient and more accurate standard for the meter, based on a particular wavelength of light, has been devised. The meter, incidentally, is equal to 39.37 inches, which is just a little longer than a yard.

Mass

The third of the fundamental quantities is mass. The standard of mass is, by definition, a platinum cylinder kept at the International Bureau of Weights and Measures in France. By international agreement this piece of metal consists of a quantity of matter equal to one *kilogram*. The weight of this cylinder of platinum is about 2.2 pounds.

As the result of Newton's work, however, we do not recognize weight as a fundamental property of matter. Although an object's weight may be independent of the observer, it does depend on the object's location in the universe. As will be shown in Chapter 5, the weight of an object on the Earth depends on the size of the Earth. A fundamental property of matter should not depend on the size of any material body such as the Earth, but should be the same whether observed on the Earth, in the stars, or between the stars. Actually, the fundamental property of matter that mass measures, described by Newton in Definition III, is *inertia*.

Inertia, the property that Newton ascribed to all matter, is somewhat intangible even though it is fundamental. Yet it is a property with which every person is familiar or which everyone has experienced, especially if he has foolishly ridden in a car with a reckless driver. Inertia makes a person push back against the seat of the car when the car quickly accelerates and makes him lunge forward when the brakes are applied too suddenly and unexpectedly. The acceleration or deceleration of every material body is resisted by the body's inertia.

Inertia is a resistance to a change in motion, whether the change be from rest to a state of motion, or from one state of motion to another. But Newton, like Oresme before him (see p. 37), pointed out that the state of rest and the state of uniform motion are relative. Consequently, inertia is that which opposes any change in motion, whether the change is from a state of motion or from a state of rest; furthermore, since any change in motion constitutes an acceleration, *we define inertia as the resistance to acceleration*. The quantitative aspect of inertia, *mass,* we define as the *numerical expression of resistance to acceleration*.

Our use of mass as a fundamental quantity is slightly altered from Newton's definition of "quantity of matter" (Definition I). The slight alteration is a shift only of emphasis, not of purpose. Newton specified that the quantity of matter in a piece of material is derived from or depends on both the object's density and its volume. Today we find it more convenient to start with mass and then define density in terms of mass and volume. In fact, we define density as the ratio of the mass to the volume

$$\rho = \frac{m}{V}$$

where ρ (Greek rho) represents the density, m the mass, and V the volume. In effect, we use Newton's Definition I, slightly altered, as the numerical expression of his Definition III. Inertia, we claim, is a fundamental property of matter; mass is the corresponding fundamental quantity.

If mass is the numerical expression of inertia, we must have some way to measure it. In Chapter 5 it is shown that weight is proportional to mass, and so if the weight of an object on the surface of the Earth can be determined, its mass can be easily calculated.

These three quantities, time, length, and mass, are fundamental in that all other mechanical quantities are dependent on or derived from them. Yet the size or magnitude of the units adopted to measure them, the second, the meter, and the kilogram (or in their more usual order, meter, kilogram, and second, the mks system of measurement), are entirely arbitrary. The standard units could have been made any size, but those in use have been accepted by international agreement.

DERIVED QUANTITIES

All the other mechanical quantities used in physics are derived from the three fundamental quantities, time, length, and mass; for example, velocity = length/time; acceleration = (change in velocity)/time.

Force

Force is one of the many derived quantities, and although it is as common a term as velocity and acceleration, its meaning is not nearly as obvious. In fact, it is partly because the term is so common that its meaning is indefinite. For example, it is used extensively to denote actions that are not physical by any stretch of the imagination: "the force of his personality . . ." "the law is still in force," etc. Part of the difficulty in defining "force" stems from the fact that in adopting the term from Aristotelian physics, its definition was changed.

Because the concept of force is of great importance to physics, we shall consider the term in more detail by analyzing its effect upon mass. Our analysis for the present will be restricted to linear motion, or motion in a straight line. We can perform a thought experiment in which a block of some material, say wood, is placed on a fairly smooth, horizontal table. The block, if firmly positioned, will remain stable and at rest. To move the block we must apply a force in the direction in which we want it to move. But what is a force?

In Definition IV Newton claimed that a force is an action exerted on a body (our block in this situation) to change its state, either of rest or of uniform motion in a straight line. A change of motion, however, is an acceleration, so *a force is what is required to accelerate an object*. The block will, in fact, start to move only if a force is applied to it, and then only if that applied force is greater than the resistance offered by the contact between the block and the table.

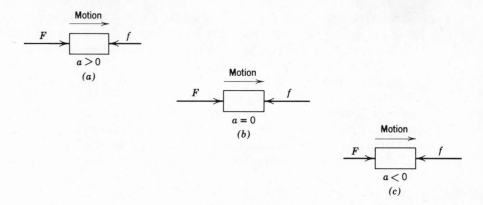

FIGURE 4-2

(a) If the impressed force *F* is greater than the force of friction *f*, the block will accelerate in a positive sense. *(b)* If the impressed force just equals the force of friction, the block's acceleration will be zero. *(c)* If the impressed force is less than the force of friction, the block will accelerate in a negative sense.

This resistance to motion, as was seen in Chapter 1, played a major role in Aristotelian physics. Aristotle recognized resistance as an impediment to motion, but only as an impediment. Newton considered it a force; today we call it the force of friction.

Therefore, according to Newton, the block has two forces acting on it, the applied force (perhaps applied by a hand) and the force of friction (Figure 4-2). Let us represent the applied force by the symbol *F* and the force of friction by the symbol f. If the applied force is greater than the force of friction (in symbolic form $F > f$), the block will start to move (Figure 4-2*a*).

Once the block is in motion, it will continue to move with the uniform motion only if the applied force is equal to the force of friction ($F = f$) (Figure 4-2*b*). If the applied force is greater than the force of friction ($F > f$), the speed of the block will increase—it will have a positive acceleration. If the applied force is less than the force of friction ($F < f$), the speed of the block will decrease—it will have a negative acceleration (Figure 4-2*c*).

Momentum

Once the block has been accelerated by the applied force, it attains a velocity and, consequently, a quantity of motion (Definition II). Newton defined quantity of motion as the product of the velocity of the object and the quantity of matter. Today we call this same property of moving matter *momentum*. This term had been in use before Newton's time, but he was the first to define it exactly. *Momentum is the product of the mass and the velocity*

$$p = mv$$

where *p* represents momentum, *m* the mass, and *v* the velocity.

The momentum with which a body travels is of great importance should that body collide with another body (for example, the collision of two pucks on an ice surface). Many men of earlier centuries observed that something in a moving body is conserved (see Galileo's observation, p. 68), and they had also observed that something in a collision between two bodies is conserved. In the second half of the seventeenth century, Newton (and some contemporaries) determined that the property conserved both in a moving body and in a collision is momentum. Momentum is an important concept to the physicist for this exact reason, because in a collision it is conserved. Hence, in addition to the principle of conservation of matter (we could now call it *conservation of mass* since conservation in this sense is quantitative), there is also *the principle of conservation of momentum: in any collision free of the influence of external forces, the momentum is conserved.* Even if both objects (ice pucks in our example) are moving before collision, the sum of the momenta before collision equals the sum of the momenta after collision, despite the fact that the momentum of each puck will change. For simplicity, let us name our pucks *A* and *B*, since these symbols will help us express the principle of conservation of momentum very neatly in an equation,

$$(mv)_A + (mv)_B = (mv')_A + (mv')_B$$

where the primes refer to the velocities after collision. If an external force is applied during the collision, the momenta and thus the velocities will, of course, be altered. Before discussing momentum further, however, we must introduce the concept of a vector quantity.

VECTOR QUANTITIES

In Figure 4-2, each of the two forces is represented by an arrow. The direction of the force is indicated by the direction of the arrow, and the magnitude of the force is indicated (to scale) by the length of the arrow. Force is one of many quantities in physics that are clearly defined only when *both the magnitude and direction are given.* Such quantities are called *vectors*.

If two or more forces operate on the block, these forces must be "added" in a manner different from the arithmetical addition of numbers. Newton was the first to present the concept of a vector and to describe this technique of addition. He discussed what we today call vectors in a series of corollaries to his three laws of motion. In effect, these corollaries introduce the concept of *vector addition* and the concept of *vector components*.

If a force acts on an object, that object will react in some manner; the force can be represented by the arrow shown in Figure 4-3a. The magnitude of the force, 10 pounds for example, is drawn to scale as the length of the arrow; for example, 1 inch equals 5 pounds. The direction of the vector is indicated by the angle α (Greek alpha) it makes with some reference axis, such as the *x*-axis.

Newton showed that one vector can be replaced by two vectors. These

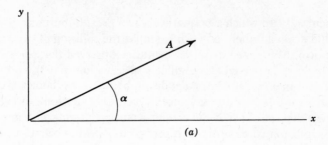

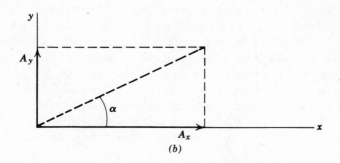

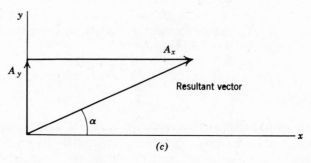

FIGURE 4-3

The vector represented by the arrow A in *(a)* is equivalent to the two vectors A_y and A_x in *(b)*. *(c)* The resultant vector is the summation of vectors A_x and A_y.

two vectors are then called *components*. It is convenient if the components are at right angles to one another, for then the original vector forms the diagonal of a rectangle and the components form two adjacent sides (Figure 4-3*b*). Hence vector A has the components A_x along the *x*-axis and A_y along the *y*-axis. The object upon which these vectors act cannot distinguish between the action of vector A alone and that of the combination of vectors A_x and A_y. Vector A is completely equivalent to the combination of vectors A_x and A_y.

We can also refer to vector A as the vector sum of vectors A_x and A_y. The vector sum is the net vector, sometimes called the resultant. This vector sum is obtained by using the *superposition principle*. Let us move the vector component A_x parallel to itself, without changing its magnitude, so that the tail of the arrow joins the head of the vector A_y. The vector sum of these two vectors is the vector that joins the tail of A_y with the head of vector A_x (Figure 4-3c). Vector addition applies to any number of vectors, and in each addition the resultant vector has a particular magnitude and direction.

The concept of the vector, vector components, and vector addition is extremely useful in physics. There are many kinds of vectors: velocity, acceleration, and force are three examples; momentum is another.

When moving pucks strike head on, the velocities continue in the same straight line. If they strike off-center, the direction of travel of each puck will change, which complicates the analysis of the collision, but momentum is still conserved. That a head-on collision is simple and an off-center collision more complex illustrates the fact that momentum is also a vector quantity.

Some examples will help illustrate the conservation of momentum and its vector nature. Let us assume that the two ice pucks have the same mass and that they are hard. Puck A, moving in a direction which for convenience we shall call the positive x-direction, strikes head on puck B initially at rest (Figure 4-4). After the collision we observe that puck A remains at rest at the scene of the collision while puck B scoots away over the ice. What are we left to conclude? We can make the following statements from our observations:

$$(mv)_B = 0$$

$$(mv')_A = 0$$

If puck B were to move away with a velocity that is less than the initial velocity of puck A, we would have to assume that some momentum had been lost during the collision; if its velocity were greater than the initial velocity of puck A, some momentum would have been gained during the collision.

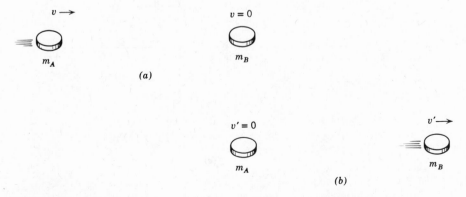

(a)

(b)

FIGURE 4-4

Two pucks on an ice surface: *(a)* before head-on collision, *(b)* after collision.

Measurements of the velocities both before and after collision reveal, however, that momentum is conserved. Since puck *B* has a mass equal to puck
A, puck *B* leaves the scene of the collision with the same velocity that puck
A had when entering it. The momentum before collision equals the momentum after collision. Using the mathematical form of the principle of conservation of momentum, we have

$$(mv)_A + (mv)_B = (mv')_A + (mv')_B$$

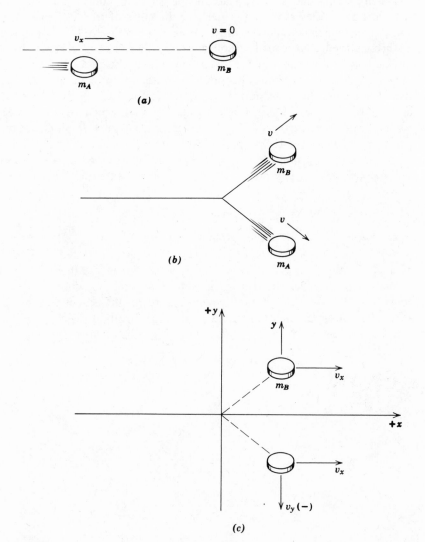

FIGURE 4-5

Two pucks on an ice surface: *(a)* before off-center collision, *(b)* after collision. *(c)* Momentum
is a vector quantity, and is conserved.

and given the values in this example, $(mv)_B = 0$ and $(mv')_A = 0$, we are left with

$$(mv)_A = (mv')_B$$

If the collision is off-center, however, only some of the initial momentum is imparted to puck B, and puck A maintains some of that initial momentum as it travels off at an oblique angle from its original direction, the x-direction (Figures 4-5 and 4-6). The momentum in the x-direction is not altered, for no outside force acted in that direction. Therefore,

$$(mv)_{A_x} = (mv')_{A_x} + (mv')_{B_x}$$

After collision, the two pucks together possess an amount of momentum in the x-direction equal to the initial momentum of puck A.

FIGURE 4-6

The puck with momentum moves from the left to collide with the puck initially at rest. The relative speeds of the pucks may be estimated by the distance they move between photographs. (Courtesy of the Ealing Corporation.)

There is no outside force acting in the y-direction either, and since the momentum in the y-direction was zero before collision, the momentum in the y-direction must be zero after collision. Since both pucks possess a component of momentum in the y-direction after collision, it is clear that one of the two momenta after collision must be a negative quantity. Hence

$$0 = (mv')_{B_y} - (mv')_{A_y}$$

or

$$(mv')_{A_y} = (mv')_{B_y}$$

Momentum is a vector quantity. How about mass? Mass is the measure of resistance to acceleration. Do you feel any different in an auto rapidly accelerating from rest in a northerly direction than you do in one accelerating at the same rate but traveling east or west or south? No, you do not; mass does not depend on the direction of acceleration. Quantities that do not have directions are called *scalar quantities*. Other examples of scalar quantities are time and temperature.

NONLINEAR MOTION

Although Galileo had indeed arrived at the concept of inertia, he expressed the idea only imperfectly. He used such terms as momentum and impetus to express this property of matter but never indicated clearly whether inertia was a linear or a circular property.

At one point he conceived of inertia as being linear, as in his example of a falling object (see pp. 71 f) that reaches a velocity at which the resistance of the air balances the natural tendency to fall. The body thereafter continues to fall straight down at a constant speed. It should be noted that Galileo did not consider resistance a force.

But in another Galilean thought experiment inertia is circular. A smooth ball continues rolling on a perfectly smooth surface which is neither inclined up nor declined down (see pp. 66 ff). Under such conditions it would roll with a constant speed; but the only condition that permits this is if that ball rolls on the surface of a sphere concentric with the Earth. Only on such a surface will the ball maintain a constant distance from the center of the Earth. The question Galileo unwittingly posed is simply this: Will a moving object upon which no forces act travel in a straight line or in a circle concentric with the Earth?

Galileo's difficulty lay in his inability to liberate his imagination from the confines of the Earth. Newton's wild imagination ventured out among the stars. Galileo was a Neoplatonist; he applied Plato's decree of circular motion in the heavens to objects on the surface of the Earth. Newton was a universalist; his imagination pictured how objects should move between the stars, beyond any possible influence of the Earth.

Centripetal Force

The French philosopher-mathematician René Descartes (1596–1650), presented the first clear idea of inertia in his book *Principles of Philosophy* (1644). Descartes attempted to describe a world system based on self-evident axioms and an irrefutable logical extension of these axioms. Had not Euclid done so with geometry? Descartes' final picture, however, did not correspond to the physical world of our everyday experiences (but this did not disturb him). His colliding objects, for example, did not behave as do those in everyday life.

Nevertheless, Descartes made a contribution to the physics of circular motion that went beyond Galileo's explanation. He realized that in order for an object to maintain circular motion, such as a stone on the end of a string or the moon moving about the Earth, the object must be forced away from linear motion. *Circular motion is in fact constrained motion.* If no unbalanced forces act on an object, its natural state is either that of rest or that of *linear motion with a constant speed.* Newton recognized that Descartes was correct in this particular view but thoroughly rejected his world system in general.

In accepting the idea that circular motion is constrained motion, Newton defined a new term, *centripetal force* (Definition V). As he defined it, the term was to apply to the Earth, to magnets, to whirling stones on strings, and the like—to all forces directed toward a center. His purpose for defining the term was simply to differentiate between the force that causes the motion of an object to deviate from straight-line motion and the force that causes the speed of an object to increase or decrease. Newton's purpose is retained in our modern definition: *centripetal force is that force which causes the motion of an object to deviate from straight-line motion.*

Just as inertia makes the rider push back against the seat of a car when it starts out suddenly and makes him lunge forward when the brakes are suddenly applied, inertia also makes the rider lean to the right when the car turns to the left. The rider has inertia; a force is required to cause his motion to deviate from straight-line motion. If the car turns to the left suddenly and the rider strikes the door, it is because no force is supplied to make him follow the curved path of the car until he hits the door—or rather the door hits him and constrains his motion away from a straight line. The door (along with the friction of the seat) supplies the centripetal force necessary to overcome inertia and alter his motion from straight-line motion.

Christian Huygens (1629–1695), a Dutch contemporary whom Newton greatly admired, was the first to determine that centripetal force is directly proportional to both the quantity of matter and the square of the velocity, and inversely proportional to the radius of the circular path followed. Expressed in the terminology of today,

$$F_c = \frac{mv^2}{r}$$

where F_c is the centripetal force, m the mass, v the velocity, and r the radius of curvature. If the force exerted on an object is exactly this amount and is continuously directed perpendicular to the direction of travel, the object will travel in a circular path.

NEWTON'S THREE LAWS OF MOTION

With the aid of these definitions, Newton was then able to state his axioms or laws of motion in the second section of *The Principia*. This section containing his laws of motion is the heart of Newtonian mechanics.

Law I

Every body continues in its state of rest, or of uniform motion in a straight line, unless it is compelled to change that state by forces impressed upon it.

Projectiles continue in their motions, so far as they are not retarded by the resistance of the air, or impelled downwards by the force of gravity. A top, whose parts by their cohesion are continually drawn aside from rectilinear [straight-line] motions, does not cease its rotation, otherwise than as it is retarded by the air. The greater bodies of the planets and comets, meeting with less resistance in freer spaces, preserve their motions both progressive and circular for a much longer time (2).

The first law of motion is often called the law of inertia, for Newton originated the idea that there is no observable difference between the state of rest and the state of uniform motion in a straight line; any body persists in that state unless acted on by a force.

For Aristotle, the state of rest was the natural state of all matter, but Newton proved that to this state must be added the state of uniform motion in a straight line, since the two states are indistinguishable. Newton further explained this concept in one of his corollaries (Corollary V) to the three laws of motion.

The motions of bodies included in a given space are the same among themselves, whether that space is at rest, or moves uniformly forwards in a straight line without any circular motion.

A clear proof of this we have from the experiment of a ship; where all motions happen after the same manner, whether the ship is at rest, or is carried uniformly forwards in a straight line (3).

Passengers playing shuffleboard or billiards aboard a ship find the games no different from what they found them ashore. On a ship the cue ball at rest on the table is at rest with respect to the ship; it moves with respect to the Earth. If the motion of the ship is uniform and in a straight line, no action or motion of the cue ball will reveal that the ship is in motion. A passenger on the ship sees the cue ball at rest; a farsighted "landlubber" observing from the shore sees the cue ball moving. Both are correct, for each is using a different *frame of reference;* that is, each is referring their motion to a different object. Motion has meaning only in relation to a frame of reference. To say that a car is traveling 70 miles per hour assumes that the Earth is the frame of reference. For the sailor the ship is the best frame of reference; objects and men aboard move with respect to the ship. The sailor could choose the moon or some imaginary reference frame in space, but motion will seem simplest in reference to his ship.

If, however, the motion of the ship is abruptly altered because it either turns sharply or is struck by a wave or another ship, the cue ball will retain its state of uniform motion with respect to the Earth until that motion is altered by a force; the same is true for the passengers and the dishes on the tables in the ship's dining room. The ball, the passengers, and the dishes all move in relation to the ship. That is to say, the ship moves out from under the objects that are not tied down.

Law II and Acceleration as a Vector Quantity

The change of motion is proportional to the motive force impressed; and is made in the direction of the straight line in which that force is impressed.

If any force generates a motion, a double force will generate double the motion, a triple force triple the motion, whether that force be impressed altogether and at once, or gradually and successively. And this motion (being always directed the same way with the generating force), if the body moved before, is added to or subtracted from the former motion, according as they directly conspire with or are directly contrary to each other; or obliquely joined, when they are oblique, so as to produce a new motion compounded from the determination of both (4).

In Newton's second law he explains how the motion of a body may deviate from rest or from the state of uniform motion with a constant speed. Any alteration in the motion of a body can result only from an applied force.

The alteration in motion is in the same direction as that in which the applied force acts. If the body is initially at rest, it will move in the direction of the force. If the body is initially moving with uniform motion in a straight line, there are three alternatives, as Newton pointed out.

1. If the applied force is in the direction of motion, the speed of the object will be increased.

2. If the applied force is in the direction opposite that of motion, the speed of the object will be decreased.

3. If the applied force is at right angles to the direction of motion, the direction of motion will change. The body will follow a curved path for as long as the force is so applied. This force, then, is a centripetal force.

We previously defined acceleration as any change in velocity (p. 60), but our discussion did not include nonlinear motion. Newton's second law makes it extremely convenient, and indeed essential, to extend our definition of acceleration to include nonlinear motion. Physicists consider acceleration as any change in motion, whether it be an increase or decrease in speed, or a change in the direction of travel.

Since the change in direction of travel is a consequence of applying a centripetal force, the resulting acceleration is called *centripetal acceleration and is defined as the acceleration in which the direction of travel is altered.* But in order to combine this with our mathematical definition of acceleration,

$$a = \frac{v_2 - v_1}{t_2 - t_1}$$

we must again redefine velocity.

Up to now, the terms "speed" and "velocity" have been used interchangeably, as indeed they are used in everyday speech even though each word may have its special connotations. But in physics speed and velocity have distinct and different meanings. *Speed is defined as the rate of change of position with respect to time but without regard to direction.* The object may be moving in any direction and along any line, curved or straight; the speed, if the motion is uniform, will not be changed. Speed *is not* altered by a centripetal force. The speed of a car may be read on the speedometer.

Velocity is defined as the rate of change of position with respect to time and with regard to the DIRECTION in which that change is made. Velocity *is* altered by a centripetal force. To know a car's velocity we must read both the speedometer and the magnetic compass.

From the definitions just given, it may be seen that speed has only magnitude; velocity has both magnitude and direction. Speed is a scalar quantity, velocity a vector quantity.

A car traveling due north at a speed of 50 miles per hour (Figure 4-7a) has the same speed as a car traveling due south at 50 miles per hour, but their velocities from the standpoint of physics are different. The speed of each car with respect to the Earth is 50 miles per hour, but the velocity of the first car with respect to the second is 100 miles per hour due north, and the velocity of the second with respect to the first is 100 miles per hour due south.

If the two cars each with a speed of 50 miles per hour with respect to the Earth travel in the same direction, they also have the same velocity (Figure 4-7b). Their velocity with respect to each other is zero, and they are at rest in relation to one another.

Since velocity is a vector quantity and includes both magnitude and direction, we can now simplify the definition of acceleration. *Acceleration is the rate of change of velocity with respect to time.* Acceleration is also a vector quantity, so that this complete definition corresponds to the mathematical definition; it includes both the alteration of speed and the change in direction of travel.

With this definition of acceleration, we observe that an automobile really as three accelerators. The throttle is the positive accelerator, the brake is the negative accelerator, and the steering wheel is the centripetal accelerator.

This clarification of the meaning of acceleration is a necessary adjunct to

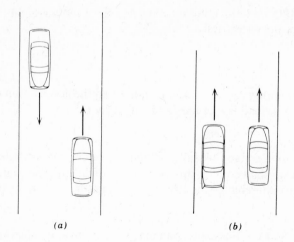

(a)   (b)

FIGURE 4-7

(a) The two cars have the same speed but different velocities. (b) The two cars have the same velocity.

the discussion of Newton's second law, for it is this law that demands the more complete definition. A force must be applied to accelerate an object, and the acceleration is directly proportional to that force. In his second law, Newton used the words "change of motion" which meant more than simply a change in velocity. That "motion" implies more than "velocity" is evident from his discussion of the third law (see p. 90 f): "The changes [are] not in the velocities but in the motion. . . ." An applied force therefore causes a change in the quantity of motion, a change in the momentum.

Furthermore, this change of momentum is a *time* rate of change: "the uniform force . . . acting equally, impresses, in equal intervals of time, equal forces upon that body, and therefore generates equal velocities. . ." (5). A constant force acting on a body will produce twice the change of motion if it operates for twice the time, three times the change of motion in thrice the time, and so on. Consequently, the *ratio* of the change of momentum to the time interval over which the force acts must be proportional to that force, or in modern mathematical notation,

$$\frac{mv_2 - mv_1}{t_2 - t_1} \propto F$$

The change in momentum, $mv_2 - mv_1$, can be written more simply by taking advantage of the delta notation introduced in Chapter 3. With the change of momentum written as $\Delta(mv)$, and the interval of time as Δt, we write Newton's second law of motion as

$$\frac{\Delta(mv)}{\Delta t} \propto F$$

For the sake of simplicity we shall consider only examples in which the mass of a system is constant; then the delta does not apply to the mass, and it may be removed from the parenthesis:*

$$\frac{m\,\Delta v}{\Delta t} \propto F$$

We note that the quantity $\Delta v / \Delta t$ is nothing more than the acceleration of the body, so Newton's second law of motion finally becomes

$$ma \propto F$$

Although proportions convey useful information, equations are much easier to handle, and we shall convert this proportion to an equation. First, for convenience, let us interchange the left- and the right-hand sides of the proportion:

$$F \propto ma$$

Then, including a constant K, this expression may be written as an equation:

$$F = K(ma)$$

As in Chapter 3, we now have the job of evaluating this constant of proportionality. In the equation relating the circumference of a circle to its diameter, $C = \pi D$, one simple method of evaluating the constant π is to measure the diameter and the circumference of a circle and then determine their ratio, $\pi = C/D$. But how do we evaluate the constant of proportionality in Newton's second law of motion?

This constant cannot be determined until the unit of force is established, or rather vice versa. If we, as Newton, define force as what is required to produce an acceleration, it becomes a simple matter to establish the unit of measure of force by using the second law of motion. In order that we may keep calculations as simple as possible, the unit of force in the mks (meter-kilogram-second) system has been defined so that the constant of proportionality in the second law of motion,

$$F = K(ma)$$

becomes unity, that is, $K = 1$. This particular unit of force is (for obvious reasons) called the *newton* (abbreviated nt). *By definition, a newton of force is the force that is required to give a mass of one kilogram an acceleration of one meter per second per second.* Newton's second law of motion now becomes

$$F = ma$$

Law III

To every action there is always opposed an equal reaction: or, the mutual actions of two bodies upon each other are always equal, and directed to contrary parts.

Whatever draws or presses another is as much drawn or pressed by that other. If you press a stone with your finger, the finger is also pressed by the stone. If a horse

---

* Recall that $mv_2 - mv_1 = m(v_2 - v_1)$.

draws a stone tied to a rope, the horse (if I may so say) will be equally drawn back towards the stone; for the distended rope, by the same endeavor to relax or unbend itself, will draw the horse as much towards the stone as it does the stone towards the horse, and will obstruct the progress of the one as much as it advances that of the other. If a body impinge upon another, and by its force change the motion of the other, that body also (because of the equality of the mutual pressure) will undergo an equal change, in its own motion, towards the contrary part. *The changes made by these actions are equal, not in the velocities but in the motions of bodies;* * that is to say, if the bodies are not hindered by any other impediments (6).

When no unbalanced force is applied, Newton's second law of motion, in effect, becomes his first law; the object is said to be in equilibrium and continues in its state of rest or of straight-line motion with constant velocity. But whenever a force, balanced or unbalanced, is applied, the third law of motion explains that this force is not the only one: *for every action there is an equal and opposite reaction.* Another way to look at the third law of motion is that in this universe of ours *forces are always exerted in pairs; a single force cannot be exerted.*

If you push against the wall of your room, the wall pushes back against you; you can feel it. This force must be equal and oppositely directed to the force you apply. If the wall cannot support this great a force, it collapses, and the force you exert causes acceleration of parts of that wall.

When you push the block of wood on the table with your finger, the block too pushes back against you. You can feel it. You push the block in one direction (Figure 4-8*a*), and in reaction it pushes back against your finger; these two forces are what we call *action-reaction pairs.* The block pushes against the table, and in reaction the table (by means of friction) pushes back against the block; these two forces are also action-reaction pairs (Figure 4-8*b*).

The block has two forces operating on it, the force of your finger and the force of friction between the table and the block (Figure 4-8*c*). These two forces are not action-reaction pairs, for action-reaction pairs never operate on the same body. Two bodies must be involved. One force (the action) operates on the first body; the other force (in reaction) operates on the second body.

The magnitude of the frictional force with which the table pushes back on the block depends both on the surfaces involved and on the force pushing these surfaces together and is thus variable up to a certain limit. If the frictional force operating on the block (and opposing the force of your finger) is large enough, it will prevent the block from slipping; the block will remain at rest with balanced forces acting on it. But if you push on the block with a force exceeding the largest possible or limiting frictional force, the block will accelerate, for there is then an unbalanced force operating on the block. Even though the block accelerates as the result of your pushing, it still pushes back against your finger with a force equal to that with which your

* These italics are mine.

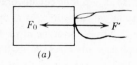

(a)

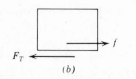

(b)

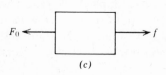

(c)

FIGURE 4-8

(a) The force F_0 which the finger exerts on the block equals the force F' which the block exerts on the finger. These two forces are action-reaction pairs. (b) The force F_T which the block exerts on the table equals the force of friction f which the table exerts on the block. These two forces are action-reaction pairs. (c) The force F_0 equals the force of friction f; but since these both act on the block, they are not action-reaction pairs.

finger pushes it. Some examples will help illustrate Newton's laws of motion.

THE HORSE. Newton's example of the horse pulling the stone is an appropriate one. The horse pulls the stone and the stone pulls the horse; each pulls with an equal but opposite force. Why then does the stone move in response to the horse and not vice versa? This question is easily answered by posing another question. What sort of luck would the horse have in pulling the stone if the stone were in dirt and the horse standing on ice? Obviously the horse would not be able to move the stone because ice does not furnish the necessary "traction," that is, frictional force. Clearly, then, it is the force of friction that pushes the horse forward. The horse pushes backward on the ground and the ground reacts (if it is able to do so) by pushing forward on the horse; we have an action-reaction pair. If the ground is not able to supply a frictional force to the horse that is greater than the frictional force between the ground and the stone—and it cannot when the horse is on ice and the stone in dirt—the stone will not be moved.

Whether it is the horse that moves the stone or the stone that moves the horse depends not on their pulling each other but on the forces of friction. In the first place, a frictional force can never initiate an action; and in the second place, if the horse is to pull the stone, the frictional forces between the horse and the ground must be larger than those between the stone and the ground.

THE CAR. The modern automobile also illustrates Newton's law of motion. An automobile starts to move only because of the frictional forces between the rear wheels and the road. The rear wheels push backward on the road, and the road in turn pushes forward on the rear wheels. If the tires are smooth and if the road is covered with a sheet of ice, the rear wheels simply spin and the car remains at rest.

If the car is traveling along the road with a constant velocity and the driver wishes to alter that velocity by either slowing down or speeding up, he again relies on the frictional forces between the road and the tires. The driver will not be able to stop the car suddenly when traveling along a road covered with ice.

Should the driver want to change direction (i.e., alter his constant velocity) by turning the steering wheel of thé car, he again is accelerating by supplying a force, but this time the force is at right angles to the direction of travel. The frictional forces between the tires and the ground turn the car. When the driver by means of the steering wheel turns the front wheels, the front wheels roll in the new direction because they roll more easily than they slide. Hence the front wheels assume the new direction of motion; the friction between the front tires and the road exerts the centripetal force that causes the centripetal acceleration. If the ground is unable to supply sufficient frictional force, the car will continue traveling in a straight line. The driver will not be able to turn the car suddenly when traveling on smooth ice.

As an example of the application of Newton's second law of motion, we might ask what force is required to accelerate a car of mass $m = 1000$ kilograms (it would weigh about 2200 pounds) at a rate of $a = 2.0$ meters per second^2?

$$F = ma$$

$$F = (1000 \text{ kg})(2.0 \text{ m/sec}^2)$$

$$F = 2000 \text{ nt}$$

A force of 2000 newtons is the same as a force of about 440 pounds, that is 1 newton $= 0.22$ pound. The acceleration of a falling stone is about 10 meters per second^2.

THE ROCKET. The second and third laws of motion are illustrated by the rocket. If the horse and the car push against the ground and the ground in reaction pushes them forward, what does the rocket—in empty space—push against? Indeed, what do motorboats and airplanes push against to propel themselves forward?

The propeller-driven craft, whether a boat or an airplane, simply pushes (or throws) a fluid (water or air) backward. The propeller, in pushing the fluid backward, pushes the craft forward—a pair of action-reaction forces. The air (or water) accelerates to the rear and the craft accelerates forward. Consequently, it is not a matter of *pushing back* against something, as tires on the road, but of *throwing* something to the rear.

A rocket that travels beyond the Earth's atmosphere, however, cannot throw air to the rear. The rocket must carry its own propellent to throw in that direction. The exhaust of the rocket (Figure 4-9) is the result of a continuing explosion in the rocket motor. The exploding gases push the rocket forward and the exhaust to the rear. If the rocket is to go forward, the exhaust must go to the rear. If the rocket is to turn to the left, the exhaust must go to the right. The rocket will neither accelerate forward nor accelerate to the left unless the exhaust is expelled and accelerated in the opposite direction, that is, to the rear or to the right.

You can easily demonstrate the rocket principle by sitting on a small wagon and throwing rocks (not pebbles) to the rear. The wagon will move forward. If you want to go faster, you can mount a machine gun on the wagon and, taking all precautions, fire it to the rear.

In each of these examples of propulsion, the total amount of momentum

FIGURE 4-9

The hot exhaust gases blasting out from the rear of this Saturn rocket during lift-off are easily discernible and are evidence in support of Newton's third law of motion. (Courtesy NASA.)

is conserved. The change of momentum of the material thrown to the rear, for example, equals the change of momentum of the craft as it is accelerated forward. If the change of momentum of the material thrown to the rear is considered negative, the change of momentum of the craft is positive; their sum, the total change of momentum, is zero.

CRITICISM OF THE LAWS OF MOTION

It is surprisingly easy to describe inertia in animistic terms (in the manner of Aristotle) because our language appears to be structured that way. In physics, however, we try to avoid saying, "The block continues in a straight line because it *wants* to." It is more common but not much better to say, "The block *tends* to travel in a straight line." Newton's statement of the first law is clear; an object *will* travel with uniform motion in a straight line unless acted on by a force. It is simply a statement of fact; yet the statement does not describe a commonly observed occurrence. How many people have actually seen an object with no net forces acting on it travel in a straight line with a constant velocity? Is the first law really a statement of observed fact or is it a pronouncement of faith?

It so happens that this statement, whether of fact or of faith, enables physicists to explain most of the phenomena in the universe about us. Newtonian mechanics works and, therefore, we use it. With Newton's laws we can, for example, predict the motions of the moon, but they do not really help much in our understanding of the relationship between the Earth and the moon.

Literally thousands of volumes have been written about Newton's laws of motion. Some have been written in criticism of these laws and some in confident support of them. Although some books extend the use of these laws in an attempt to make them truly universal, other books demonstrate their limitations. Newton's laws established a distinctive view of and an attitude toward nature that allowed men to predict rationally and quantitatively the outcome or results of events, such as the application of a force to a block resting on a table. It enabled men to predict what happens when a stone is whirled on the end of a string and what happens when that string is released. Today, all the physical sciences and engineering are based largely on these three laws of motion (as well as Newton's law of gravity, the subject of Chapter 5). But are these laws so well established that they are beyond criticism? Certainly not. Criticism is necessary if we are to understand the universe better. We must not relegate ourselves to the level of those who followed Aristotle without question.

No one denies that an object will accelerate if an unbalanced force is applied to it, and that the acceleration will be proportional to and in the same direction as the applied force and inversely proportional to the mass of the object. But what really is a force? What is mass? What, indeed, is a physical law? A statement becomes a physical law if that statement describes some action in the physical world and describes that action invariably whenever the conditions exist for which that statement is intended.

Suppose that an observer is riding on a large, smoothly rotating merry-go-round and that he is not aware of his location. This observer cannot experimentally verify Newton's laws of motion; he finds that those laws do not describe the actions of his experiments. For example, if a constant force is applied to a block on the floor of the rotating merry-go-round, the observer finds that the resulting acceleration depends on the direction in which the force is applied. A force applied in one direction (directed toward the outside of the merry-go-round as seen by an outside observer) results in a greater acceleration than that same force applied in the opposite direction, toward the center of the merry-go-round.

If that same force is applied at right angles to the radius of the merry-go-round (tangent to the circle in which the experimenter is traveling), he finds that the block accelerates the same in both directions (i.e., in both the forward and backward directions). Hence, in this rotating system, acceleration and force are not related by a simple proportionality $a \propto F$. A factor Q specifying the direction has to be included in this statement to make it a good description of the experiment: $a \propto QF$. Furthermore, if the speed of rotation of the merry-go-round varies, the resulting experiment is even more difficult and the resulting description even more complex. If the speed of rotation varies regularly with time, a time factor T as well as a direction factor has to be included, which further complicates the description: $a \propto TQF$. If the speed changes in a random fashion, it is impossible to describe the actions of the experiment with any simple expression.

We must therefore specify exactly the conditions under which Newton's laws of motion apply. They were first observed on the Earth, and the Earth, since it rotates much more slowly than the merry-go-round, approximates a system in which the laws of motion apply. Newton felt that the "fixed stars" formed an absolute system in which his laws were invariant. Today we call such a system an *inertial system and define it as any system in which an object will not accelerate if it is released and is free to move*. The surface of the Earth is *not* such a system. An object released and free to move does accelerate downward. By Newton's second law, then, a force must be acting on that object. This force, the force of gravity, is the subject of our next chapter.

What is mass? If we define mass as the numerical expression of resistance to acceleration, are we not defining mass in terms of the second law of motion? Actually we are not. Mass can be defined and measured under conditions in which there is no concern whatever for the forces that may be involved. This definition was first presented by Ernst Mach (1838–1916), almost two centuries after Newton.

When two objects collide, their velocities change that is, each is accelerated. The velocities before and after collision can be determined by measuring the distances traveled and the periods of time required to cover a certain distance. If *one* of the two bodies in the collision is arbitrarily selected as a standard mass (one kilogram), and if all the velocities are measured, the mass of the second body can be determined. This determination is based on the principle of conservation of momentum (see p. 82 ff.):

$$(mv)_A + (mv)_B = (mv')_A + (mv')_B$$

If object *B* is chosen as the standard mass, the mass of object *A* can be determined by measuring the velocities of the two objects before and after collision.

If the two masses are equal, the accelerations will be equal. If one mass is greater, it will have the smaller acceleration; the ratio of accelerations will be in an inverse proportion to the ratio of the masses. Hence, mass may be defined and determined by the changes of velocities resulting from such a collision.

Thus, the principle of conservation of momentum may be used to define mass as well as to establish its numerical value without relying in any way whatsoever on the second law of motion. Indeed, the word force never even entered our discussion; there simply is no need to consider forces, much less to measure the value of any force.

It is true that we may apply the second law to a collision between two bodies and say that since the bodies were accelerated, there must have been forces acting. The application of the second law, however, is not only unnecessary in this situation but is, in effect, a distraction. The reason for selecting the situation of a two-body collision to define mass in the first place was to be able to establish its properties and numerical value without resorting to any derived units such as a force. The only measurements required are those of the two fundamental quantities, distance and time, which we need for calculating the velocities.

Furthermore, it is equally true that the third law of motion states that in every collision between two bodies, two equal and opposite forces are exerted. Because they are equal and oppositely directed, however, these forces may be neglected. They cancel each other out; the net force on the system is zero.

Hence, mass may be defined and established without recourse to the second law of motion. Can the same be said for the concept of force?

If we define force as that which produces an acceleration, it appears that the mathematical statement

$$F = \frac{\Delta(mv)}{\Delta t} = ma$$

does not describe the actions of nature but is simply a mathematical definition of force. And so it may be. But the fact of the matter is that the definition works with all the various kinds of forces operating in the world of man, as long as he examines forces that are neither "too big" nor "too small." In the next two chapters we show that the only forces included in this statement are gravitational and electrical forces, and that Newton's second law of motion does work admirably (and almost completely) for both. In fact, since the law does work for both kinds of forces, it offers us two independent methods of measuring forces. Measurements of forces by both methods have been shown to be completely consistent with each other.

The second law of motion can therefore be considered more than just a definition of force. It can be considered a law, even though the concept of force is not as clear as some would like it to be.

References

1. Isaac Newton, *Principia*, University of California Press, Berkeley (paperback), Volume 1, 1962, pp. 1 ff.
2. *Ibid.,* p. 13 f.
3. *Ibid.,* p. 20.
4. *Ibid.,* p. 13.
5. *Ibid.,* pp. 21 f.
6. *Ibid.,* p. 13 f.

Questions

1. Define (*a*) mass, (*b*) force, (*c*) momentum, and (*d*) centripetal force.

2. Discuss the need for a system of fundamental quantities in physics.

3. Discuss the need for definitions in physics, as well as in any field of study.

4. Differentiate between a scalar and a vector quantity, and give several examples of each.

5. Read the "Rules of Reasoning in Philosophy" on p. 398 of Newton's *Principia*, Volume 2, University of California Press paperback, and answer the following questions.

(*a*) Apply Rule I and demonstrate a preference between the Ptolemaic and the Copernican systems.

(*b*) Show that Rule II eliminates the Aristotelian "two worlds" universe.

(*c*) Discuss Rule III and the experimental method. (Gravitational forces are discussed in Chapter 5.)

(*d*) Rule IV establishes the relationship between the theoretical scientist and the experimental scientist. Explain.

Problems

1. What is the density of a chunk of matter with a mass of 8.9 g if it displaces 3.3 cm³ of water when submerged?

2. What force is required to accelerate an automobile with a mass of 1500 kg at a rate of 3 m/sec²?

3. What will be the acceleration of a 6.0-kg block sliding on a horizontal table if the only unbalanced force acting on it is a force of friction of 4.0 nt?

4. The force impressed upon and moving an object is 17 nt; the force of friction opposing the motion is 3 nt.

(*a*) What is the resultant force acting on the object?

(*b*) If the mass of the object is 3.0 kg, what is its acceleration?

(*c*) What is the acceleration if the impressed force is reduced to 2.0 nt, and if the force of friction remains the same?

5. By means of a scale drawing, determine the magnitude of the x- and y-components of a vector whose magnitude is 15 units, if it makes an angle of 37 degrees (counterclockwise) with the positive x-axis?

6. Calculate the momentum of an automobile with a mass of 2000 kg (weight about 4500 lb) if it is traveling with a velocity of (*a*) 15 m/sec (about 33 miles/hr), (*b*) 35 m/sec (about 77 miles/hr). (*c*) If an average force of 8×10^3 nt is applied to change the momentum from that of part *a* to that of part *b*, over how long a period of time was the force exerted?

7. What is the change in momentum if a force of 22 nt is exerted on an object for 3 sec?

8. A ball of mass 450 g (weight about 1 lb) traveling with a velocity of 150 cm/sec collides head on with a second ball initially at rest whose mass is 225 g. The more massive ball continues on after the collision with a velocity of 50 cm/sec. What is the velocity of the less massive ball after collision?

9. A 0.5-kg rock is whirled on the end of a string 1.2 m long. The rock makes 15 revolutions in 10 sec.

(*a*) What is the speed of the rock?

(*b*) What centripetal force must the string exert on the rock to maintain it in a circular orbit?

10. Determine the mass m_x of an object if it collides with a standard 1-kg mass m_0 ; the unknown mass is accelerated 4.5 times the acceleration of the standard mass.

Newton and the Law of Gravity

W e have already seen how Newton's second law of motion did not entirely clarify the meaning of the term force; the enunciation of his law of gravity only aggravated the situation. The difficulty lay with the moon.

We cannot reproduce Newton's reasoning in arriving at his three laws of motion and his law of gravity; we can be reasonably sure, however, that these four laws were derived in the same logical process, even if that process took some time. Therefore, we shall not attempt to reconstruct Newton's thinking; let us, instead, use his three laws of motion as discussed in Chapter 4 to demonstrate that if we accept them, a law of gravity is essential. We shall then discuss that law, the universal law of gravitation.

THE MOON AND NEWTON'S LAWS OF MOTION

According to Newton's first law of motion, the moon, made of the same chemicals as the Earth, should travel in a straight line. But it does not! The moon moves in a curved path; it is continually being accelerated toward the center of its nearly circular orbit. Therefore, according to the second law of motion, a force must be continually acting on the moon, a centripetal force. The center of the moon's orbit lies nearly at the center of the Earth, and the only plausible explanation Newton could offer for this motion was that the Earth exerts an attractive force on the moon. Newton's third law of motion explains that there are no single forces exerted in the universe; since, if the Earth exerts an attractive force on the moon, the moon must exert an equal and opposite attractive force on the Earth.

Herein lies the problem; the moon travels in a curved path and consequently a force must be acting on it. But how can a force be exerted across such great distances? How can a force be exerted by any means other than physical contact? What kind of force is it? Does its strength decrease with increasing distance?

To answer these questions to his own satisfaction, Newton found it necessary not only to propose a mighty principle, the universality of physical laws, but also to formulate an entirely new kind of mathematics, the calculus of infinitesimals.

If the questions Newton was asking about the moon and the Earth were valid, they should also be asked about other members of the solar system, for example, about the Earth and the sun, about Jupiter and the sun, in fact, about any of the planets and the sun. Indeed, the laws of physics holding true for the surface of the Earth should be universally applicable. The same laws should operate on the moon, on the sun, on any planet, and between the planets and the stars. This is one universe (the term *one universe* is really a redundancy), and we can best describe it by formulating laws that are applicable throughout its vast extent. This principle gave the *coup de grâce* to Aristotle's two-world system.

FIGURE 5-1
Johannes Kepler (1571–1630).

KEPLER'S LAWS

Newton was a Copernican in that he accepted the heliocentric concept of the universe. But more than half a century earlier, during the time of Galileo, Johannes Kepler (1571–1630), a German astronomer, had proposed a heliocentric system much superior to that of Copernicus. It will be recalled that Copernicus did little more than interchange the position of the Earth and the sun in the Ptolemaic theory (which was indeed quite a bit). Copernicus still utilized an extremely complex system of epicycles, deferents, and eccentrics to account for the celestial motions. Kepler discovered a vastly simpler and more accurate method of explaining them. Because science strives for the simplest accurate explanation of any phenomenon, Kepler's laws have replaced the more complex Copernican theory.

Kepler had acquired the fairly precise observations of the positions of the planet Mars that had been made by Tycho Brahe (1546–1601). These observations were crucial in the history of astronomy because they were the first meticulous and systematic observations of planetary positions. Before the time of Brahe, the positions of planets in the sky had been observed at only a few critical places in their epicyclic motions. Brahe made observations of all the visible planets night after night, tracing the course of each through the sky with large, exact instruments which he had constructed himself and which measured with an accuracy never before achieved.

Working with Brahe's observations of Mars, Kepler, after years of time-consuming and grueling calculations, discovered a much simpler, and thus preferable, shape for the planetary orbits. Furthermore, Kepler realized how the planets must move in these orbits. He made this discovery in spite of himself, for he too had the preconceived Platonistic notion that celestial

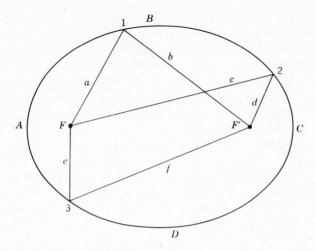

FIGURE 5-2

The figure *ABCD* is an ellipse, so the distances $a + b$, $c + d$, and $e + f$ are all equal.

motions were all circular and were performed at a uniform rate. But he found that their motions can best be described if the planetary orbits are considered *ellipses* and if the motion of the planets is not regarded as uniform.

An ellipse is a figure formed by all the points so located that the sum of the distances from each point along the curve to two given points, called *foci*, is always constant (Figure 5-2):

$$a + b = c + d = e + f$$

If the two foci F and F' coincide, the ellipse becomes a circle.

Kepler stated, in what is called his *first law*, that the orbits of the planets are indeed ellipses with the sun at one of the foci. Kepler's *second law* states that the line joining any planet with the sun (for example, SA of Figure 5-3),

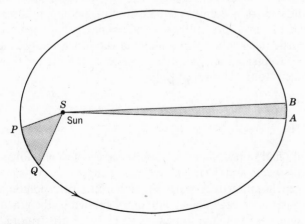

FIGURE 5-3

If a planet travels from P to Q in the same length of time it takes to travel from A to B, then the shaded areas are equal.

called the *radius vector*, sweeps out equal areas in equal periods of time. For example, if the planet moves from *A* to *B* in 28 days and from *P* to *Q* in the same interval of time, the shaded areas will be equal.

In effect, Kepler's second law makes it clear that when a planet is in the part of its orbit closest to the sun (the point *P*, called perihelion), it moves with the greatest speed. As the planet travels farther from the sun, its speed decreases, reaching a minimum when it is at its greatest distance (the point *A*, called aphelion). Modern observations indicate that at perihelion the Earth has an orbital speed of 18.9 miles per second, at aphelion a speed of 18.3 miles per second.

Kepler's *third law* relates the periods of time any two planets require to complete their orbits to their average radius vectors

$$\frac{p^2}{P^2} = \frac{r^3}{R^3}$$

where *p* and *r* represent the period and average radius vector of one planet, and *P* and *R* the period and average radius vector of any other planet. In effect, this law states that the greater a planet's average distance from the sun, the more slowly it travels in its orbit. Modern observations indicate that Mercury, the planet closest to the sun, has an average orbital speed of about 30 miles per second. The Earth has an average orbital speed of 18.6 miles per second, and Pluto, the most distant known planet, a speed of about 3 miles per second.

It is of interest that Kepler's three laws were published during the lifetime of Galileo (the first two laws were published in 1609 and the third law in 1619), and that he and Galileo had been in correspondence. Galileo, however, never gave the slightest recognition to Kepler's three laws. There may be several reasons for this. First, Kepler was a mystic, and like other mystics filled his works with numerological distinctions between numbers, considering some "more beautiful" than others and by that token more appropriate to use in describing the world. Perhaps Galileo was so repelled by this mysticism that he did not study Kepler's works carefully enough to discern the three laws, which are never directly stated and are nearly hidden among the numbers and mystic reflections. Or possibly Galileo was actually aware of the formulation of Kepler's laws, but simply preferred the perfect circles of Plato and Copernicus to the ellipses of Kepler.

THE LAW OF GRAVITY

In any event, Newton had to postulate and describe a gravitational force of attraction between the Earth and the moon in order to supply the centripetal force necessary to keep the moon in its nearly circular orbit. Correspondingly, the centripetal force acting on each planet would be supplied by the gravitational force of attraction between the sun and each planet. Furthermore, in order that this gravitational force be truly universal, it would also have to account for the force of attraction between the Earth and, let us say, an apple falling from a tree. Kepler's laws are restricted; they apply only to planetary-like motion. Newton wanted his laws to apply universally to all matter in all kinds of motion.

These are rigorous requirements. In the first place, Newton had to assume that matter exerts a force of attraction upon other matter and, furthermore, that this force is exerted over great distances with no apparent physical contact. He also had to assume that if the distances are large in comparison to the size of the objects, as, for example, the distance between the planets and the sun, the sizes of the individual objects themselves may be neglected.

By Kepler's second law, the speed of each planet in its elliptical orbit decreases with increasing distance from the sun, and by Kepler's third law the speed of each successive planet also decreases with increasing distance from the sun. These two laws of Kepler led Newton to conclude that the gravitational force acting on each planet must decrease with increasing distance from the sun. In fact, Newton proved that the gravitational force is inversely proportional to the square of the distance from the sun:

$$F \propto \frac{1}{d^2}$$

That is, if the distance from the sun is doubled, the force becomes one-fourth as great; if the distance is tripled, the force becomes one-ninth as great; if the distance is made four times as great, the force becomes one-sixteenth as great.

Kepler's third law gives a clue to the relation between the force of gravity F_g acting on each planet and the planet's mass m_p. This law describes the motion of any planet about the sun in relation to the motion of any other planet. Since even the very early telescopic observers saw that the planets clearly differ in size, it was reasonable for Newton to assume that they differ in mass as well. But if they differ in mass, why does Kepler's third law ignore the masses of the planets? How can a planet's motion, its centripetal acceleration, be independent of its mass?

The answer, once grasped, is easy enough. If the force of gravity F_g is proportional to the planet's mass, m_p, then by Newton's second law, $F = ma$, the acceleration will have to be independent of the mass. For example, assume that the force of gravity acting on a given planet is proportional to its mass; then if that mass were double its present value, the force of gravity would also be double. But if both the mass of the planet m_p and the force of gravity F_g acting on it were double, the acceleration would be the same as it now is:

$$F_g = m_p a$$
$$(2F_g) = (2m_p)a$$

The accelerations in these equations are equal to each other, and each is equal to F_g/m_p, because the two's in the equations (showing that both the mass and the force are doubled) cancel out. If the acceleration is independent of the mass, the motion of any planet would not change even if its mass were double, triple, or quadruple its present mass. Thus, by assuming that the force of gravity on any planet is proportional to its mass, Newton explained why Kepler's third law ignores the mass.

By Newton's third law of motion, however, the force of gravity between any planet and the sun acts on both the planet and the sun. The two forces constitute an action-reaction pair; they must be equal in magnitude and opposite in direction. The force of gravity acting on the Earth, for example, must therefore depend on the mass of both the Earth m_E and the sun m_s. The force of gravity might possibly depend on the sum of the two masses, $m_E + m_s$, or it could depend on the product of the two masses, $m_E m_s$. However, if the force of gravity is to be doubled by doubling the mass of the Earth *alone*, the force of gravity must depend on the product of the masses.

The same argument must apply for any planet and the sun, and to be universal it must also apply for any two bodies of mass m_1 and m_2; consequently, the mutual force of gravity F_g acting between two bodies must be proportional to the product of their masses:

$$F_g \propto m_1 m_2$$

That is, if the mass of one object is doubled, the force of attraction is doubled; if the mass of each object is doubled, the force becomes four times as great. Accordingly, the speed of a planet in its orbit depends only on the mass of the sun and on the planet's distance from the sun.

By combining these two proportionalities, we obtain

$$F \propto \frac{m_1 m_2}{d^2}$$

Newton demonstrated that the same relationship exists between the moon and the Earth. But does it apply to the Earth and an apple? When these two bodies are considered together, the size of the Earth certainly cannot be neglected in the same way that the sizes of the sun, the moon, and the planets were neglected when considering forces among them.

Newton realized that if matter attracts other matter according to the proportion just given, the apple is actually attracted by all the mass of the Earth—by each and every particle that goes to make up the Earth. He was therefore faced with the tremendous task of adding up vectorially all the forces exerted on the apple by each particle in the Earth. The vector sum of all these forces gives the resultant force which, Newton hoped, would conform to the description of gravitational forces for the planets and the moon.

In order to perform this vector addition, Newton had to devise an entirely new method of addition, a method based on the addition of extremely small elements, so small indeed that we consider them infinitesimally small. Newton called his technique the theory of fluxions; today we call it calculus. By this technique Newton was able to add the gravitational forces of attraction exerted on the apple by each particle in the Earth. He determined that the resultant force is the same as the force that would be exerted if all the mass of the Earth were concentrated at its center. That is, mathematically we may consider the Earth a point and the apple another point. These two points are separated by a distance equal to the radius of the Earth. Once we consider the Earth and the apple as points, the force exerted by the Earth on the apple does indeed conform to Newton's description of the forces in the solar system, and this description thus became the universal law of gravitation. This proportion

$$F \propto \frac{m_1 m_2}{d^2}$$

becomes an equality by the insertion of a constant

$$F = G \frac{m_1 m_2}{d^2}$$

The symbol G is called the *universal gravitational constant*. Its numerical value was not determined by Newton but by Henry Cavendish (1731–1813) in the eighteenth century. Cavendish actually measured the force of attraction between two objects in the laboratory. Since gravitational forces are weak forces, this achievement marks a high point in the history of man's experimental investigation (1).

The law of gravity is usually stated as: *Every particle in the universe attracts every other particle with a force that is proportional to the product of their masses and inversely proportional to the square of the distance between them.*

THE THREE STRANDS FROM ARISTOTLE

If Newton's laws of motion and his law of gravity are truly comprehensive, they certainly should be able to explain fully the three strands of thought considered in Chapter 2. These three concepts, free fall, projectile motion, and planetary motion, will be discussed in turn.

Free Fall

Any object on or near the surface of the Earth is subject to the force of gravity exerted by the Earth on that object. By the law of gravity that object also attracts the Earth with an equal but oppositely directed force (action-reaction forces). Let us, for the moment, place an object (shall we again consider it to be an apple?) on a table. We then have two objects, the Earth and the apple, attracting each other with equal forces. But these objects are held apart by the table, that is, the table pushes up on the apple and down on the Earth. When, however, we remove the apple from the table and release it, we have for a brief period of time two objects exerting a force on each other and each free to move. Newton's laws must describe the resulting motion.

From the law of gravity the force exerted on the apple F_1 (Figure 5-4) must equal the force exerted on the Earth F_E:

$$F_1 = F_E$$

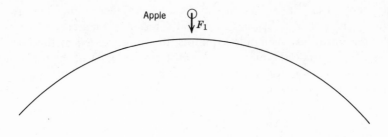

FIGURE 5-4

Since the Earth and the apple exert a force upon each other, and since the motion of each is described by Newton's second law, they fall toward each other.

The apple, according to the second law of motion, accelerates after it is released, for an unbalanced force is acting on it:

$$F_1 = m_1 a_1$$

But the Earth, too, must accelerate according to the force acting on it and its mass:

$$F_E = m_E a_E$$

Because the two forces are equal,

$$m_1 a_1 = m_E a_E$$

Hence the apple appears to fall when it is released, but according to Newton's law both the apple and the Earth accelerate toward each other. Since the mass of the Earth ($m_E = 6 \times 10^{24}$ kilogram*) is so much greater than that of an apple ($m_1 = \frac{1}{4}$ kilogram or less), the acceleration of the Earth is, by the same ratio, that much smaller than the acceleration of the apple. If the apple falls for one second, it will cover a distance of about 5 meters (16 feet). In the same second, the Earth will have moved a distance of only 2×10^{-25} meter, certainly a smaller distance than we can even hope to perceive, much less measure.

Not all apples have the same mass, yet Galileo showed that they all fall with the same acceleration. How can this be? Let us, for example, consider

---

* See Appendix A for a description of the scientific notation for very large and very small numbers.

two apples, one with a mass of ¼ kilogram and a very large one with a mass of ½ kilogram. By the law of gravity the force acting on the larger apple will be twice the force acting on the smaller apple, for the larger one has twice the mass of the smaller one. The law of gravitation implies what Aristotle maintained—the larger the mass, the greater the force causing it to fall. Aristotle, however, did not realize that *acceleration and not velocity is the important aspect of motion.* Newton's second law makes it clear that if the two apples are to have the same acceleration, the larger one must have twice the force acting on it, for it has twice the mass. We may express this relationship between mass and acceleration mathematically.

Since the mass of the larger m_2 is twice the mass of the smaller m_1,

$$m_2 = 2m_1$$

the force of gravity on the larger F_2 must be twice that on the smaller F_1,

$$F_2 = 2F_1$$

Now each acceleration equals the ratio of force to mass:

$$a_1 = \frac{F_1}{m_1}$$

$$a_2 = \frac{F_2}{m_2}$$

If, in the equation for the acceleration of the second apple, we substitute $2F_1$ in place of F_2 and $2m_1$ in place of m_2 (equals may be substituted for their equals in any equation), we obtain

$$a_2 = \frac{2F_1}{2m_1}$$

which reduces to

$$a_2 = \frac{F_1}{m_1}$$

which is precisely the acceleration of the first apple.

By extending this reasoning, we conclude that the acceleration of all objects at or near the surface of the Earth is constant, and the results of Galileo's measurements are accounted for nicely by the application of Newton's universal laws.

Since all objects at or near the surface of the Earth have the same acceleration, we refer to that acceleration by the special symbol g (see page 63). Numerically, it is very nearly equal to 32 feet per second^2, or 10 meters per second^2 (more precisely 9.8 meters per second^2). For objects in free fall the statement of the second law of motion becomes

$$F = mg$$

This force is equal to the weight of an object:

$$weight = mg$$

Hence the mass of an object is indeed proportional to the object's weight:

$$m = \frac{\text{weight}}{g}$$

We can therefore determine the mass of an object simply by weighing it. Weighing is a much simpler procedure than measuring the velocities of colliding objects, as is necessary with our previous definitions of mass (see p. 100). But before we rely on weighing, the acceleration produced by gravity must be determined for each place on the Earth; since the Earth is rotating and is not exactly spherical, the value of g varies slightly over the surface of the Earth.

Weight is therefore a force. In the English system of units weight is expressed in pounds. In the metric system it is expressed in newtons. One pound is equal to 4.4 newtons.

The Projectile

Projectile motion is the second strand of Aristotle's description of motion to be explained by Newton's laws. Galileo discovered (see p. 66f) that the motion of a projectile can best be understood if it is considered to perform two motions simultaneously, a horizontal motion and a vertical motion. For example, if the projectile is fired horizontally, its horizontal velocity will not change, since no force operates in the horizontal direction (neglecting the resistance of the air). Newton's first law indicates that there is no horizontal acceleration.

The force of gravity, like all forces, is a vector quantity, and it operates only in the vertical direction. The acceleration produced by this force must therefore be only in the vertical direction. Consequently, the vertical velocity continues to increase while the horizontal velocity remains constant (Figure 5-5). As Galileo pointed out, the curved path followed by the projectile is a parabola. The velocity is directed along the curve and is represented by an arrow tangent to the curve. This velocity is the resultant velocity of the horizontal and vertical motions of the projectile. The resultant velocity of an object traveling in a curved path is often called the tangential velocity. The velocity of a falling projectile increases in magnitude and changes its direction to become more nearly vertical.

If an object is thrown straight up in the air, its horizontal velocity is zero and remains so; but according to Aristotle the object is still a projectile. Since the force of gravity is the only force acting on it and, furthermore, since that force (near the surface of the Earth) is constant, the acceleration is constant. Its value is g and is acting in the same direction in which the force acts, that is, downward.

For convenience, let us call the upward direction positive. The downward direction then becomes negative; hence the acceleration is negative. To throw the object up we must give it an initial velocity upward of v_0. As it travels up, its velocity decreases because its velocity is positive (directed

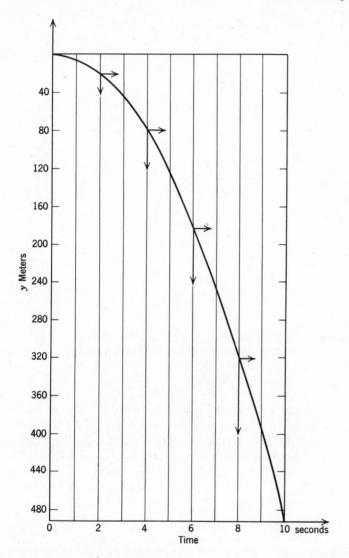

FIGURE 5-5

Since the only force acting on a projectile (in ideal motion) is its weight, the horizontal velocity remains constant while the vertical velocity increases.

upward) and its acceleration is negative (directed downward). The graph of *velocity* versus *time* slopes downward to the right (Figure 5-6).

When the velocity decreases to zero, the object no longer travels up; it comes to rest. Since the constant force of gravity continues acting on this projectile, it still has a constant acceleration directed downward. The object's velocity therefore changes from zero to negative values, and the object starts to fall. During the object's fall, the velocity increases in a negative sense, for both its velocity and acceleration are negative. The negative acceleration increases the negative velocity of the projectile.

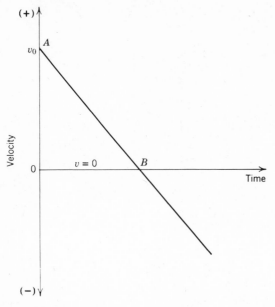

FIGURE 5-6

The velocity-time graph of an object thrown straight up. That object is continually accelerated downward: on the way up, its positive velocity decreases; on the way down, its negative velocity increases.

The slope of the curve shown in Figure 5-6 is constant, that is, the curve is a straight line. This slope represents the change of velocity with respect to time—the acceleration. The steeper the slope, the greater the change of velocity. This curve shows how a projectile can travel upward with decreasing speed, stop, and then travel downward with increasing speed, all with a constant acceleration.

The Planets

Newton's laws also resolve the problems connected with the third strand of Aristotelian motion. His *Principia* describes the motions of planets and comets about the sun and, indeed, of any satellite about its mother planet.

If a projectile is given a horizontal velocity, it will (neglecting air resistance) maintain that horizontal velocity. If it is given a small horizontal velocity, it will travel only a short distance horizontally before striking the ground. If the projectile is given a very large horizontal velocity, that is, if it is thrown very hard indeed, it will travel quite a distance before finally striking the ground. Whatever the original horizontal velocity, the projectile falls at the same rate, for as it falls gravity is the only force acting on the projectile. The downward acceleration is constant and independent of any horizontal motion (see Galileo's arguments on p. 66 ff).

This analysis of projectile motion, in which a greater and greater horizontal velocity is imparted to a projectile, was extended to its ultimate and logical conclusion by Newton.

That by means of centripetal forces the planets may be retained in certain orbits, we may easily understand, if we consider the motions of projectiles; for a stone that is projected is by . . . its own weight forced out of the rectilinear path, which by the initial projection alone it should have pursued, and made to describe a curved line in the air; and through that crooked way is at last brought down to the ground; and the greater the velocity is with which it is projected, the farther it goes before it falls to the earth. We may therefore suppose the velocity to be so increased, that it would describe an arc of 1, 2, 5, 10, 100, 1000 miles before it arrived at the earth, till at last, exceeding the limits of the earth, it should pass into space without touching it [the earth].

Let *AFB* [Figure 5-7] represent the surface of the earth, *C* its center, *VD*, *VE*, *VF*

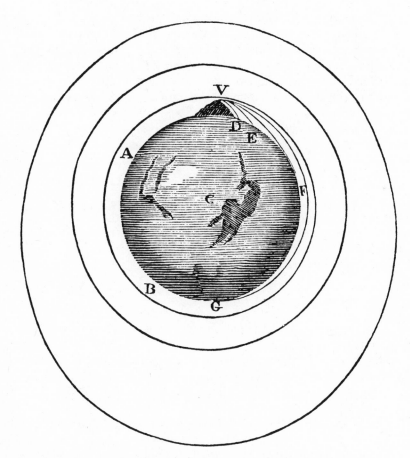

FIGURE 5-7

Newton realized that if he could throw a projectile fast enough, that projectile would assume an orbit about the Earth (Newton's *Principia*).

the curved lines which a body would describe, if projected in a horizontal direction from the top of a high mountain successively with more and more velocity; and, because the celestial motions are scarcely retarded by the little or no resistance of the spaces in which they are performed . . . let us suppose either that there is no air about the earth, or at least that it is endowed with little or no power of resisting; and for the same reason that the body projected with a less velocity describes the lesser arc *VD*, and with a greater velocity the greater arc *VE*, and, augmenting the velocity, it goes farther and farther to *F* and *G*, if the velocity was still more and more augmented, it would reach at last quite beyond the circumference of the earth, and return to the mountain from which it was projected.

And since the areas which by this motion it describes by a radius drawn to the center of the earth are proportional to the times in which they are described, its velocity, when it returns to the mountain, will be no less than it was at first; and retaining the same velocity, it will describe the same curve over and over, by the same law (2).

The correctness of Newton's conclusions has been amply demonstrated since October 1957, when the Russians launched the first artificial satellite. A rocket must both lift the satellite above the Earth's atmosphere and at the same time gradually alter its direction of motion so that it travels more or less parallel to the local horizon. At the same time the rocket's velocity increases to at least the minimum value required for the satellite to maintain the proper orbit.

As long as the force of gravity continues to act on the satellite, it continues to accelerate, for it is the force of gravity, let us call it F_g,

$$F_g = G \frac{m_s m_E}{d^2}$$

which supplies the centripetal force necessary to maintain the rocket in orbit about the Earth. In this equation, m_s is the mass of the satellite, m_E is the mass of the Earth, and d is the distance from the center of the Earth to the satellite.

Centripetal force (see p. 89) is given by

$$F_c = \frac{mv^2}{r}$$

where r is the radius of the satellite's circular orbit and v is the orbital velocity or speed of the satellite.

If the force of gravity operates in a direction perpendicular to the direction of travel and, furthermore, if it is the only force acting, then

$$F_g = F_c$$

To achieve a circular orbit, the rocket must be traveling parallel to the local horizon and with the correct speed at the instant the rocket motors are turned off. Its distance from the center of the Earth must equal the radius r of the circular orbit, and its speed v must be such that

$$G \frac{m_s m_E}{d^2} = \frac{m_s v^2}{r}$$

which, since $d = r$ and $m_s = m_s$, reduces to

$$G \frac{m_E}{d} = v^2$$

Therefore,

$$v = \sqrt{\frac{Gm_E}{d}}$$

This equation gives the velocity that must be attained by a rocket in order to place an artificial satellite in a circular orbit, at a distance d from the center of the Earth or at a distance d' from the surface of the Earth where $d' = d - 4000$ miles (4000 miles is the radius of the Earth). The circular orbit is orbit b of Figure 5-8a.

If the velocity is greater than the value given by the previous equation, the satellite will travel in a curved path with a larger radius of curvature (orbit c of Figure 5-8a). The force of gravity (for the same height, i.e., the same height at which the satellite goes into orbit) remains the same, and thus the centripetal force remains the same. Consequently, if the velocity v is increased, by the equation for centripetal force, the radius of its orbit must also be increased. In the resulting elliptical orbit the satellite travels farther from the Earth.

If the velocity is less than that required for a circular orbit, the radius of the orbit will also be smaller and the satellite will fall closer to the Earth, traveling in elliptical orbit a in Figure 5-8a. In any event, the plane formed by the satellite's orbit must pass through the center of the Earth.

If the speed of the space vehicle is greater than that producing elliptical orbit c, the vehicle will travel in a parabola (orbit d in Figure 5-8a). The vehicle will not move in a closed curve about the Earth, that is, it will not become a satellite. The parabola is an open curve and the vehicle will leave the immediate vicinity of the Earth and assume an elliptical orbit about the sun. It is possible, of course, to give a space vehicle such a large velocity that it assumes a parabolic orbit about the sun; it would then leave the entire solar system and become an interstellar vehicle.

The Merging of the Three Strands

The three strands of thought, the concepts of free fall, projectile motion, and planetary motion, were considered separately by Aristotle: freely falling bodies constituted natural motion; projectiles were considered in violent motion; planetary motions were treated by a different law, the law of the heavens. Gradually, through the passing centuries, the strands were brought closer and closer together, until finally Newton wove them into one strong cable that is the strength of his physics.

According to Newton, all motion is governed by the same set of physical laws. Freely falling bodies, projectiles, artificial satellites, planets, all have the same basic motion. They are all freely falling bodies with different horizontal velocities. An object falling vertically toward the Earth (or any

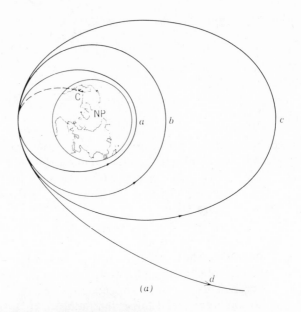

FIGURE 5-8

    (a) The shape of the satellite's orbit depends on its velocity, that is, its speed and its direction of travel at the moment the rocket motors cease firing. (C represents Cape Kennedy and NP represents the North Pole.) *(b)* The astronaut and the space capsule are each in an orbit (courtesy NASA).

planet) has no horizontal motion; the horizontal velocity of a projectile is small compared to the tremendous velocity of 17,000 miles per hour (5 miles per second) necessary to realize Newton's wild example of centripetal force and launch an artificial satellite.

Planets are freely falling bodies; they fall toward the sun. The force of gravity acting on a planet depends on three factors: the mass of the sun, the distance of the planet from the sun, and the mass of the planet. Since the mass of a planet and its distance from the sun are unique quantities, scientists have found it convenient to refer to the *force per unit mass* which varies inversely as the square of the distance from the sun:

$$\frac{F}{m_p} = G\frac{m_S}{d^2}$$

By Newton's second law, a force per unit mass is nothing more than acceleration:

$$\frac{F}{m_p} = a$$

Hence we may speak of the force per unit mass or the acceleration produced by the sun's gravitational influence and mean the same thing.

The inner planets have very large tangential velocities. Since the force per unit mass decreases with increasing distance from the sun, the closer the planet is to the sun, the faster it must travel tangentially to maintain its orbit. As we learned earlier, Mercury, the planet closest to the sun, travels with an orbital speed of 30 miles per second. The Earth's orbital speed is 18.6 miles per second, and Pluto, the most distant known planet, moves at only 3 miles per second. The force per unit mass acting on Pluto is relatively small, and therefore it is not falling very fast.

Newtonian physics furnished the basis for explaining these apparently different kinds of motion with one set of fairly simple laws. This simplicity is the supreme advantage of the Newtonian system and has made it the foundation for nearly all the physical sciences and engineering.

The Elevator

Newton's second law of motion is illustrated very well by the elevator problem (Figure 5-9a). What does Newtonian mechanics tell us about the forces acting on the elevator as it first starts from rest, travels up, stops, starts back down, and finally stops once more?

In the analysis of such problems it is convenient to isolate the body under consideration; we therefore draw the elevator, indicating all the forces acting on it in the direction of motion. There are two forces, as shown in Figure 5-9b: the weight of the elevator is represented by *mg* and the tension in the cable supporting the elevator by *T*. The cable prevents the elevator from assuming free-fall motion.

According to Newton's second law of motion, the vector sum of these two forces must account for the resultant motion of the elevator. Since these

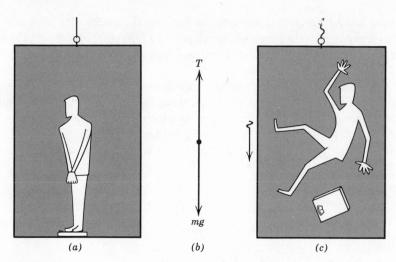

FIGURE 5-9

The two forces acting on the elevator *(a)*, are its weight *mg* and the tension in the cable *T* *(b)*. Should the elevator cable break, the elevator, the occupant, and the bathroom scales would all fall freely together *(c)*.

two forces act in opposite directions, we need to choose one as the positive direction. Let us choose upward. The tension *T* in the cable is then acting in the positive direction and the weight *mg* in the negative direction. The net force ΣF^* is equal to the resultant of these two, $T - mg$, so Newton's second law

$$F - ma$$

may be written as

$$T - mg = ma$$

If the cable does nothing more than support the elevator, the acceleration *a* is zero and

$$T - mg = m(0)$$
$$T - mg = 0$$
$$T = mg$$

As we would expect, the tension in the cable just equals the weight of the elevator. Under these conditions the elevator does not accelerate; but can it be moving? Yes, of course. The elevator may be moving either up or down with a constant velocity, that is, with zero acceleration; if it is, the tension in the cable will just equal the weight of the elevator. Alternately, the elevator may be at rest with a zero (a constant) velocity. Unless the elevator jiggles, an occupant cannot determine whether the elevator is at rest or moving with a constant velocity.

Suppose that the operator wishes to accelerate the elevator upward; he must increase the speed of the motor that lifts the elevator. By our sign con-

---

*Σ, Greek sigma, means the summation of any number of items, in this instance forces.

vention an upward acceleration is positive; therefore *ma* must be positive. If *ma* is positive, then *T* must be greater than *mg*, for

$$T - mg = ma$$

On the other hand, if the operator decreases the speed of the motor, or allows it to stop, then by our sign convention the acceleration will be negative. Consequently, *T* will be less than *mg*.

Let us illustrate the problem by using some numbers, since this often makes the problem clearer for the student. In addition, the numbers may help to clarify the difference between weight and mass. Suppose the elevator weighs 10,000 newtons (2250 pounds). To determine the elevator's mass we recall that

$$\text{weight} = mg$$

and using $g = 10$ meters per second^2, we substitute these numbers into the equation

$$10,000 \text{ nt} = m \, (10 \text{ m/sec}^2)$$

$$m = \frac{10,000 \text{ nt}}{10 \text{ m/sec}^2}$$

$$m = 1000 \text{ kg}$$

If the operator accelerates the elevator upward at the rate of 2.0 meters per second^2, we can find the tension in the cable by adding the forces acting on the elevator and setting that sum equal to *ma*:

$$T - mg = ma$$

Therefore,

$$T = mg + ma$$

This second expression tells us that not only must the tension in the rope support the elevator by exerting a force *mg*, but it must also accelerate the elevator by exerting an additional force *ma*. Substituting our numbers in the equation, we find that

$$T = (1000)(10) + (1000)(2.0)$$
$$T = 10,000 + 2000$$
$$T = 12,000 \text{ nt}$$

A force of 12,000 newtons is equivalent to 2700 pounds, which is greater than the weight of the elevator.

If the elevator is accelerated downward at a rate of 2.0 meters per second^2, by our sign convention the acceleration is negative, -2.0 meters per second^2.

$$T = mg + ma$$
$$T = (1000)(10) + (1000)(-2.0)$$
$$T = 10,000 - 2000$$
$$T = 8000 \text{ nt}$$

The tension in the cable during the downward acceleration will be less than the weight of the elevator, not a surprising result.

If a passenger in the elevator were to stand on a bathroom scales while the elevator moves up and down, what would he observe? When the elevator is at rest or moving with a constant velocity, the bathroom scales exerts a force upward equal to the weight of the passenger. The passenger also exerts the same force on the bathroom scales, and the reading of the scales is therefore his weight. When the elevator accelerates upward, the passenger is also accelerated upward, and the bathroom scales exerts a greater force on him, a force equal to his weight plus an additional force required to accelerate him. The bathroom scales indicate some value greater than the passenger's weight. When the elevator accelerates downward, the bathroom scales indicates a value somewhat less than his weight. The passenger's stomach agrees with the bathroom scales. If the cable breaks and the elevator, bathroom scales, and occupant fall freely, the passenger is no longer able to stand on the bathroom scales and successfully observe a reading; he then appears to be weightless (Figure 5-9c). If he were truly weightless, however, he would not fall!

To clear up this apparent contradiction, we must define weight and then speak of true weight and apparent weight. *The true weight of an object is the force of gravitational attraction between that object and the Earth* (or between that object and the moon or Mars). The force expressed by Newton's law of gravity is the weight of any object. Only when an object is entirely free of all gravitational influences is it truly weightless.

FIGURE 5-10
A beam balance.

The apparent weight of an object is the reading of a spring balance, such as a bathroom scales, which uses a spring to measure the force. The apparent weight will depend on the motion of an object; it will increase if the object is accelerated upward and decrease if it is accelerated downward. The true weight of the object depends only on the astronomical body to which it is attracted and how far the object is from that body.

It is interesting to compare the operation of a beam balance (Figure 5-10) with that of a spring balance. If an object, for example, a rock, is balanced on a beam balance against a standard mass of, let us say, 1.5 kilograms, will the two objects remain in balance if the instrument is transferred to the moon or Mars? Yes, the operation of a beam balance is independent of the gravitational field in which it resides; the beam balance is a mass determiner. If the rock is hung from a spring balance and this balance is transferred to the moon or to Mars, the reading on the spring balance scale will be changed (actually, the reading on both the moon and Mars will be lower than the reading on the Earth). A spring balance is a weight determiner; its reading depends on gravitational attraction of an astronomical body for an object.

CRITICISM OF THE LAW OF GRAVITY

Newton's law of gravity was not immediately accepted without criticism. In fact, beyond the confines of the British Isles it was not well received. The French remained loyal to the ideas of their national hero, Descartes, who earlier in the seventeenth century had developed his own world system. The remainder of Europe looked skeptically upon the Newtonian system while favoring that of Descartes.

Critics everywhere asked the same questions. How can this force of gravity act across countless miles of empty space? How can the sun exert a force on all the planets to maintain each in its orbit? The force must be an occult force! Is science returning to magic?

Newton, in part, anticipated these criticisms and concluded the third book of his *Principia* with the following statement:

Hitherto we have explained the phenomena of the heavens and of our sea [the tides] by the power of gravity, but have not yet assigned the cause of this power. This is certain, that it must proceed from a cause that penetrates to the very centers of the sun and planets, without suffering the least dimunition of its force; that operates not according to the quantity of the surfaces of the particles upon which it acts (as mechanical causes used to do) but according to the quantity of the solid matter which they contain, and propagates its virtue on all sides to immense distances, decreasing always as the inverse square of the distances. . . . But hitherto I have not been able to discover the cause of those properties of gravity from phenomena, and I frame no hypotheses; for whatever is not deduced from the phenomena is to be called a hypothesis; and hypothesis, whether metaphysical or physical, whether of occult qualities or mechanical, have no place in experimental philosophy. In this philosophy particular propositions are inferred from the phenomena, and afterwards rendered general by induction. Thus it was that the impenetrability, the mobility, and the impulsive force of bodies, and the laws of motion and of gravitation, were discovered. And to us it is enough that gravity does really exist, and act according to the laws which we have explained, and abundantly serves to account for all the motions of the celestial bodies, and of our sea (3).

In effect, Newton said that if we accept the straight line as described by Euclid to be a fundamental property of space, the first law of motion, with its "uniform motion in a straight line," stipulates a property of matter, that of inertia. Furthermore, it follows that the second law explains all motion that deviates from the inertial state defined by the first law. If we accept the linearity of space and the three laws of motion, we have no alternative but to accept the law of gravity as a description of one aspect of Nature, even if we cannot understand the process by which it acts.

MECHANICAL CONTACT VERSUS FORCES AT A DISTANCE

The eighteenth century was the century of Newton. The scientific world studied and digested the ideas of Newton's *Principia* and of his second book, *Opticks*. The universal laws of motion and the law of gravity repeatedly demonstrated their amazing ability to order and describe the actions of Nature.

The new mathematical insight provided by the invention of calculus opened the way for a vastly more penetrating study of all things that undergo change.

The invention of calculus was, incidentally, not Newton's alone. The German philosopher-mathematician Gottfried Wilhelm Leibniz (1646–1716) also developed the idea, but with quite different notation. This independent and nearly simultaneous development of calculus was the cause of a long-standing bitterness between these two great men. History has shown, however, that independent and nearly simultaneous inventions or discoveries are quite common. It appears that when the scientific world advances to the point that a particular step forward is required, men in various parts of the world, quite free of any influence upon each other, may independently take that step.

Through the new mathematics and Newton's laws, the motions of the planets were examined more and more carefully. It was found that the planets do actually influence one another as they revolve around the sun. It became quite clear that a force is exerted at a distance and that all matter in the universe does exert a force of attraction on all other matter. The motion of Mars, for example, is determined principally by the sun, but that motion is also influenced by the gravitational attractive forces of Jupiter, Saturn, and the Earth. In fact, the motion of Mars can be predicted with precision only when all these forces are taken into consideration.

But the mechanism by which a force is exerted at a distance was never made clear. Moreover, Roger Boscovich (1711–1787), born in what is now Yugoslavia, began to question how a force is exerted even by physical contact.

The atomistic view of nature, originating with Leucippus and Democritus, survived until the time of Newton. Indeed, Newton gave the view his full support. Much ground work had to be done, however, before experiments supporting the idea of atoms could be devised and carried out, and before mathematicians could describe the behavior of atoms.

It had been assumed that atoms were the ultimate minute particles of matter, that they could not be broken up into smaller particles, and that they were hard and impenetrable. Hence a mechanical force can be exerted by the collision of these hard impenetrable atoms. Boscovich asked, "What happens when two such particles collide?" Their velocities change, because of the force exerted by contact. But for how long a time are they in contact?

The change in velocity depends not only on the force exerted but also on the length of time the force is applied. This is described clearly by Newton's second law of motion,

$$F = m \frac{\Delta v}{\Delta t}$$

which may be rewritten, in the form presented by Newton, as

$$F \Delta t = m \Delta v$$

The left-hand side of the equation, $F \Delta t$, is called the *impulse* and includes all the factors affecting any change in momentum—or, as long as the mass remains constant, any change in velocity. Expressing the second law of motion in this form reveals that the same change in velocity may be achieved by a small force acting over a long period of time, or by a large force acting over a small period of time. For example, a force of 4 newtons acting on a 1-kilogram mass over a period of 4 seconds will produce a change of velocity of 16 meters per second. If the object is initially at rest, at the end of 1 second its velocity will be 16 meters per second. But a force of 8 newtons acting on the same object over 2 seconds will produce the same change in velocity, as will a force of 16 newtons in 1 second, or a force of 32 newtons in $\frac{1}{2}$ second, or of $\frac{1}{2}$ newton in 32 seconds. It is the product of the force and the time interval over which the force acts that determines the change of momentum.

Boscovich wrote:

In the year 1745 [18 years after the death of Newton] . . . I . . . began to investigate somewhat more carefully that production of velocity which is thought to arise through impulsive action, in which the whole of the velocity is credited with being produced in an instant of time by those, who think, because of that, that the force of percussion is infinitely greater than all forces which merely exercise pressure for single instants. It immediately forced itself upon me that, for percussions of this kind, which really induce a finite velocity in an instant of time, laws of their actions must be obtained different from the rest.

However, when I considered the matter more thoroughly, it struck me that, if we employ a straightforward method of argument, such a mode of action must be with-drawn from Nature, which in every case adheres to one and the same law of forces, and the same mode of action. I came to the conclusion that really immediate impulsive action of one body on another, and immediate percussion, could not be obtained, without the production of a finite [change in] velocity taking place in an indivisible instant of time, and this would have to be accomplished without any sudden change or violation of what is called the Law of Continuity; this law indeed I considered as existing in Nature . . . (4).

Boscovich concluded that the universality of Newton's laws of motion was the best way to approach the problem, but he soon came to recognize that during the collision either the principle of continuity is not correct or the collision is not "mechanical."

The principle of continuity stipulates that any change in velocity must be made gradually, there being no such thing as a change in velocity from one speed to another without going through all the intermediate speeds. Calculus relies on this law, for the technique of calculus is based on the infinitesimally small change, which corresponds to a continuous change as opposed to a discontinuous or sudden change.

For example, the impulse $F \Delta t$ causes a change in momentum $m \Delta v$. If the force acting on an object is applied over a fairly long interval of time and measurements of the velocity are made every second, the graph of these measurements might look like that shown in Figure 5-11a. We might next ask what the graph would look like if we had made measurements every $\frac{1}{2}$ second, or every 1/10 second, or every 1/1000 second? According to the principle of continuity, no matter how small the interval of time, the graph must continue smoothly from one point to the next as in Figure 5-11b. If the force is constant, the curve will be a straight line. But in any event, both the change in time and the change in velocity will be accomplished continu-ously, not in discontinuous jumps as in Figure 5-11c or 5-11d.

But if the changes in velocity are continuous, that is, if they are made in infinitesimally small steps, then, as Boscovich pointed out, the collision between two ultimately minute, hard, and impenetrable particles of matter is not well understood.

He considered two such particles of the same mass traveling in the same direction along the same line. The one in front travels with a speed of 6 degrees (Boscovich refers to arbitrary units of speed), and the other one travels with a speed of 12 degrees. Since the faster one is behind, it will catch up and collide with the slower one. By the law of conservation of momentum, if the particles of equal mass are completely hard so that neither is distorted during the collision, the resulting velocity of each particle will be 9 degrees of speed, each continuing in the original direction.

Boscovich then pointed out that such a collision is not possible. If both particles are absolutely hard and impenetrable, and if the changes in velocity occur only when the particles are in contact, such changes must be *discon-tinuous*. The faster one would have to change instantaneously from a speed of 12 to 9 degrees without going through any intermediate speeds, and the

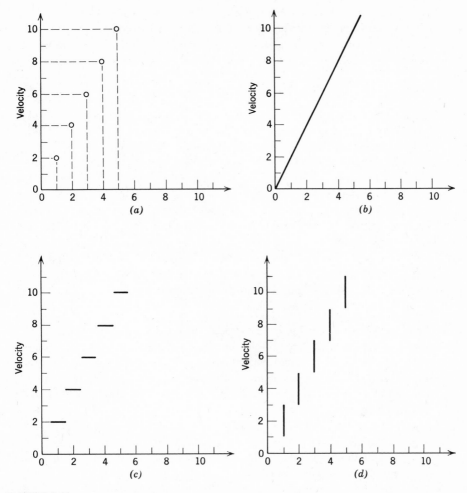

FIGURE 5-11

(a) A plot of measurement of velocity made every second. *(b)* If the acceleration is uniform, the law of continuity stipulates that the velocity changes through all intermediate values. All changes are smooth. *(c)* The law of continuity does not permit velocities to change abruptly from one value to another. *(d)* The law of continuity does not permit time to change abruptly from one instant to the next. Time flows continuously.

slower one would have to change instantaneously from 6 to 9 degrees without going through any intermediate speeds.

For it cannot possibly happen that this kind of change is made by intermediate stages in some finite part, however small, of continuous time, whilst the bodies remain in contact. For if at any time the one body then had 7 degrees of velocity, the other would still retain 11 degrees; thus, during the whole time that has passed since the beginning of contact, when the velocities were respectively 12 and 6, until the time at which they are 11 and 7, the second body must be moved with a greater [average] velocity than the first; hence it must traverse a greater distance in space than the other. It follows that the front surface of the second body must have passed beyond the back surface of the first body; and therefore some part of the body that follows behind must be penetrated by some part of the body that goes in front (5).

It is apparent that something is wrong; impenetrable bodies do not penetrate one another. Boscovich chose to stand by the second law of motion, the principle of continuity, and the principle of conservation of momentum. He rejected the belief that hard, impenetrable particles are the ultimate minuteness of matter. He proposed that the change of velocity takes place before actual contact, in fact, that "real contact" never actually happens. In place of hard, impenetrable particles, Boscovich suggested that atoms of matter are nothing but geometric *points* which act as centers *of force*. When these atoms come extremely close to one another, this force becomes repulsive; the particles repel each other gradually, and so the concept of "real contact," and the dilemma of the collision, vanishes.

As the particles approach each other more closely, the force of repulsion increases, resulting in an increase in the rate of change of momentum. Mechanical "collision" becomes nothing more than the interaction of forces at a distance. To this Boscovich says:

They call forces like those I propose nonmechanical, and reject them, just as they also reject the universal gravitation of Newton, for the alleged reason that they are not mechanical, and overthrow altogether the idea of mechanism which the Newtonian theory had already begun to undermine. Moreover, they also add, by way of a joke in the midst of a serious argument derived from the senses, that a stick would be useful for persuading anyone who denies contact.

It will be enough just for the present to mention that, when a body approaches close to our organs, my repulsive force (at any rate it is that finally) is bound to excite in the nerves of those organs the motions which, according to the usual idea, are excited by impenetrability and contact . . . (6).

Boscovich simply invokes Newton's third law of motion; the particles in both the stick and the nerve feel a force exerted on them. Hence the sensation is the same.

We are left, then, with atoms of matter that are nothing more than points from which force originates. When the atoms are very close to one another, the force is repulsive; when the atoms are farther apart, it is the attractive force of gravity described by Newton.

We previously observed (see p. 113) that Cavendish, in 1798, was able to measure the force of attraction between two objects in his laboratory. The objects he used were two pairs of lead spheres. In essence, his experiment involved the suspension of two spheres A and B of equal mass from a very fine wire (Figure 5-12a). By bringing two much larger lead spheres M and N up to these as shown in Figure 5-12b, the spheres A and M would attract each other and tend to twist the wire in a counterclockwise direction as seen from

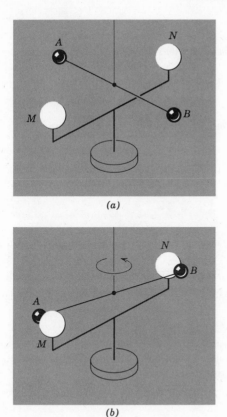

(a)

(b)

FIGURE 5-12

(a) When the large weight M and N are some distance from the smaller weight A and B, the supporting wire is not twisted. *(b)* As the weights M and N are moved closer, the attraction between A and M, and between B and N causes the supporting wire to twist a measurable amount.

above. So too would spheres B and N attract each other and twist the wire in a counterclockwise direction. By measuring the number of degrees through which the wire was twisted, Cavendish was able to measure the force of attraction between the two pairs of spheres.*

This experiment demonstrated without question that the law of gravity was indeed a valid description of one aspect of nature. Hence the scientific world was obliged to accept it, even if they did not understand how such a force was exerted. But then if Newton's law of gravity muddled the meaning of the term force, Boscovich only made it worse, for he denied that objects ever come into "mechanical contact" with each other.

*The torsion balance is treated in more detail on p. 144 ff. in the discussion of Coulomb's experiment.

References

1. Morris H. Shamos, ed., *Great Experiments in Physics*, Holt-Dryden, New York (paperback), 1959, p. 75.

2. Isaac Newton, *Principia*, University of California Press, Berkeley (paperback), 1962, pp. 551 f.

3. *Ibid.*, pp. 546 f.

4. Max Jammer, *Concepts of Force*, Harvard University Press, Cambridge, Mass., 1957, p. 171.

5. *Ibid.*, p. 172.

6. *Ibid.*, p. 174.

Questions

1. Indicate any similarity between Kepler's laws and Ptolemy's use of the eccentric and the equant.

2. Read Query 21 of Newton's *Opticks*, p. 350, Dover paperback, and comment.

3. Discuss the similarity between Newton's third law of motion and the principle of conservation of momentum.

4. The moon is constantly accelerating. Explain.

5. Read the section "The Master-stroke: Universal Gravitation" in *The Birth of a New Physics*, by I. B. Cohen, pp. 170–180, Doubleday Anchor paperback, to learn in more detail how the law of gravity derives from Kepler's laws.

6. Explain the statement that a spring balance is a weight determiner and a beam balance a mass determiner.

Problems

1. The period of revolution of Jupiter is 11.9 years and its distance from the sun is 5.20 times that of the Earth's. Show that Kepler's third law holds true if Jupiter is compared with the Earth.

2. A 1-kg object is placed at the distance of Jupiter from the sun; another 1-kg object is placed at the distance of the Earth from the sun. What is the ratio of the gravitational force of attraction between the sun and each object?

3. Find the force of gravitational attraction between the Earth and the moon. The mass of the Earth m_E is 6.0×10^{24} kg, and the mass of the moon m_m is 7.4×10^{22} kg. The radius of the moon's orbit is 3.8×10^8 m. ($G = 6.6 \times 10^{-11}$ nt-m^2/kg^2)

4. The moon makes one complete revolution about the Earth in 27.3 days (2.4×10^6 sec). Determine (*a*) the speed (tangential velocity) of the moon in its orbit; (*b*) the centripetal acceleration of the moon; (*c*) the centripetal force exerted on the moon to cause this centripetal acceleration. Compare this force with the gravitational force determined in problem 3.

5. Calculate the weight (in newtons) of a 10-kg mass.

6. Determine the mass (in kilograms) of a 50-lb weight. A 1-kg mass weighs about 2.2 lb.

7. An elevator with a mass of 1800 kg descends with an acceleration of 2.0 m/sec². What is the tension in the supporting cable?

8. A man who weighs 180 lb on the ground ascends in an elevator which accelerates at 2.2 ft/sec². What is the man's apparent weight during this acceleration?

9. Calculate the velocity necessary to place a satellite in a circular orbit 800 miles above the surface of the Earth.

10. A particle receives an impulse of 150 nt-sec. What is its change of momentum?

Chapter Six

Faraday, and Electricity and Magnetism

Oersted demonstriert die Ablenkung der Magnetnadel durch den elektrischen Strom
Nach einer Zeichnung von K. Storch

The observation of forces acting at a distance was troublesome with phenomena other than gravitation. The lodestone, a rock with naturally acquired properties of a magnet, attracts a small piece of iron even if the two are not in physical contact. A lodestone extends its "powers" beyond physical contact. The Greeks (first Thales and later Aristotle) endowed the lodestone with the animate powers of a soul to account for this action at a distance.

Amber* has similar attractive powers, for when rubbed it will attract small pieces of paper, "bits of straw and the chaff of grain." Again since early times men had observed that the amber causes the chaff to move, even though the two are not touching each other.

ATTRACTION AND REPULSION

Although known for centuries, little use was made of these phenomena until the beginning of the seventeenth century. In the year 1600, William Gilbert, an English physician and scholar, published his work *On the Lodestone and Magnetic Bodies, and on the Great Magnet the Earth*. This work was one of the very first to introduce the concept of experimentation into the study of nature. Hence Gilbert, along with Galileo and Bacon, should be given some recognition as one of the "fathers of experimental science." In fact, Gilbert had a direct influence on both the other men.

Gilbert was the first to recognize clearly the difference between magnetic and electrical phenomena. A magnet does not have to be rubbed to attract bits of iron; amber will attract small pieces of matter *only* if it is rubbed. A magnet attracts only pieces of iron; rubbed amber will attract many different substances, even drops of water. Amber, however, does not affect all substances; for example, it has no effect on a candle flame. A magnet, placed on a piece of wood floating in water, will align itself nearly north and south; amber, even if rubbed, will not so align itself. The end of the magnet that points north is called the north-seeking pole, or simply the north pole; the other end is called the south pole.

To explain the properties of the magnet, Gilbert relied on the Greek idea endowing the magnet with animate powers of a soul. To explain the ability of amber to exert a force at a distance, however, he formed a rather elaborate theory incorporating a substance he called *effluvia* (plural). He described effluvia as leaving the rubbed amber and drawing the bits of chaff to it.

No action can be performed by matter save by contact [and yet] these electric bodies do not appear to touch; ... of necessity something is given out from theone to the other to come into close contact therewith, and be a cause of incitation to it (1).

But ... effluvia lay hold of the bodies with which they unite, enfold them, as it were, in their arms, and bring them into union with the electrics ... (2).

---

* Amber is fossilized pitch of the pine tree which, fortunately for paleontologists, often retains insects millions of years old in an undamaged state.

His description of the effluvia as having arms is merely his way of saying that the effluvia could exert a force on bits of chaff and draw them to the electric which had been rubbed. (Gilbert classified any substance as an *electric* if it had these amberlike properties; those that did not were called nonelectrics. The Greek word for *amber* is *elektron*.)

But how could Gilbert suggest that effluvia was strong enough to pick up chaff yet so rarefied that it did not disturb the flame of a candle? All the physical evidence was there for Gilbert to draw this conclusion. The amber *must* "of necessity . . ." emit something in order to attract a bit of chaff without physically touching it. So he called the things emitted effluvia and endowed them with the properties necessary to account for all that he observed.

Gilbert, like Newton, was faced with the difficult task of explaining action at a distance. As a man of his times, Gilbert did the best he could. We cannot say that he was wrong or that he was right. His explanations are as sound as one could expect, given the intellectual environment of his time. The explanations offered were consistent with his observations; they accounted for the properties of the amber as he determined those properties, and they were used by subsequent investigators as a guide in their researches. These are, after all, the prime functions of any theory. Thus, taking these considerations into account, Gilbert was correct; he made a sizable contribution to the science of electricity and magnetism. The reason his theories do not satisfy us today is simply that we have many more observations to account for. The intellectual climate in which we work is also vastly different from that of Gilbert's day. We have benefited not only from his work but from the work of all the others during the intervening centuries.

THE ONE- AND TWO-FLUID THEORIES OF ELECTRICITY

In the closing years of the seventeenth century and the first few years of the eighteenth, men began to build machines by which they could generate quite a large amount of electric charge. The construction of these machines was based on the observation that substances such as amber, glass tubes, and other "electrics" can be electrified by merely rubbing them. The machines simply improved upon this rubbing process. For example, in several of these machines, called *electrostatic generators,* large globes, rotating at high speeds, were charged by placing a hand on them. The culmination in the development of these generators was reached in 1931 with the Van de Graaff generator, invented by the late R. J. Van de Graaff, formerly at the Massachusetts Institute of Technology.

The Two-Fluid Theory

In the early part of the eighteenth century, investigators discovered that the "electric virtue" could be conducted over large distances along metallic wires, and that not only do electrified bodies attract objects, but under certain conditions they also repel one another. Strangely, electrical repulsion was not observed until one hundred years after Gilbert. Charles Dufay (1698–1739) in France was the first to report this observation.

I discovered a very simple principle that accounts for a great part of the irregularities and, if I may use the term, caprices which seem to accompany most of the experiments on electricity. This principle is that an electrified body attracts all those that are not themselves electrified, and repels them as soon as they become electrified by . . . [conduction from] the electrified body. Thus gold leaf is first attracted by the tube [a glass tube which had been rubbed]. Upon acquiring an electricity . . . [by conduction from the tube], the gold leaf is of consequence immediately repelled by the tube. Nor is it reattracted while it retains its electrical quality. But if . . . the gold leaf chance to light on some other body, it straightway loses its electricity and consequently is reattracted by the tube, which, after having given it a new electricity or repels it a second time (3).

Dufay's single fortuitous observation led him to suggest that there are two kinds of electric fluids, a *vitreous* electric fluid and a *resinous* electric fluid. He defined resinous electricity as that which appears upon a rubbed piece of silk string; vitreous electricity appears on a piece of rubbed glass. Dufay suggested that two objects both with the same kind of fluid repel each other, whereas two objects each with different kinds of electricity attract each other.

For the first time electrical repulsion was recognized as such. Yet the phenomenon must have been fairly common and known even to Gilbert. Why had Gilbert not recognized it as a repulsion? Surely small bits of chaff must have been attracted to a piece of amber he had rubbed, only to be repelled by that "electric" after the two had come into contact.

We can only surmise that Gilbert did observe it, but that his theory of effluvia, which could not account for repulsion, had so conditioned his thinking that he could not accept these observations as the consequence of any action or property of the amber. Hence he must have passed it off as a trivial occurrence, the action of gravity perhaps. Herein lies the danger of theories—they are likely to prejudice our thinking so that we fail to recognize the significance of an observation.

By the middle of the eighteenth century, the effects of electricity had become well known and indeed famous. As better electrostatic generators were built, apparatus to store the electricity generated was devised (we now call these storage elements *capacitors*). Furthermore, public lectures on the subject of electricity were given throughout Europe and England. Each lecturer attempted to outdo the others in spectacular showmanship and startling, indeed, shocking effects.

The One-Fluid Theory

One lecturer, a Dr. Spencer of Scotland, traveled to the British colonies in America, and in 1743 he performed in Boston before an audience that included the thirty-seven-year-old Benjamin Franklin (1706-1790). Franklin was quite impressed by the demonstration, and having acquired enough money to permit free time for investigations, he proceeded to set up a very active research program. One result of this study is well known. Franklin flew a kite during a thunderstorm and identified, without question, the electric nature of lightning. Franklin's theory of electricity, a theory that differed from that of Dufay, is less familiar to the general public but is nonetheless more important to the development of science.

Franklin made famous the *one-fluid* theory of electricity. In 1747, he claimed that if object A, perhaps a piece of silk, rubs object B, perhaps a piece of glass, the glass takes on an excess amount of this one fluid, and the silk is left with a deficiency.

Hence have arisen some new terms among us. We say B is electrized *positively;* A, *negatively*. Or rather, B is electrized *plus;* A, *minus*. And we daily in our experiments electrize *plus* or *minus*, as we think proper. To electrize *plus* or *minus* no more needs to be known than this: that the parts of the [glass] tube or sphere which are rubbed do, in the instant of the friction, attract the electrical fire, and therefore take it from the thing rubbing; the same parts immediately, as the friction upon them ceases, are disposed to give the fire they have received to any body that has less (4).

As Dufay had postulated that rubbed silk thread gains *resinous electricity*, so Franklin postulated that glass when rubbed takes on the fluid and becomes *positive*. Dufay's terminology has been dropped; Franklin's is still with us. A glass rubbed by a silk cloth assumes a positive charge, the silk becomes negative.

Franklin, who was strongly influenced by Newton (whom he nearly met during one visit to England), assumed that the electrical fluid is composed of minute particles that repel each other. The most significant aspect of Franklin's work, however, was his view that the fluid leaving one body must travel to another; the fluid is neither created nor destroyed. Today we call this the *principle of conservation of electric charge*, and it stands, with the other conservation principles, as one of the pillars of physics.

Franklin's one-fluid theory, as influential as it was, did not account for all observations. For example, he could not explain why two bodies, each having a negative or minus quantity of electricity, repel each other.

To explain the repulsion of two negatively charged bodies, it was suggested that if two objects each have a deficiency of electric fluid, the particles of ordinary matter will repel one another. According to this view, if two bodies have normal quantities of electric fluid, the electric fluid is just sufficient to counterbalance the repulsion of ordinary matter, since electric fluid and ordinary matter attract each other.

To account for electrical repulsion by supposing that, in the absence of the electric fluid, particles of matter repel each other is, however, in direct contradiction with Newton's law of gravity, and this was not very acceptable. The law of gravity, after all, makes clear that particles of matter attract each other, and that the closer they are to each other, the stronger is the attractive force between them. The route of escape from this dilemma appeared in 1758 when Boscovich suggested that particles might repel each other at one distance and attract at another distance (see p. 132).

When completed, both the one-fluid theory and the two-fluid theory of electricity were able to account for all the observations of electrical phenomena known in the eighteenth century, and certain investigators felt that the two theories were not as different as they appeared to be. In each theory the particles or portions of one fluid repelled the particles or portions of that same fluid, and attracted particles or portions of the other (Figure 6-1). Consequently, the only real difference between the two theories was that both fluids in the two-fluid theory moved, whereas in the one-fluid theory only the electrical particles moved, the particles of matter remaining fixed.

Until direct evidence of such particle motion was obtained, it was impossible to say which theory was the better description of the capricious effects of electrification. In the intervening years it has become apparent that both theories are acceptable. The one-fluid theory may be used to explain

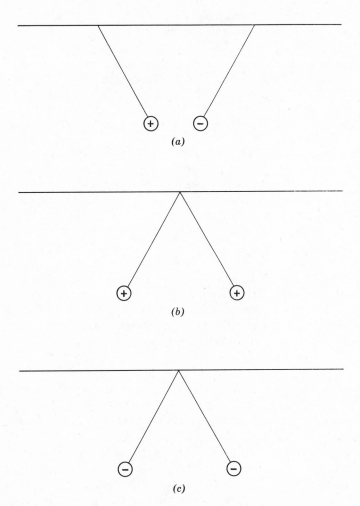

FIGURE 6-1

In *(a)* unlike charges attract. The plus sign is Franklin's designation of an excess of "electrical fluid," the negative, a deficiency. In *(b)* and *(c)* like charges repel.

the actions of electrical phenomena in solids, and the two-fluid theory may be used to explain electrical effects in both liquids and gases.

COULOMB'S LAW

Through Franklin's initiative the researches into electrical phenomena became more quantitative. One of the most pressing problems was to develop the relationship between the forces exerted by two electrically charged bodies and the distance separating them. As a consequence of Newton's law of gravity, it was natural to suspect (and perhaps hope) that this force is also an inverse square law force, but this suspicion had to be proved by actual measurements.

Since the electric force exerted between two small electrically charged objects is very small, how can this minute force be measured? Charles Coulomb (1736–1806), a French military engineer turned physicist, was the first to make the measurements leading to the establishment of the inverse square law of electric forces, now called *Coulomb's law*.

The Torsion Balance

During his studies of the mechanical properties of matter, Coulomb found that only a small force is necessary to twist a strand of wire, and by careful measurements he determined the relationship between the force causing the wire to twist and the angle through which it twists. This observation led him to the conclusion that a fine wire might be used to measure minute electric forces.

When discussing a twisting action, it is not sufficient to talk of the force alone. For example, if a bolt cannot be unscrewed with a short-handled wrench, perhaps a long-handled wrench will do the job. That is, the longer wrench offers more "leverage." Even if the same force is exerted, the longer wrench has more effect on the bolt because that force acts at the end of a longer lever arm. The twisting action of a force exerted at the end of a lever arm is called a *torque*, defined mathematically as

$$\tau = F \times R$$

where τ (Greek letter tau) is the torque, F is the force, and R is the lever arm (Figure 6-2). By definition, the lever arm is the shortest distance (the perpendicular distance) between the center or axis of rotation and the line along which the force acts.

Coulomb found that the angle through which a wire can be twisted depends on the material from which the wire is made, on the length and cross-sectional area of the wire, and on the torque. If the torque is doubled, the angle is doubled; that is, there is a direct relationship between the torque and the angle through which the wire is twisted. The relationship is

$$\theta = K\tau$$

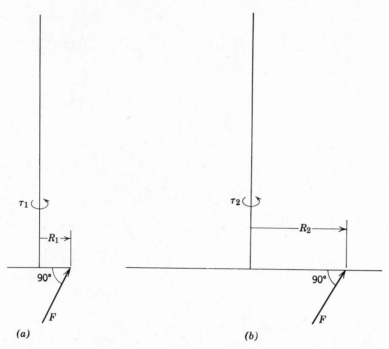

FIGURE 6-2

Given the same force F, the torque on the lever arm will increase in proportion to the length of the lever arm (a and b). Given the same lever arm, the torque will be proportional to the force F. In this figure $\tau_2 > \tau_1$.

where θ (Greek letter theta) is the angle through which the wire is twisted and K is the constant of proportionality. This constant K depends on the material from which the wire is made, on the cross-sectional area of the wire, and on its length.

The Inverse Square Law Again

If the wire is made fine, and if a small arm with a pith ball* at each end is carefully balanced and suspended from this wire (Figure 6-3a), it is then possible to measure the small forces of electrical repulsion. A third pith ball is introduced and fixed in place next to one of the suspended and resting ith balls (Figure 6-3b). Two pith balls are now in contact and a fourth pith ball with an electric charge is introduced to charge the two that are touching. These two pith balls, each with the same electric charge, now repel each other (Figure 6-3c); the angle through which the wire is twisted as the result of this repulsion can be measured.

To determine the exact relationship between the force exerted and the distance between the two pith balls, Coulomb suspended the fine wire from

---

* Pith is a substance of very low density found inside certain hollow-stemmed plants. A small pith ball is therefore very light and when charged with electricity is easily accelerated by any electric forces acting on it.

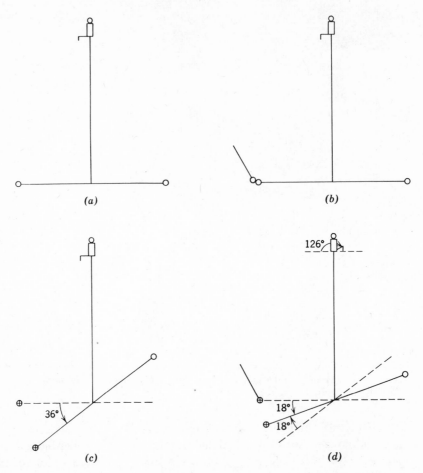

FIGURE 6-3

(a) The torsion balance rests in equilibrium. *(b)* A third pith ball is brought into contact with one of the pith balls on the torsion balance, and fixed in place. *(c)* The two pith balls in contact are both given the same charge. They then repel each other, twisting the support in a counterclockwise direction (as seen from above). *(d)* By twisting the top of the support in a clockwise direction, the two charged pith balls are brought closer together, increasing the force of repulsion.

a knob that could be rotated. A micrometer was attached to the knob to measure the angle through which the knob was turned. By rotating the knob attached to the top of the wire, the distance between the two pith balls could be controlled.

If the two pith balls repel each other so that they rotate or twist the bottom of the wire in a counterclockwise direction (as seen from above), they can be brought closer together by rotating the top of the wire in a clockwise direction (Figure 6-3*d*). Because the force of repulsion increases as the two pith balls are brought closer together, the top of the wire will have to be twisted through a larger angle than that through which the bottom turns.

An increase in the force of repulsion means that a greater torque is required to twist the wire, and consequently the wire must be twisted through a greater angle.

In his paper presented to the French Academy of Sciences in 1785, Coulomb quoted some representative measurements. When the two pith balls were first charged, the bottom of the wire was twisted through an angle of 36 degrees. Coulomb then twisted the top of the wire through an angle of 126 degrees in order to reduce the separation of the balls to only 18 degrees (one half of 36 degrees). Since the bottom of the wire was twisted 18 degrees in one direction and the top 126 degrees in the other direction, the total twist in the wire was then 144 degrees (18 + 126). But 144 = 4 × 36. Comparing these two measurements, we see that as the distance is reduced by a factor of one half (36 to 18 degrees), the force increased by a factor of four (36 to 144 degrees). This result is in agreement with an inverse square law of distance (Table 6-1).

Table 6-1

Measurement	Separation	Micrometer	Total Twist
1	36°	0°	36°
2	18	126	144
3	8½	567	576

In the third measurement the top of the wire was twisted through a total angle of 567 degrees to again reduce the separation by a factor of one-half, to 8½ degrees (nearly 9 degrees). This resulted in a total twist of 576 degrees (four times 144 degrees). Again, when the distance is reduced by one-half, the force is increased by a factor of four. The forces of electrical repulsion were *measured* to be inversely proportional to the square of the distance

$$F \propto \frac{1}{d^2}$$

Coulomb realized that the linear distance between the two charged pith balls is not really measured by the angle. He felt that the difference between the *chord* of a circle, the linear distance between the two pith balls, and the *arc*, measured by the central angle, was not significant in this experiment.

The Product of Two Electric Charges

Demonstrating that electric forces decrease as the inverse square of the distance from a charged body is only part of the law formulated by Coulomb. This much is certainly analogous to Newton's law of gravity; but the law of gravity says something about the "quantity of matter," the mass of the bodies exerting the gravitational force. Coulomb felt that the law of electric forces must contain something comparable about the quantity of electric charge. But before Coulomb's time nobody had measured an amount of electricity, so he devised his own method.

Coulomb realized that an electric charge will distribute itself evenly over the object on which it is placed. Hence, if one of two identical pith balls is charged and the other is not, then after they are brought into contact with each other, the charge will distribute itself evenly over both balls. Each ball will now hold half the charge that the first contained before contact. If a third uncharged pith ball of the same diameter is brought into contact with either of the first two, the charge will again be divided in half, so that each of these latter two will now hold one-fourth the charge on the original ball.

By keeping the distance between the pith balls constant and by repeatedly reducing the charge by a factor of one-half, first on one and then on the other, Coulomb was able to show that the force of electrical repulsion does indeed depend on the quantity of electricity on each of the pith balls in a manner analogous to the way gravitational forces depend on mass. The force of electrical repulsion is proportional to the product of the electric charges,

$$F \propto q_1 q_2$$

where q_1 and q_2 represent the quantity of electric charges. Hence the law of electrical force becomes

$$F \propto \frac{q_1 q_2}{d^2}$$

Coulomb also showed that this same law applies for attractive as well as repulsive electric forces.

This proportion may be stated as an equality if a constant of proportionality k is included:

$$F = k \frac{q_1 q_2}{d^2}$$

It now remains to establish appropriate units for the quantity of electric charge. In the mks system of measurements the unit of electricity is the *coulomb*. In practice, the magnitude of the coulomb is not determined by measuring forces between known charges but by other more accurate methods. It turns out that the constant k in Coulomb's law is not equal to one for the mks system but is measured to be nearly equal to 9×10^9 newton-meters^2 per coulomb^2. Consequently, in the mks system, Coulomb's law can be written

$$F = (9 \times 10^9) \frac{q_1 q_2}{d^2}$$

With this choice of units two similar electric charges, each of one coulomb placed one meter apart, will exert a force of 9×10^9 newtons on each other! This is a very large force, nearly 2×10^9 pounds or one million tons! Clearly, it is not easy to place two such charges that close together. In fact, it is not easy to isolate one coulomb of a single kind of charge because one coulomb is such a large charge. Nevertheless, the coulomb is retained as the unit of electric charge because it is so conveniently applied to electric currents (flow of electric charge).

Coulomb's ingenious instrument, called a torsion balance, had been used independently by Henry Cavendish who was able to measure the gravitational force of attraction between two lead masses. Cavendish announced his results in 1798. During the latter half of the eighteenth century the inverse square law became firmly established as the best method of describing how action-at-a-distance forces vary when the distance is altered. The fact that both electric and gravitational forces behave in much the same fashion has intrigued physicists from the late eighteenth century until the twentieth century. Both electric and gravitational forces are inverse-square-law forces. Electric charges, however, may attract or repel other electric charges, depending on the charge; that is, electric forces are both attractive and repulsive. There is, apparently, only an attractive gravitational force.

CHEMICAL ACTION GENERATES AN ELECTRIC CURRENT

Until the second half of the eighteenth century all investigations, all observations, and all theories in the study of electricity had been confined to what we call *static electricity*, that is, electricity which does not flow.

Galvani and Frogs' Legs

After 1780, however, the approach to studies of electricity suddenly changed. In that year Luigi Galvani (1737-1798), a professor of anatomy at the University of Bologna, fell heir to one of those fortuitous discoveries in science. He related the incident and the succeeding events some years later (in 1791). He had

dissected and prepared a frog, and laid it on a table, on which, at some distance from the frog, was an electric machine. It happened by chance that one of my assistants touched the inner crural nerve of the frog, with the point of a scalpel; whereupon at once the muscles of the limbs were violently convulsed.

Another of those who used to help me in electrical experiments thought he had noticed that at this instant a spark was drawn from the conductor of the machine. I myself was at the time occupied with a totally different matter; but when he drew my attention to this, I greatly desired to try it for myself, and discover its hidden principle. So I, too, touched one or other of the crural nerves with the point of the scalpel, at the same time that one of those present drew a spark; and the same phenomenon was repeated exactly as before (5)

A chance happening, an accident; had it occurred to someone untrained, someone without the curiosity and the ability to interpret the incident, it would have gone unnoticed. But Galvani made the most of the opportunity presented him.

He suspected that the phenomenon might somehow be related to atmospheric electricity (after all it was only twenty-eight years after Franklin's famous kite experiment had aroused so much interest in this phenomenon), and he set about to determine whether his hypothesis was correct. Galvani showed that a frog's legs when hung by a crural nerve do indeed twitch when lightning flashes nearby; he then

wished to try the effect of atmospheric electricity in calm weather. My reason for this was an observation I had made, that frogs which had been suitably prepared for these experiments and fastened, by brass hooks in the spinal marrow, to the iron lattice round a certain hanging-garden at my house, exhibited convulsions not only during thunderstorms, but sometimes even when the sky is quite serene. I suspected these effects to be due to the changes which take place during the day in the electric state of the atmosphere; and so, with some degree of confidence, I performed experiments to test the point; and at different hours for many days I watched frogs which I had prepared for the purpose; but could not detect any motion in their muscles. At length, weary of waiting in vain, I pressed the brass hooks, which were driven into the spinal marrow, against the iron lattice, in order to see whether contractions could be excited by varying the incidental circumstances of the experiment. I observed contractions tolerably often, but they did not seem to bear any relation to the changes in the electric state of the atmosphere.

However, at this time, when as yet I had not tried the experiment except in the open air, I came very near to adopting a theory that the contractions are due to atmospheric electricity, which, having slowly entered the animal and accumulated in it, is suddenly discharged when the hook comes in contact with the iron lattice. For it is easy in experimenting to deceive ourselves, and to imagine we see things we wish to see.

But I took the animal into a closed room, and placed it on an iron plate; and when I pressed the hook which was fixed in the spinal marrow against the plate, behold! the same spasmodic contractions as before. I tried other metals at different hours on various days, in several places, and always with the same result, except that the contractions were more violent with some metals than with others (6).

Galvani's discovery led directly to the development of the electric battery and the electric current; the term "current" was chosen because of the apparent fluid nature of electricity. This current was also called a galvanic current.

It is significant that these results have absolutely nothing to do with the initial observation made in Galvani's laboratory, the twitching of a frog's leg when its nerve was touched by a scalpel at the same time that a spark was drawn from an electrostatic generator. It is difficult to predict where a particular discovery will lead the development of science.

Volta and the Voltaic Cell

Allesandro Volta (1745-1827), professor of physics at the University of Pavia, studied the phenomenon that came to be known as galvanism (note: galvanized iron). During these studies he discovered a technique by which a continual supply of electricity can be produced.

FIGURE 6-4

 The first battery, made by Volta, was a series of zinc *(Z)* and silver *(A)* discs separated by a series of discs of cardboard soaked in salt water.

Thirty, forty, sixty, or more pieces of copper, or better of silver, each applied to a piece of tin, or, much better still, of zinc; and an equal number of layers of water or of some other liquid which may be a better conductor than simple water, such as brine, lye, or pieces of card, skin, etc., well soaked in these liquids, such layers being interposed between each pair or combination of two different metals, —one such alternate series, and always in the same order, of these three kinds of conductors, is all that constitutes my new instrument (7).

 This first basic "instrument" (Figure 6-4) was the forerunner of the modern electrochemical battery. The electrochemical battery is more appropriately called a *voltaic cell*, which may be defined as any instrument that produces an electric current through chemical action. Volta himself, however, did not recognize the chemical nature of the instrument he had invented, but it was soon realized that before any electric current could be produced, a chemical action had to take place. This realization led investigators to suspect that electricity and matter were really very intimately related. The discovery that electric current from a voltaic cell decomposed water into hydrogen and oxygen (see p. 167) was further evidence of this close relationship.

MAGNETIC EFFECTS OF ELECTRIC CURRENTS

Oersted: Electric Currents and a Compass

 The invention of the voltaic cell led to the discovery of yet another significant relationship, the relationship between electricity and magnetism. Investigating a possible connection during one of his lectures, Hans Christian Oersted

(1777-1851), professor of natural philosophy in Copenhagen, placed a compass needle close to and perpendicular to a wire. He then passed an electric current through the wire but could detect no response in the compass needle. After the lecture, however, he placed the compass needle parallel to the wire, that is, he placed the wire north and south, and started an electric current through the wire. The compass needle now responded with a pronounced deflection away from its position parallel to the wire. His announcement in 1820 of the discovery of what he called an "electric conflict" about a current-carrying wire was a major step in the development of physics.

The direction in which the compass needle swings is important if the apparently magnetic "region," or magnetic field (using contemporary terminology explained in Chapter 9), about the current-carrying wire is to be described. The shape of the magnetic field is readily determined by running a wire vertically through a small hole in a piece of cardboard, permitting an electric current to pass through that wire, and then sprinkling iron filings on the cardboard (Figure 6-5).

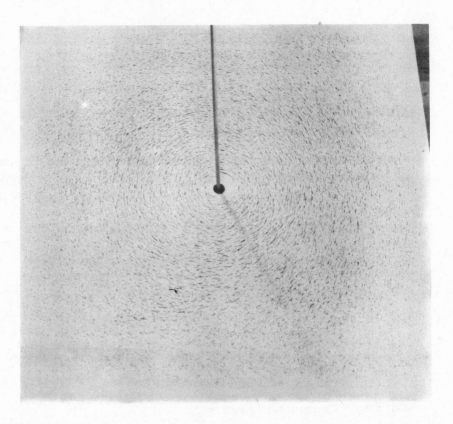

FIGURE 6-5

Iron filings sprinkled on a piece of cardboard through which a current-carrying wire passes assume a very definite pattern. (From *Physics,* Physical Science Study Group, D. C. Heath and Company, 1960.)

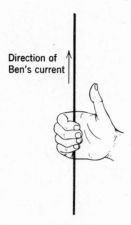

Direction of Ben's current

FIGURE 6-6

If the fingers of the right hand are wrapped about a current-carrying wire with the thumb pointing in the direction of Ben's current, the fingers will point in the direction of the north-seeking poles of compasses were they placed about the wire.

When placed in the magnetic field about a permanent magnet or a current-carrying wire, iron filings become, temporarily, small magnetic compasses. All the iron filings line up so as to form concentric circles about the current-carrying wire. In their circular pattern the north pole of each iron filing points to the south pole of its neighbor. We can therefore conclude that the magnetic field about the current-carrying wire is circular. If the iron filings were replaced by small permanent magnetic compasses, and if the direction of the current were reversed, the compass needles would then all swing around and point in the opposite direction. Therefore, to complete the description, we must relate the direction in which the compass needles point to the direction of the current in the wire. We must specify a direction for the electric current.

Early in these investigations it was established that one of the metals of each voltaic cell assumes what Franklin called a positive electric charge and the other metal a negative charge. For example, the zinc plate of a zinc-copper cell assumes a positive charge, the copper plate a negative charge. Furthermore, if we adopt Franklin's proposed direction of flow as that from the positive end of the wire to the negative end (from the zinc plate through the wire to the copper plate) and call this *Ben's current*, we are able to relate this direction to the direction in which the compass needles point.

If the *right hand* grasps the wire as shown in Figure 6-6, with the thumb pointing in the direction of Ben's current, the fingers wrapping about the wire point in the same direction that north ends of the compass needles point. If the direction of the current is reversed, the hand must also be reversed and the fingers will point in the opposite direction.

If the wire, lined up in the north-south direction, does not carry an electric current, the compass needle remains in a position parallel to the wire. If an electric current is now passed through the wire, the needle will swing

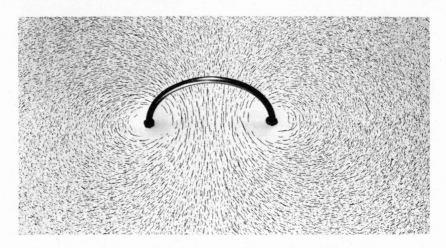

FIGURE 6-7

Iron filings outside a current-carrying loop of wire form a pattern reminiscent of the pattern about the permanent magnet (see Figure 6-12a). Inside the loop, the filings form lines; presumably they would do the same inside a permanent magnet. Many such loops placed one after another comprise a coil. (From PSSC PHYSICS, D. C. Heath and Company, A Division of Raytheon Education Company, Boston, 1965.)

one way or the other; and it was found that the larger the voltaic cell, that is, the larger the number of plates, the more vigorously the needle swings. The simple compass thus became an instrument for measuring the intensity of electric current (then called a galvanic current or galvanism). This instrument soon came to be known as a *galvanometer*, the name still applied to it today. In modern instruments, however, the needle is brought back to a central or "zero" position by a spring rather than by the Earth's magnetic field.

It was also discovered that if the wire carrying the current is wrapped in the form of a helix or coil, the magnetic region is intensified in the volume enclosed by the coil. In fact, the behavior of a coil carrying an electric current appears similar in every way to that of a regular bar magnet (Figure 6-7).

Oersted's Experiment and Newton's Third Law

Almost immediately after Oersted's announcement, a number of scientists in England and Europe were proposing a logical extension of the discovery. If a current-carrying wire can make a compass needle move, the wire must exert a force on the needle; hence by the symmetry of Newton's third law the compass needle must also exert a force on the current-carrying wire. Would it not be possible, therefore, to make an instrument in which the compass needle is fixed and the current-carrying wire free to move?

In 1821 the English scientists William Wollaston (1766-1828) and Sir Humphry Davy (1778-1829) devised a very neat instrument in which a piece of wire was placed across two knife edges and was consequently free to roll. A current could then be passed through the wire by attaching each knife to one end of a voltaic cell. The whole was placed between the poles of a magnet, such as a U magnet (Figure 6-8), so that when an electric current was passed through the wire, the wire was seen to move. The instrument gave a simple but very convincing demonstration. The outcome of the Wollaston-Davy experiment was, of course, the electric motor, but the techniques for building a usable electric motor were not developed until some years later.

Working in the same laboratory as an assistant to Davy was Michael Faraday (1791-1867), who without any formal education became one of science's leading experimentalists. He too developed a technique by which a fixed magnet could cause a current-carrying wire to move.

MAGNETICALLY INDUCED ELECTRIC CURRENTS

Seeing beyond this symmetrical relationship between electricity and magnetism, Faraday sought a more profound symmetry. If an electric current somehow creates a magnetic region about it, why should it not be possible to generate an electric current by somehow manipulating a wire in a magnetic field?

Faraday's Initial Observation

In a paper read before the Royal Society on November 24, 1831, Faraday recounted his experiment to show this symmetry, for

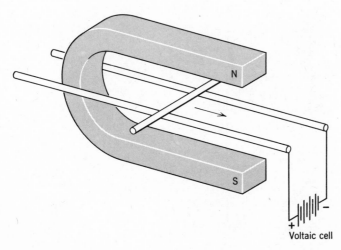

Voltaic cell

FIGURE 6-8

The current-carrying rod in this magnetic field will roll to the right. Why?

FIGURE 6-9

Michael Faraday (1791–1867). (By permission of The Royal Institute.)

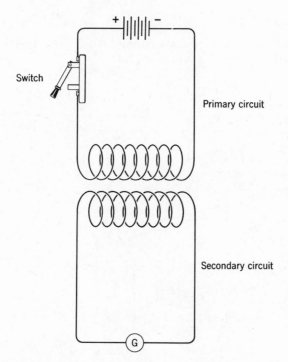

Switch

Primary circuit

Secondary circuit

G

FIGURE 6-10

As the switch in the primary circuit is either closed or opened, the changing current in the primary induces a current in the secondary.

. . . it appeared very extraordinary, that as every electric current was accompanied by a corresponding intensity of magnetic action at right angles to the current, good conductors of electricity, when placed within the sphere of this action, should not have any current induced through them, or some sensible effect produced equivalent in force to such a current.

These considerations, with their consequence, the hope of obtaining electricity from ordinary magneticism, have stimulated me at various times to investigate experimentally the inductive effect of electric currents. I lately arrived at positive results; and not only had my hopes fulfilled, . . . and also to discover a new state, which may probably have great influence in some of the most important effects of electric currents (8).

Faraday then took two coils of wire. Through one of these he sent an electric current. This coil, now called the *primary coil*, forms an electromagnet. To the *secondary coil* Faraday attached a galvanometer (Figure 6-10). He expected the magnetic compass of the galvanometer in the secondary coil to move when he directed an electric current in the primary coil, even if the two coils were not touching. Faraday felt that the magnetic field of the primary coil would induce a current in the secondary coil. But his first attempts met with failure. So he build a larger set of coils, used a larger set of voltaic cells, and found that

when contact was made, there was a sudden and very slight effect at the galvanometer, and there was also a similar slight effect when the contact with the battery was broken. But whilst the voltaic current was continuing to pass through the one helix [coil], no galvanometrical appearances nor any effect like induction upon the other helix could be perceived, although the active power of the battery was proved to be great, by its heating the whole of its own helix. . . .

Repetition of the experiments with a battery of one hundred and twenty pairs of plates produced no other effects; but it was ascertained, both at this and the former time, that the slight deflection of the needle occurring at the moment of completing the connection, was always in one direction, and that the equally slight deflection produced when the contact was broken, was in the other direction; and also, that these effects occurred when the first helices were used.

The results which I had by this time obtained with magnets led me to believe that the battery current through one wire, did, in reality, induce a similar current through the other wire, but that it continued for an instant only . . . (9).

This was the first announcement of an electric current being induced by magnetic action. When the electric current continued steadily through the primary circuit, the galvanometer did not indicate a current in the secondary. When the primary circuit was broken, stopping the electric current through the primary coil, the galvanometer was again deflected, but in the direction opposite to the deflection produced when the primary circuit was initially closed.

Continuing his researches, Faraday learned many other salient features of induced electrical currents.

1. A current can be induced in the secondary coil even while the current

in the primary coil is steady, *if the two coils are moved in relation to one another*.

2. If a small voltaic cell is placed in the secondary circuit (as well as in the primary circuit), a current can still be induced in the secondary coil and will simply add its effect to the voltaic current already there. If the two currents are in the same direction, the resulting current is stronger and is equal to the sum of the two currents. If the two are in the opposite direction, the stronger of the two overrides the other; the resulting current is equal to the stronger minus the weaker.

3. If the two coils, the primary and the secondary, are joined by a ring of iron (Figure 6-11), the induced current is greatly enhanced but still drops to zero as soon as the primary current reaches a steady or constant value.

4. An electric current can be induced in the secondary simply by thrusting a permanent magnet into the coil. The direction of the induced current depends on the direction of motion of the permanent magnet and on which end of the magnet, the north-seeking pole or the south-seeking pole, is thrust in. No current is induced while the magnet and coil are at rest in relation to each other.

FIGURE 6-11

(a) Faraday's actual iron-ring experiment. The two coils, primary and secondary, are clearly visible. *(b)* The page of Faraday's notes in which he records his experiments with the iron ring. (By permission of The Royal Institute.)

Faraday and Lines of Force

In his deliberations on the effects discussed here as well as in the many other researches he conducted, Faraday came to think of the apparent lines formed by the iron filings about a magnet as constituting *magnetic lines of force*. According to Faraday, these lines of force are continuous (Figure 6-12*a*), starting at one pole of a magnet, by convention the south-seeking pole, and passing through space to a north-seeking pole, which may not be part of the same magnet. That is, lines of force will pass from one magnet to another if the magnets are close enough. These magnetic lines of force can be traced by a compass.

Place a magnet on a large piece of paper and then place a small compass near the magnet. Mark a dot on the paper at the north and south pole of the compass needle. Then shift the compass so that the south pole lies on the dot that the north pole lay on before the compass was moved. In this step-by-step fashion a line of force can be traced (Figure 6-12*b*), and then another and another. Yet no matter where the compass is placed near the magnet, its needle will respond, even if it is placed between two of the traced lines of force. The region about the magnet is therefore not really composed of magnetic lines of force with nothing between these lines. *A magnetic field* is continuous. The concept of the field will be discussed in more detail in Chapter 9, but the description of a magnetic field in terms of lines of force will be given here. Magnetic lines of force will also serve to describe the action necessary for the formation of an induced current.

To begin with. when a magnetic compass needle. not enclosed in a case but suspended on a needlelike point, is placed in a magnetic field, the needle can be set into oscillation by rotating it slightly from its equilibrium position. Its frequency of oscillation can be determined by counting the number of oscillations that occur in a particular interval of time. If the compass needle is then moved to another part of the magnetic field, its frequency of oscillation may well change. If the frequency of oscillation is determined for many positions about the bar magnet, for which the lines of force have already been drawn, the frequency of oscillation will be greater for those regions where the lines are more closely spaced, that is, where their density is greater.

Without entering into a detailed description of the causes of oscillation, it should be clear that the needle oscillates with a higher frequency in those regions where the bar magnet exerts a greater force on each magnetic pole of the needle. In short, the frequency of oscillation gives us a measure of the strength of the magnetic field about the bar magnet. We observe that the magnetic field strength is greater in those regions where the lines of force about the magnet are closer together. It should be pointed out that the magnetic field strength does not increase by merely drawing more lines of force. The strength of the field is determined by the magnet. The number of lines of force drawn is arbitrary; a number crossing through a certain cross-

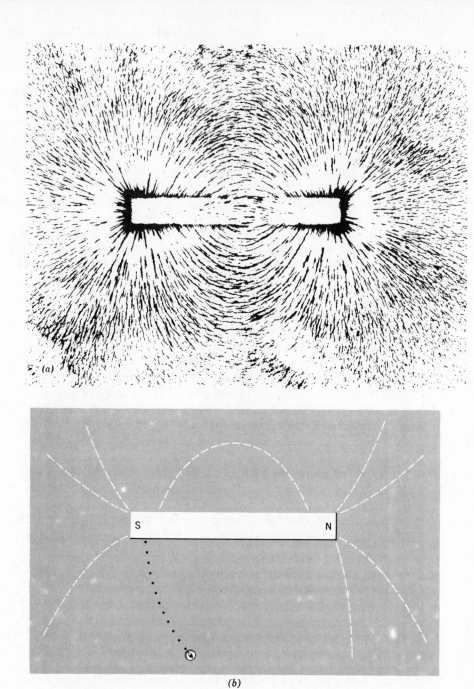

FIGURE 6-12

(a) Iron filings on top of a piece of paper under which a bar magnet rests. (From *Physics,* Physical Science Study Committee, D. C. Heath and Company, 1960.) (b) A magnetic compass can be used to trace out lines of force. The line indicated by dots terminating at the compass shows how this is done; other lines have been sketched in.

sectional area does, however, constitute one unit of measure for magnetic field strength.

According to Faraday, *if a wire is formed into a closed loop of any shape, a current will be induced in that loop whenever the number of lines of force passing through that loop changes.* If the area enclosed by the loop remains the same, we can say that a current is induced in a wire loop whenever the number of lines passing through that loop changes.*

If a magnet, either electrical or permanent, is moved in relation to the secondary coil, the magnetic lines of force will follow the magnet. If these lines of force move so that the number passing through the secondary coil changes, a current will be induced in that coil.

When the primary and secondary coils are held stationary, a current is induced *only* when the current in the primary circuit is varied, for example, when it is started or stopped. Before the current passes through the primary coil, there are no magnetic lines of force about it. As the current starts, it increases from zero to some steady value. During this time when the primary current increases, the number of magnetic lines of force are building up from zero to some constant or steady value which they achieve and maintain only when the current reaches a steady value.

The increase in the number of magnetic lines of force constitutes the motion of these lines; the number of lines intersecting the windings of the secondary coil changes, and an electric current is induced in that coil.

When the primary circuit is broken, the current stops, but not instantaneously. During the period in which the current changes from a steady value to zero, the magnetic field diminishes in strength and the magnetic lines collapse toward the primary coil; in so doing, the number of lines passing through the area enclosed by the windings of the secondary coil changes and a current is induced in the secondary coil.

The direction of the induced current depends on whether the number of lines of force intersecting the secondary coil increases or decreases. If the current in the primary is alternately started and stopped, the current induced in the secondary surges back and forth, producing what we call an *alternating current* (abbreviated ac). (The current produced by a voltaic cell is called a *direct current*, abbreviated dc.) An alternating current is also generated in the secondary coil if a magnet, either permanent or electrical, is thrust back and forth inside the secondary coil. The electricity transmitted from the large generators to the cities and towns of our technological world is alternating current.

The consequences of Faraday's work are difficult to assess; it may be said that essentially all electromagnetic devices stem either directly or indirectly from his work. And there are indeed many applications for electromagnetic devices.

---

* The term *flux* describes the same property of magnetic fields, but it refers not to the number of lines of force per unit area but to the magnetic field strength times the unit area.

Lenz's Law

One very important question remains to be answered, however. In which direction does the electric charge flow in the induced current? Faraday offered some suggestions, but the most profound answer was formulated by the Russian scientist Emil Lenz (1804-1865) and published in 1834.

Lenz's law can be derived by observing the direction of the induced current through a galvanometer, while noting the action that induced the current. The direction of that current is then related to the inducing action by some general statement applying to all conceivable situations. It must first be recognized that two magnetic fields are involved: (1) the magnetic field causing the induced current, and (2) the magnetic field formed by that induced current. The interaction of these two magnetic fields forms the basis for *Lenz's law: the induced current sets up a magnetic field that opposes the inducing action*.

For example, let us imagine a magnetic field set up by a straight section of current-carrying wire which is actually part of a large wire loop. This loop constitutes the primary circuit of our example, and the current passing through the wire loop sets up a magnetic field illustrated in Figure 6-13. The number of lines of force per unit area decreases with increasing distance from the current-carrying wire.

Adjacent to and parallel to the straight section of wire in the primary circuit is the straight section of another large wire loop. This section does not initially have a current passing through it and is therefore the secondary circuit of our example. If this secondary loop is moved away from the primary loop (i.e., the distance between the straight sections is increased), the

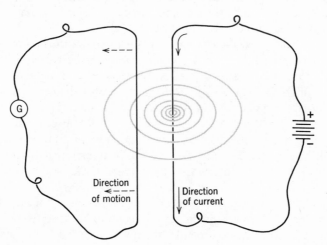

FIGURE 6-13

A current will be induced in the wire on the left when it is pulled away from (or pushed closer to) the current-carrying wire on the right.

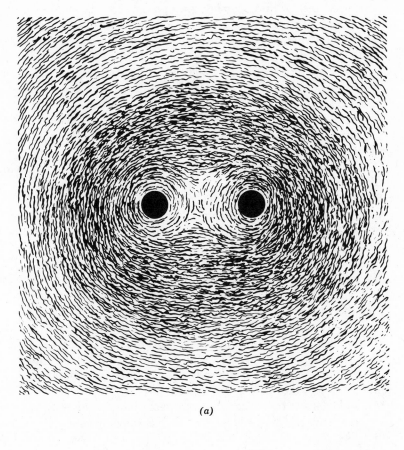

(a)

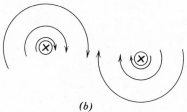

(b)

FIGURE 6-14

(a) The magnetic lines of force about two parallel wires carrying currents in the *same* direction. (Reprinted with permission from *Physical Science for the Liberal Arts Student,* by Swenson and Woods, John Wiley & Sons, Inc., 1957.) *(b)* Since the currents in the two parallel wires both flow into the paper (the tail feathers of the arrows indicate flow into the paper), the fields between the wires are in the opposite direction. These two fields tend to annul each other, so the overall electric field is weaker in the region between the two wires than in the region outside the wires.

number of lines of force passing through the secondary loop will decrease and a current will be induced. In which direction will that induced current be traveling when it passes through the straight section of the secondary circuit? Will it pass in the same direction as the current in the parallel straight section of the primary circuit, or in the opposite direction? Before we can answer this question, we must determine how the magnetic fields of two parallel wires, each carrying a current, interact with each other.

The magnetic fields of two parallel wires carrying currents in the same direction form a magnetic field, shown in Figure 6-14a. The field in the region between the two wires is weakened, which can be verified by the right-hand rule (Figure 6-14b). In both wires the direction of the current is into the page (we see the "tail feathers" of the arrow indicating the direction of flow). The right-hand rule stipulates that the compass needles point in a clockwise direction, but since each field points in a clockwise direction, the two fields oppose each other in the region between the wires. The field is therefore weaker in the middle.

Experiments show that two parallel wires carrying currents in the same direction *attract each other*. Therefore, we conclude that a current-carrying wire in a magnetic field will have a force exerted on it which moves the wire from the stronger to the weaker portion of that field.

Can we apply this observation to the previous example of a current induced in the straight section of the secondary wire loop in Figure 6-13 when that loop is forcibly moved in a magnetic field? The straight section of the secondary wire is being moved away from the straight section of the primary wire which carries the voltaic current. The movement of the wire is the inducing action; according to Lenz's law, the induced current must set up a magnetic field which opposes the inducing action. To oppose the inducing action—the movement of the secondary wire away from the primary—the two wires must attract each other. Consequently, the induced current must pass in the same direction as the voltaic current, for then and only then will these two wires attract each other and the inducing action be opposed.

Should the secondary wire be moved toward the primary wire (the one with the voltaic current), the induced current would pass in the direction opposite that of the voltaic current. The two currents, passing in opposite directions, set up a magnetic field that is stronger in the region between the wires (Figure 6-15), and the two wires will repel each other. The repulsion opposes the inducing action, the movement of the secondary wire toward the primary wire.

Lenz's law can be extended to any situation in which an electric current is induced, whether the inducing action is the mechanical motion of a wire or a changing magnetic field. If it were possible for the induced current to set up a magnetic field that favored the inducing action, the induced current could perpetuate itself, and it would be very inexpensive to generate electricity. Obviously, Lenz's law excludes this possibility; we do not get something for nothing.

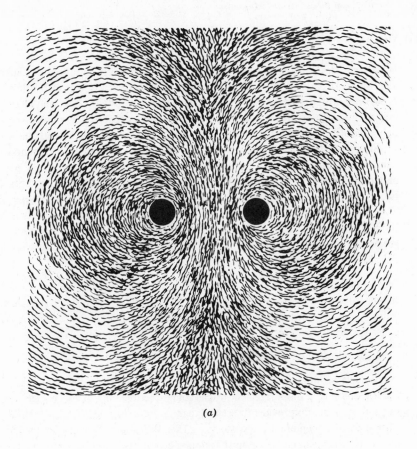

(a)

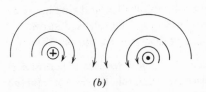

(b)

FIGURE 6-15

(a) The magnetic lines of force about two parallel wires carrying currents in *opposite* directions. (Reprinted with permission from *Physical Science for the Liberal Arts Student* by Swenson and Woods, John Wiley & Sons, Inc., 1957.) *(b)* Since the currents in the two parallel wires flow in opposite directions (the point of the arrowhead indicates flow out from the paper), the fields between the wires are in the same direction. These two fields tend to strengthen one another, so the overall field is stronger in the region between the wires than in the region outside the wires.

CHEMICAL EFFECTS OF ELECTRIC CURRENTS

Electricity does not stand as an isolated or completely independent subject. We have already discussed the similarities between Coulomb's law and Newton's law of gravity (see p. 147 f) and the relationship between electricity and magnetism as investigated by Faraday. This great experimentalist established another important relationship, that between electricity and chemistry.

Shortly after the invention of the voltaic cell, it had been learned that chemical compounds can be decomposed by an electric current, a process called *electrolysis*. This discovery was again one of those accidents that occur in science but apparently only in the presence of the "right person." Two Englishmen were involved, Sir Anthony Carlisle (1768-1840) and William Nicholson (1753-1815). In 1800 Carlisle read of Volta's discovery and set about to make a voltaic cell for himself. Nicholson, working with Carlisle, wrote of the incident.

On the 30th of April, Mr. Carlisle had provided a pile consisting of seventeen half-crowns [silver], with a like number of pieces of zinc, and of pasteboard, soaked in salt water. These were arranged in the order of silver, zinc, card, etc. . . . the silver was . . . under the zinc (10).

In the course of subsequent experiments,

the contacts being made sure by placing a drop of water upon the upper plate, Mr. Carlisle observed a disengagement of gas round the touching wire. This gas, though very minute in quantity, evidently seemed to me to have the smell afforded by hydrogen when the wire of communication was steel. This, with some other facts, led me to propose to break the circuit by the substitution of a tube of water between two wires.

After several trials using silver and zinc in the tube of water, the two men found that hydrogen was indeed released at one of the wires while the other became blackened, presumably by the reaction with oxygen. Consequently,

two wires of platina . . . were inserted into a short tube of a quarter of an inch inside diameter. When placed in the circuit, the silver side [of the voltaic cell] gave a plentiful stream of fine bubbles, and the zinc side also a stream less plentiful. . . . It was natural to conjecture, that the larger stream from the silver side was hydrogen, and the smaller oxygen. . . .

By changing the apparatus so that it would capture larger quantities of the gases, they were able to test and weigh these gases.

The process was continued for thirteen hours, after which the wires were disengaged, and the gases decanted into separate bottles. On measuring the quantities, which was done by weighing the bottles, it was found that the quantities of water displaced by the gases were respectively, 72 grains by the gas from the zinc side, and 142 grains by the gas from the silver side. . . . These are nearly the proportions in bulk, of what are stated to be the component parts of water (10)

We see that the gases were not weighed, but rather the water displaced by those gases. Nearly twice as much water was displaced by oxygen.

The idea that a chemical compound could be dissociated (broken down) into its component elements by an electric current was carried to its logical conclusion by Faraday. He established the fact that there is a definite ratio between the total amount of electricity that passes through the decomposing cell and the amount of material that is dissociated.

What Faraday called the law of definite action (now extended and called Faraday's laws of electrolysis) is simply the statement of his observation that a given quantity of electricity will always decompose a specific and definite amount of each chemical substance, the amount being different for each substance. The amount decomposed he called the equivalent weight of each substance. From this observation Faraday drew the far-reaching conclusion that

the equivalent weights of bodies are simply those quantities of them which contain equal quantities of electricity, or have naturally equal electric powers; it being the ELECTRICITY which *determines* the equivalent number, *because* it determines the combining force. Or, if we adopt the atomic theory or phraseology, then the atoms of bodies which are equivalents to each other in their ordinary chemical action have equal quantities of electricity naturally associated with them. But I must confess I am jealous of the term *atom*; for though it is very easy to talk of atoms, it is very difficult to form a clear idea of their nature, especially when compound bodies [molecules] are under consideration (11).

To the reader of today the passage appears to be a very clear statement that electricity comes in little bundles, each bundle containing the same quantity of electricity. The atoms of matter accept electricity only in that same quantity or multiples of that quantity. And when the particulate nature of electricity was ready to be accepted, this statement of Faraday's was indeed quoted as authority.

Faraday himself, however, did not subscribe to the theory that electricity is a "substance" composed of minute particles. His concept of electricity was not so clear-cut. He could not keep from thinking in terms of the lines of force that had served him so eminently. He imagined that the passage of an electric current was somehow related to a temporary alignment of the atoms of matter. The atoms were aligned, in his view, by the process that generated the electricity itself. Again preconceptions misled a truly great scientific investigator. This erroneous view, however, does not in the slightest detract from the greatness of Michael Faraday.

Today we recognize that elements differ in their electrical nature. In decomposing hydrochloric acid into hydrogen and chlorine gas, each atom of hydrogen gains one electron and each atom of chlorine gives up one electron. In the decomposition of hydrogen sulfide, each atom of hydrogen again takes on one electron but each atom of sulfur gives up two electrons. Hence the molecule of hydrogen sulfide is composed of one atom of sulfur and two atoms of hydrogen, H_2S.

An important result of Faraday's work with electrolysis was that it established a means for measuring a definite amount of electricity. The basic unit of electricity is the coulomb (see p. 148). In the study of electric currents we are concerned with the quantity of electricity flowing per unit time interval; the unit for the rate of flow of electricity is the *ampere*, defined as the passage of 1 coulomb per second. To liberate 1 gram of hydrogen in electrolysis, 96,500 coulombs of electricity are required. The same amount of electricity is required to liberate 23.0 grams of sodium and 35.5 grams of chlorine. The numbers 1, 23.0, and 35.5 are in proportion to the atomic masses of hydrogen, sodium, and chlorine respectively. Because this quantity of electricity is so fundamental, it has been given a special name, the *faraday*.

References

1. William Gilbert, *De Magnete*, Dover Publications, New York (paperback), 1958, p. 92.
2. *Ibid.*, pp. 89 ff.
3. James B. Conant and L. K. Nash, eds., *Harvard Case Histories in Experimental Science*, Vol. II, Harvard University Press, Cambridge, Mass., 1957, p. 585.
4. *Ibid.*, p. 598.
5. Sir Edmund Whittaker, *A History of the Theories of Aether and Electricity*, Harper Torchbook, New York, 1960, pp. 67 f.
6. *Ibid.*, pp. 68 f.
7. A. Wolf, *A History of Science, Technology, and Philosophy in the 18th Century*, Harper Torchbooks, New York, 1961, p. 264.
8. Morris H. Shamos, ed., *Great Experiments in Physics*, Holt-Dryden, New York (paperback), 1959, p. 131 f.
9. *Ibid.*, p. 133.
10. A. Wolf, *op. cit.*, pp. 266 f.
11. Michael Faraday, *Experimental Researches in Electricity*, Great Books of the Western World, Encyclopaedia Britannica, Chicago, Vol. 45, pp. 389 f.

Questions

1. Why is it that the one-fluid theory can account for the electric actions in solids, and the two-fluid theory is required to account for the actions in fluids (both gases and liquids)?

2. Why should electricity spread itself uniformly over a conducting spherical surface?

3. In reading the portion of Cavendish's experiment and the portion of the experiment performed by Coulomb contained in *Great Experiments of Physics*, M. Shamos, ed., Holt-Dryden Press, 1959, it becomes evident that Cavendish had to take far greater precautions against extraneous effects from the surroundings and to make much more delicate readings than Coulomb did. Explain.

4. Comment on the probability of an important discovery, such as Galvani's, being made by someone relatively untrained in the sciences.

5. Faraday sought to induce an electric current by magnetic action, since an electric current produces a magnetic effect. His reasoning was one of symmetry. If A produces B, why should not B produce A? Comment on the concept of symmetry when applied to a field of study such as physics. Is it desirable? Is it to be relied on? Is it a man-made crutch?

6. If a wire moves parallel to magnetic lines of force, will a current be induced in that wire? Explain.

7. When an alternating current is directed through a primary coil, it will appear to pass through a secondary coil as well, if the two coils are placed close together but are not touching. Such an arrangement is called a transformer. Will a voltaic current appear to pass through a transformer? Will a pulsed direct current (on-off, on-off, . . .) appear to pass through a transformer? Explain.

8. A fairly loosely coiled light spring will, if permitted, contract when it conducts an electric current. Explain.

9. A positively charged particle moves northward through a horizontal magnetic field directed to the east. In which direction will the force exerted on that particle act?

10. A negatively charged particle moves northward through a vertical field directed down. In which direction will the force exerted on that particle act? (Use the left-hand rule rather than the right-hand one for negatively charged particles.)

Problems

1. A force of 6.0 nt acts on a lever arm of 0.42 m. What is the resulting torque?

2. In a hypothetical repetition of Coulomb's experiment incomplete measurements were made. Predict what the missing measurements should be (fill in the blank spaces in the following table).

Measurement	Separation	Micrometer Reading	Total Twist
1	48°	0°	48°
2	24		192
3		756	

3. Calculate the electric force of repulsion between two positive charges of 0.01 coul each when placed 2 m apart.

4. At what distance should an electric charge of 1.0 coul be placed from an equal charge of the same kind if the electric force of repulsion is to be equal to 1.0 nt?

5. If an electric current of 5 amp passes through an electric heater for 10 minutes, how many coulombs of electricity pass through the heater?

Heat, Work, and Energy

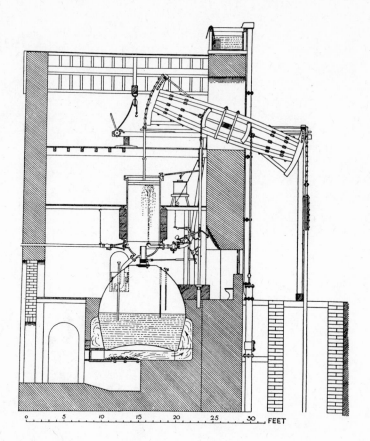

M an's investigations of the phenomenon of heat provide a striking example that the development of a field of study in physics depends almost entirely upon the ability to express ideas quantitatively. Clearly, mankind has always been aware of the sensations of "hot" and "cold," and certainly, the theories of the early natural philosophers accounted for heat. Until the nineteenth century, two theories of heat competed for acceptance by the scientific community: the first theory held that heat is a substance that flows something like water; the second maintained that heat is a form of vibratory motion. Those who professed heat to be a form of vibration were of necessity believers in the particulate or atomistic nature of matter. According to this view, atoms vibrate, and the faster they vibrate, the hotter the substance becomes. In the seventeenth century influential men like Bacon, Descartes, and Newton supported the idea that heat is a form of motion, a vibratory motion of atoms. Yet until the nineteenth century all theories of heat were qualitative and speculative and were not subject to verification.

THE THERMOMETER AND TEMPERATURE

The invention of the thermometer was the *first step* quantifying the study of heat. That invention ultimately dates back to Galileo and an observation made sometime between 1592 and 1603. Galileo noticed that if a glass stem with a bulb on one end is filled with air and the stem is emersed in water (Figure 7-1), heating the bulb lowers the level of the water in the stem. If the glass bulb is cooled, the level of the water in the stem rises.

The arrangement of the glass bulb and stem did not constitute a thermometer, and although Galileo attached a piece of paper to the stem marked off in "degrees," he apparently did not realize the significance of his observations. The instrument Galileo used was only semiquantitative, for the markings were not standardized in any way. Had he written of his experiments, using the numbers read from the markings on the paper, other investigators would not have known what those numbers meant. What Galileo really demonstrated is that air expands noticeably when heated, even by the hand, and contracts when cooled.

Galileo's air "thermometer" has been replaced by the modern thermometer, which consists of a small glass bulb on the end of a short glass tubular stem. A fluid that expands when heated and contracts when cooled is sealed inside. The fluids commonly used are alcohol (colored red) and mercury. Sealing the tube makes the modern thermometer insensitive to changes in the pressure of the Earth's atmosphere; Galileo's thermometer was subject to such changes. Yet the modern thermometer still does nothing more than indicate the same property that Galileo demonstrated was characteristic of air: mercury and alcohol expand when heated and contract when cooled.

173

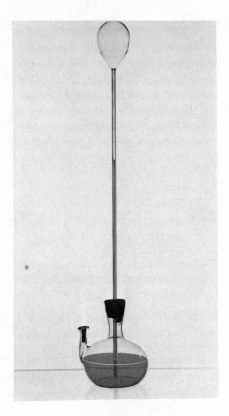

FIGURE 7-1

Galileo's "thermometer" was only a means of demonstrating that air expands when heated and contracts when cooled. His "thermometer" was also sensitive to changes in the atmospheric pressure.

TEMPERATURE SCALES

Investigators learned that when the thermometer is immersed in a mixture of melting ice and water, the level of mercury is always located at the same height in the stem; similarly, when the thermometer is immersed in boiling water, the mercury level always rises to the same height—or nearly the same height. Locating these two points, appropriately marked on the glass stem, constituted the *second step* in quantifying the studies of heat. They are points of reference that can be duplicated in any laboratory.

The *third step* in quantifying the study of heat was to establish the number of evenly spaced marks to be included between these two quite arbitrarily chosen points on the glass stem. The selection of two reference points and the evenly spaced divisions between them creates a temperature scale.

A temperature scale can also be set up by selecting only one reference point and then choosing an arbitrary division for the scale markings drawn on the thermometer stem above and below the reference point. In the early eighteenth century, scientists, choosing one method or the other, established a number of temperature scales; those we use today derive from them.

The most important thermometer scale for the physicist is now called the *Celsius* scale, in honor of the inventor, Anders Celsius, a Swedish astronomer. (In the United States this scale has been called the centigrade scale.) The two reference points on this scale are the freezing point and the boiling point of pure water at sea level; the former was arbitrarily designated as 0°, the latter as 100°.

The second scale, the one used in English-speaking countries (but not by the scientists in those countries), is the Fahrenheit scale, named in honor of its originator, G. D. Fahrenheit, a German-born maker of meteorological instruments. This scale uses the same points of reference but designates them as 32° and 212° respectively. The zero point on this scale was originally the temperature of a mixture of ice, water, and table salt. Between the freezing and boiling points of pure water there are 180 divisions on the Fahrenheit scales, whereas on the Celsius scale there are only 100 divisions.

It is fairly simple to convert one temperature reading to another by referring to an intermediate scale, which allows the conversion from 100 divisions to 180 (or from 180 to 100) to be done on scales with the same zero point (Figure 7-2). Not only does the intermediate scale graphically indicate the meaning of the two equations commonly given for the transformation of temperatures from one scale to the other, but it also eliminates the necessity of memorizing these equations (which are usually forgotten when they are most needed).

Anticipating our needs in later chapters, we shall now describe a third temperature scale, the Kelvin scale, named in honor of Lord Kelvin. The zero points on the Celsius and Fahrenheit scale were chosen arbitrarily, but the zero point on the Kelvin scale is fundamental and is dictated by observational and theoretical evidence of a wide variety. Called absolute zero, it is, according to contemporary theory and practice, the lower limit of temperature and therefore colder than any temperature men can achieve. It is in this sense that absolute zero is of fundamental importance.

Although absolute zero was established from observational evidence, the scale itself, that is, the size of the divisions, was chosen arbitrarily to be the same as the Celsius scale. The zero point and the divisions of the Kelvin scale are such that water freezes at 273°K and boils at 373°K; there are a hundred divisions between the freezing and boiling points of pure water. Consequently, to transform a reading from the Celsius scale to the Kelvin scale, it is merely necessary to add 273; absolute zero is, therefore, at $-273°$C (or $-460°$F).

We have introduced three widely used thermometer scales and it would seem appropriate at this point to define temperature. After all, physics depends heavily on precise definitions. What is a precise definition of temperature? Is temperature a measure of something? Yes. As read on ther-

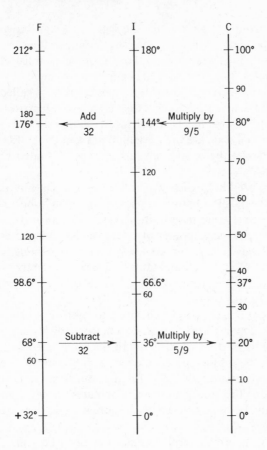

FIGURE 7-2

To change a Fahrenheit reading to a Celsius reading, it is necessary to go through an Intermediate scale whose zero is the same as the Celsius but whose divisions are the same size as Fahrenheit. For example, from 68°F, 32 must be subtracted to arrive at 36° on the Intermediate scale. To change from the Intermediate scale to Celsius, 36° must be multiplied by 5/9. The result is 20°C. To change a Celsius reading into a Fahrenheit reading the process is simply reversed: 80°C × 9/5 = 144°; 144° + 32 = 176°F.

mometers of the type just discussed, it is a measure of the expansion and contraction of mercury or alcohol. Temperature is a common word in our language and yet it is a word that is difficult to define. Most people would be likely to define it as "degrees of hotness or coldness," and this is a reasonable definition, although not very precise. A more precise definition is related to the motion of molecules; the faster the molecules of a substance move, the hotter it is.

THE QUANTIFICATION OF HEAT

Despite the difficulty of defining temperature, the thermometer does enable us to establish the concept of a specific quantity of heat and to measure this quantity. Joseph Black (1728-1799), a Scottish physicist, is notable for his pioneering efforts in this field of measurement. Black was the first to make a clear distinction between heat, as a measurable physical quantity, and temperature. Although we must wait until the end of this chapter to define heat, Black's distinction between heat and temperature can be readily understood.

Joseph Black and Thermal Equilibrium

Black's *Lectures*, a compilation of the lectures he delivered at the University of Edinburgh, published posthumously in 1803, begins with a discussion of distribution of heat:

An improvement in our knowledge of heat, which has been attained by the use of thermometers, is the more distinct notion we have now than formerly of the *distribution* of heat among different bodies. Even without the help of thermometers, we can perceive a tendency of heat to diffuse itself from any hotter body to the cooler ones around it, until the heat is distributed among them in such a manner that none of them is disposed to take any more from the rest. The heat is thus brought into a state of equilibrium.

This equilibrium is somewhat curious. We find that, when all mutual action is ended, a thermometer applied to any one of the bodies undergoes the same degree of expansion. Therefore the temperature of them all is the same. No previous acquaintance with the peculiar relation of each body to heat could have assured us of this, and we owe the discovery entirely to the thermometer. We must therefore adopt, as one of the most general laws of heat, the principle that *all bodies communicating freely with one another, and exposed to no inequality of external action, acquire the same temperature, as indicated by a thermometer* (1).

A number of aspects of Black's ideas should be commented on. First, we recognize that he understood temperature to be what is indicated by the expansion of the fluid in his thermometers. Second, we see that he was aware that only the thermometer could have given him the information needed to formulate his *principle of thermal equilibrium*—the italicized portion of the final sentence in the quotation from Black's *Lectures*. The human senses do not permit us to draw the same conclusion.

If a piece of iron and a piece of wool are left in the same room (let us, for the sake of argument, assume that the room is cold) for 24 hours and are then touched by the hand of an observer, the observer will say that the iron is cold, the wool warm. A thermometer, however, would indicate that they have the same temperature. Hence it is only through his use of the thermometer that Black was able to arrive at his principle of thermal equilibrium.

The human senses tell us that the iron is cool, because upon touching, the hand and the iron become part of a new thermal system. The hand is warmer than the iron and so loses heat to that iron. Our senses, perceiving a drop in temperature because of this loss of heat, tell us that the iron is cold. The heat that the iron gains is quickly distributed throughout the piece of iron, for heat is readily conducted by metals. Additional heat is then drawn from the body in order to establish a new thermal equilibrium between the hand and the iron.

When the hand touches the wool, it feels much warmer. The wool, too, draws heat from the body, but not as much. Since wool does not conduct heat very well, the heat drawn from the hand of the observer remains close to the region of contact. Thermal equilibrium is quickly established because only a small amount of heat is needed to bring this small portion of wool up to body temperature. The body actually loses little heat, and the temperature of the hand does not drop noticeably; our senses tell us that the wool is warm.

Apparently, the bodily senses cannot be trusted for even reasonable guesses at quantities of heat. Furthermore, and much more important, we recognize that the experimenter always runs the risk of disturbing what he is attempting to observe—in this example the temperature of the iron in our cold room. If we want to measure the temperature of an object, it is not fruitful to use an instrument (the human body in this situation) that will alter this temperature. Certainly, a thermometer brought from a warm room will also alter the temperature of the steel, but to a much smaller degree. In other words, it is difficult to carry out Galileo's Platonistic wish that man become an entirely detached and objective observer (see p. 78). The problem remains even today. How do we observe without disturbing that which we observe?

Heat Exchange

Black's *Lectures* continue with his thoughts on the measurement of heat and ultimately reach the conclusion that all substances do not undergo the same change of temperature upon receiving (or giving) the same quantity of heat. For example, when two liquids of different temperatures are mixed and thermal equilibrium is established, Black assumed that the heat flowing from the hotter liquid to the colder one is conserved. That is, the heat lost by the hotter substance equals the heat gained by the colder one. But the temperature change will not be the same if the substances mixed are different. In discussing this aspect of the study of heat, Black refers to work done and reported by Hermann Boerhaave (1668-1738), a Dutch scientist and teacher of great influence in his day. Boerhaave apparently worked with Fahrenheit.

After relating an experiment on the mixing of hot and cold water which Fahrenheit made at his desire, Boerhaave also tells us that Fahrenheit agitated together quicksilver [mercury] and water of initially different temperatures. From the Doctor's account, it is quite plain that the quicksilver, though it has more than 13 times the density of water, had less effect in heating or cooling the water with which it was mixed than would have been produced by an equal volume of water. . . .

To make this plainer by an example in numbers, let us suppose the water to be 100°F and that an equal volume of warm quicksilver at 150°F is suddenly mixed and agitated with it. We know that the temperature midway between 100° and 150°F is 125°F, and we know that this middle temperature would be produced by mixing cold water at 100°F with an equal volume of warm water at 150°F, the temperature of the warm water being lowered by 25 degrees while that of the cold is raised just as much. But when warm quicksilver is used in place of warm water, the temperature of the mixture turns out to be only 120°F, instead of 125°F. The quicksilver, therefore, has cooled through 30 degrees, while the water has become warmer by 20 degrees only; and yet the quantity of heat which the water has gained is the very same as that which the quicksilver has lost (2).

Finding it more fruitful to work with equal *masses* (rather than equal volumes as Boerhaave had done), Black was led to the concept of heat capacity: equal *masses* of different substances will undergo a different change of temperature upon acquiring or losing the same quantity of heat. Hence, different substances have a different "capacity" for heat. By expressing this capacity in algebraic notation, we can develop Black's idea more clearly.

The observable quantity in such experiments is the change in temperature which we designate as ΔT. The quantity of heat lost or gained we designate as ΔQ. Black discovered that when two different quantities of water at different initial temperatures are mixed, the changes in temperature are directly proportional to the amount of heat lost by one and gained by the other; that is,

$$\Delta T \propto \Delta Q$$

He also showed by this technique that the change in temperature is inversely proportional to the mass of the object:

$$\Delta T \propto \frac{1}{m}$$

Hence,

$$\Delta T \propto \frac{\Delta Q}{m}$$

This proportion can be rewritten so that we may eventually equate the amount of heat lost by one substance to that gained by the other:

$$\Delta Q \propto m \, \Delta T$$

Or, put into the form of an equation with a constant of proportionality, it finally becomes

$$\Delta Q = cm \, \Delta T$$

The ratio of the amount of heat lost or gained by an object to the resulting change in temperature, $\Delta Q/\Delta T$, is called the *heat capacity* of that object. The heat capacity per unit mass is known as the *specific heat* and is the constant c in this equation. Actually, the specific heat of a substance is not a constant quantity; its value varies with the temperature of that substance. As long as the temperatures considered are moderate, however, the specific heat can be considered constant without introducing much error.

Units of Heat

We can now define the unit for measuring heat. Because it is desirable that the unit be readily and accurately duplicated in any laboratory, the specific heat of water is chosen as one. The unit of heat, called a *calorie, is that amount of heat required to raise the temperature of one gram of water one degree Celsius* (the abbreviation of calorie is cal). Since specific heat does depend on the temperature, the standard calorie is defined for a change of temperature of water from 14.5 to 15.5°C. The kilocalorie, of course, is the amount of heat required to raise one kilogram of water one degree Celsius. This is the "calorie" used by those who, in watching their weight, must count their calories.

In engineering the unit of heat is called the *British thermal unit* (abbreviated Btu). It is the amount of heat required to raise the temperature of one pound of water one degree Fahrenheit.

As with temperature, our discussion of heat has progressed to a fairly quantitative level without anything having been said about what "heat" is. Early work led many researchers of the eighteenth century to adopt a fluid theory of heat, a theory akin to the fluid theories of electricity then only recently proposed. Certainly the substance, heat or *caloric* as it came to be called, was a very subtle substance. Many attempts were made to see whether substances did not gain weight when heated, or did not gain weight when melted.

Melting was considered a process of chemically uniting heat and material particles, and vaporization was explained in a similar way. Attempts were therefore made to determine the difference between the weight of ice and that of water; ice was observed to absorb so much heat upon melting that scientists thought the difference in weight should certainly be measurable. But the evidence was contradictory; the experiments had to be conducted in cold rooms, and the balances did not always work properly. Moreover, convection currents from the flame used to melt the ice disrupted the experiment.

It should be noted at this point that experiments on the melting of ice depended entirely on the availability of that substance. These experiments were not performed in Italy. In fact, Black comments at one point that an idea occurred to him during the summer of 1761, but "as there was no ice house in Glasgow . . . ," he had to wait patiently until winter to perform the experiment to test the idea. Indeed, the early work on heat, which of necessity involved experiments with ice, was done by Black in Scotland, Boerhaave,

Fahrenheit, and Pieter van Musschenbroeck in Holland, J. C. Wilcke in Sweden, and other Northerners!

The caloric theory of heat was actually a very successful theory in the development of physics. It guided Black through his work and also stimulated many other first-rate scientists to venture into new areas of study that have led to our modern ideas of thermodynamics. Thus the caloric theory gave men a sturdy framework upon which they could build. Even today, the basic mathematical equation for the flow of heat through a piece of metal is of the same type or form as the equation for the flow of water through a pipe. Yet today we do not consider heat to be fluid; we consider it to be the result of the motion or vibration of atoms.

JAMES JOULE

Electrical Prelude

Of those who rejected the caloric theory of heat, James Joule (1818-1889) was one of the most influential. Born near Manchester, England, Joule was raised in an atmosphere dominated by manufacturing interests. His father was a manufacturer and was deeply engrossed in the problems of the newly developing mechanization. Joule's early work was therefore directed toward the solution of some of these problems.

Joule, however, developed interests encompassing many fields of endeavor in physics. He eventually showed that these fields are all interrelated

FIGURE 7-3

James Joule (1818–1889). (Reprinted from *Joule's Scientific Papers,* Vol. I, Taylor and Francis, Ltd., 1884.)

by making one of the important contributions leading to the recognition and adoption of the principle of conservation of energy. Since Joule's work was so inclusive and since it illustrates so well the interrelationships of many fields of study, it is instructive to look more closely at his experimental work and his evolving ideas that led to the concept of energy and its conservation.

Joule's first researches began in 1837, when he was only nineteen years old and only six years after Faraday first reported electromagnetic induction. Because of Joule's interests in manufacturing, it is no great surprise to discover that he first attempted to build an electric motor that would be competitive with the steam engine (which had been developed in the eighteenth century).

Because the principle of the electric motor had not been discovered until 1821 (see p. 154f), the motor itself was not very well developed by the time Joule began his work. Therefore, in order to learn as much about the electric motor as he could, Joule studied the relationships of all the factors involved in the motor's production of power. The number of voltaic cells used was a measure of what he called "electric force" and what we now call *electromotive force*. He measured the current (with a galvanometer of his own design), the velocity of rotation of the rotor (the rotating coil of the motor), and the "duty" of the motor (the number of pounds that a motor could raise to a height of one foot by the consumption of one pound of zinc in the voltaic cells). Joule even considered the friction of the bearings in adjusting this "duty." (Engineers developing the steam engine, which consumed pounds of coal rather than zinc, first applied the word duty to the work done by a machine.)

This very thorough research into the operation of the electric motor proved enlightening to the young Joule. He had hoped that as the number of voltaic cells placed across the motor was increased, the electric current through the motor would also increase. With this increase in current, Joule felt that the motor ought to turn faster and consequently deliver *more power at less expense*.* He discovered, however, that as the speed of rotation *increased*, the current through the motor *decreased*. This decrease in the current indicated to Joule that the electric motor has limitations that he had not suspected. Within a few months of Joule's discovery, the work of Lenz was published. After reading the principle we now call Lenz's law, Joule realized why the voltaic current decreased as the speed of the motor increased.

Voltaic cells are connected to a motor in such a way that the current runs through both the stationary coils and the rotating coil. The current in the stationary coils sets up a magnetic field. Thus the current in the rotating coil (which is made of wire) is actually a current in a wire situated in that magnetic field. Wallaston and Davy had previously shown (see p. 155) that a current-carrying wire placed at right angles to a magnetic field has a force

---

* Joule had learned that the magnetic force of his motors was proportional to the *square* of the electric current. Hence he reasoned that doubling the current would increase the magnetic force by a factor of four.

exerted on it. This force acting on the rotating coil causes it to move or rotate and the motor runs (Figure 7-4). But once the rotating coil starts to rotate, the number of lines of force passing through the coil changes. A current is induced in that coil. In which direction is that current induced? According to Lenz's law, the induced current must oppose the inducing action.

The inducing action is the voltaic current in those same wires of the rotating coil, the current that started the coil rotating in the first place. Therefore, the induced current must oppose that voltaic current. The induced current passes through the same wires as the voltaic current, but in the opposite direction, thus reducing the net current. The net current equals the voltaic current minus the induced current. The voltaic current is always greater than the induced current; otherwise the motor would run backward, and of its own accord! The opposing action induced in the rotating coil is now called *back electromotive force*, or simply *back emf*.

Once Joule understood the limitation placed upon the operation of the motor by Lenz's law, he was convinced that further work on the motor at that time was not worthwhile. His interests turned elsewhere, to the principle behind Lenz's law rather than to its applications. What is the basic principle that explains why he could not run an electric motor with greater electric current at less expense? In effect, why was it impossible to get something for nothing and outdo the steam engine?

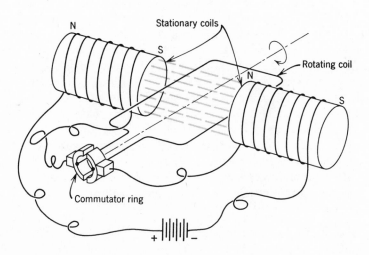

FIGURE 7-4

Joule connected his electric motor to a voltaic cell in such a way that the same electric current passed through both the stationary coil and the rotating coil. The current is passed to the rotating coil by means of the commutator ring.

Electricity and Heat

In March 1841, less than two months after his comments on Lenz's law appeared, Joule's next paper was published. The man who wrote it no longer wished to build a better electric motor; he had become concerned with the more profound aspects of nature. The paper, which reported on the heating effect of an electric current, an effect now called Joule's law, opened with this comment:

There are few facts in science more interesting than those which establish a connection between heat and electricity. Their value, indeed, cannot be estimated rightly, until we obtain a complete knowledge of the grand agents upon which they shed so much light (3).

James Joule, who at age nineteen began to build a better electric motor, had, by age twenty-three, embarked on a search for the "grand agents" of nature. Joule the engineer had become Joule the scientist, a research scientist on a par with his older contemporary in London, Michael Faraday.

Electricity, Chemical Action, and Heat

Joule studied the heat produced by the electric motor as well as the relationship between an electric current in a simple wire and the heat produced by that current. He also studied the chemical action of the voltaic cell (acid solution acting on metals) and the relationship between that action and the heat produced. Some voltaic cells form bubbles during their operation, others do not, and Joule recognized that when other considerations are made equal, a voltaic cell in which *bubbles form* produces *less heat* than a cell in which bubbles do not form.

Joule realized that chemical action, bubble formation, electric currents, and a motor lifting a weight are all related to each other, and that *each is related to heat*. He became convinced that all these actions have something in common. Such a conviction was based on two beliefs: (1) the idea that heat is a vibratory motion of atoms, a motion that can be produced mechanically by rubbing or pounding, and (2) the conviction that nature is both consistent and conservative.

In a paper titled *On the Heating Effects of Magneto-Electricity, and on the Mechanical Value of Heat* published in July 1843, Joule stated his now famous conclusion that every mechanical effort that is converted into heat is converted at a constant ratio. So much mechanical effort, so much heat—no more, no less.

It is pretty generally, I believe, taken for granted that the electric forces which are put into play by the magneto-electrical machine [electric generator] possess, throughout the whole circuit, the same . . . [heating] properties as currents arising from other sources [such as voltaic cells and electrostatic generators]. And indeed when we consider heat not as a *substance*, but as a *state of vibration*, there appears to be no reason why it should not be induced by an action of a simply mechanical character, such, for instance, as is presented in the revolution of a coil of wire before the poles of a permanent magnet. . . .

The general plan which I proposed to adopt in my experiments . . . was to revolve a small compound electromagnet, immersed in a glass vessel containing water, between the poles of a powerful magnet, to measure the electricity thence arising by an accurate galvanometer, and to ascertain the . . . [heating] effect of the coil of the electromagnet by the change of temperature in the water surrounding it (4). [The water jacket is not shown in Figure 7-5.]

With this equipment, Joule was able to show that by mechanically turn-ing the crank he could generate an electric current, which in turn generated heat. The amount of electricity was measured by a galvanometer; the amount of heat was measured (as Black had done) by the rise in temperature of a known amount of water. But could the amount of mechanical effort be measured? Yes. Joule's experience of measuring the "duty" of an electric motor led him to reverse the process in this experiment. If lifting a weight is a measure of the power of a motor, then the falling of a weight through a certain distance at a constant velocity is a measure of the mechanical effort expended in the generation of heat by means of an electric current (Figure 7-5c).

Mechanical Equivalent of Heat

Joule then performed the necessary experiments and calculations to find the ratio between the mechanical effort of the falling weight and the heat gen-erated, a ratio which we now call *Joule's equivalent* (the mechanical equiva-lent of heat). These calculations are difficult for the modern reader to inter-pret because they are in Joule's own system of units. The result of the first experiment as stated by Joule was that

1° [Fahrenheit] of heat per lb of water is therefore equivalent to a mechanical force capable of raising a weight of 896 lb to the perpendicular height of one foot (5).

Repetitions of the same experiment determined the weight to be 838 pounds rather than the 896 pounds first obtained.

Joule soon realized that the intermediate step of generating electricity was not necessary, and he proceeded to convert the action of the falling weights directly into heat. In his first attempt, he let water fall through narrow tubes, thus increasing the temperature of the water. Measuring this increase in temperature, and having established the amount of water and the height through which it had fallen, he found that he

obtained one degree of heat per pound of water from the mechanical force capable of raising about 770 lb to the height of one foot, a result which will be allowed to be very strongly confirmatory of our previous deductions. I shall lose no time in repeating and extending these experiments, being satisfied that the grand agents of nature are, by the

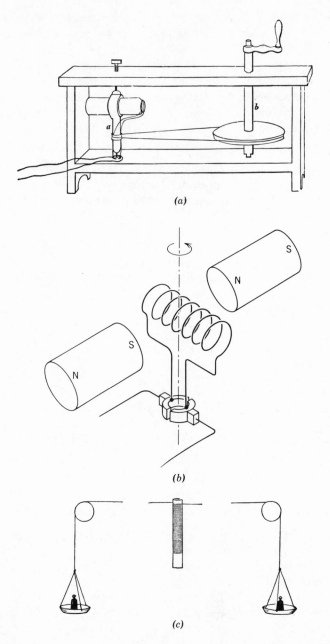

FIGURE 7-5

(a) By turning the crank, Joule could rotate the coil that was placed between the poles of a large magnet and inside of a water jacket (neither the magnet nor the water jacket are shown in this drawing). (b) This schematic drawing shows the coil in place between the two large magnets. The rotation of the coil generated an electric current. The electric energy so generated was converted into heat. (c) Joule replaced the crank with a pulley arrangement so that a known weight falling with a constant velocity through a measured distance could rotate the coil.

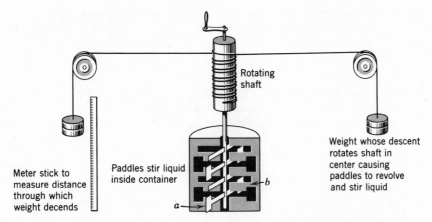

Rotating shaft

Weight whose descent rotates shaft in center causing paddles to revolve and stir liquid

Paddles stir liquid inside container

Meter stick to measure distance through which weight decends

b

a

FIGURE 7-6

Joule's famous "paddle wheel" experiment in which falling weights turning paddles immersed in the water caused the temperature of water to rise a measurable amount. The container had vanes *(b)* built in to prevent the water from simply moving with the paddles *(a)*.

Creator's fiat, *indestructible*; and that where ever mechanical force is expended, an exact equivalent of heat is *always* obtained (6).

Joule's excitement at this point is obvious; he would "lose no time" in verifying and extending these results. And he did not lose any time. He determined the mechanical equivalent of heat by a number of different methods, and by 1850 he finally produced a paper describing his now famous "paddle wheel" experiment. The paper was read before the Royal Society in London by Michael Faraday. In his previous experiment with the electromagnet, the weight (falling at constant velocity) turned a wire coil between the poles of two electromagnets. In the paddle wheel experiment the weights (again falling at constant velocity) turned paddles immersed in water (Figure 7-6). The water was placed in a cylindrical container that had vanes projecting into the water to prevent it from circulating with the rotating paddles. The stationary vanes thus created more friction so that the falling weights moved more slowly. From the experiment he obtained the following value for the mechanical equivalent of heat: 772 pounds falling a distance of one foot will generate enough heat to raise the temperature of one pound of water one degree Fahrenheit. The value used today is 778 pounds.

THE PRINCIPLE OF CONSERVATION OF ENERGY

Joule demonstrated that mechanical effort, electricity, magnetism, chemical reactions, and heat all have something in common. But what is it? That "something" is called *energy*, a term that came into use in the latter part of the nineteenth century. To understand the concept of energy, however, it is necessary to understand the concept of work, which forms the all-important connecting link between Newtonian physics and energy.

The concept of work stems directly from the concept of "duty" that Joule had inherited from the engineers who developed the steam engine. Joule rated a motor by how great a weight can be lifted a certain distance while one pound of zinc is consumed in the voltaic cells. The motor must exert a force on that weight and as a result cause the weight to move. Furthermore, Joule recognized that work done on an object changes some aspect of that object. The particular aspect that changes depends on which form the energy takes. Recognition of this change in energy led to the formulation of *work-energy theorem: work done on an object changes the energy of that object.*

To make this theorem applicable, however, we need to express both the work done and the change in energy in mathematical notation, and in such a way that measurements can be made to verify the statements we develop. During our discussion we will be guided by the work of Joule which led him to propose what we now call *the principle of conservation of energy: the energy of a system upon which no work is done must remain constant.* We can define a system in any way we please. It should, however, include all the important aspects of the problem we are dealing with. For example, Joule might have included his motor, the weight it lifted, and the voltaic cells in a system.

If no work is done on the system, its energy content should not change. The energy can, however, be transformed from one form into another. If energy is lost to or gained from the surroundings, the total energy of the system will, of course, change accordingly. We will classify such energy loss under a very loose definition of work. That is, it may not always be immediately obvious which force is acting over what distance. It remains, then, to develop the mathematical statements of work and energy. Let us first consider the concept of work.

Work

Joule's motor did work on the weight in lifting it. The amount of work it did on that weight depended on two things: the force exerted by the motor and the distance through which that force acted. We can, therefore, define work W as the product of the force F and the distance s over which that force operates:

$$W = Fs$$

We should make clear that the force and the distance moved must be in the same direction.

If a force of 10 pounds acts on an object that moves a distance of 5 feet while that force acts on it, the work done will then be

$$W = Fs$$
$$W = (10 \text{ lb})(5 \text{ ft})$$
$$W = 50 \text{ ft-lb}$$

The unit of work-energy in the engineering system of units is the foot-pound. The units are different in the mks system. For example, suppose that a force of 8 newtons is exerted on an object and acts over a distance of 3 meters. How much work will have been done on that object?

$$W = Fs$$
$$W = (8 \text{ nt})(3 \text{ m})$$
$$W = 24 \text{ nt-m}$$

The newton-meter, however, is given another name, the *joule*:

$$W = 24 \text{ joules}$$

The joule is defined as the amount of work done by a force of one newton acting over a distance of one meter.

So far we have been able to express work in a straightforward mathematical manner. It now remains to develop mathematical expressions for the various forms of energy. We hope that there are not too many forms of energy, and Joule's work indicates that there are a limited number.

Kinetic Energy

Let us suppose that a force acts on a block resting on a horizontal table and let us, for a moment, neglect the force of friction. You can push the block with your finger and make it accelerate (Figure 7-7). The amount of acceleration will depend on the mass of the block and on the force you exert. By exerting this force on the block, work is done, and by the work-energy theorem energy is given the block. What form does this energy take? To select the particular form of energy that best describes this action, we should look for some aspect of the block that changes while the force is being exerted on it. The most obvious change is the block's velocity, which increases while the force is acting. We propose that as the velocity increases, so does the amount of this form of energy. Therefore, we can call this energy of motion, or *kinetic energy*.

If we can express the kinetic energy mathematically and equate it to the work done in a way that satisfies all the measurements we might make of the motion of this block, we may feel assured that our choice of kinetic energy is a good one. If it also satisfies measurements of motions of all kinds of objects, we may be confident that we have selected an expression that describes very well one aspect of the physical world.

We have defined the work done on an object as

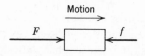

FIGURE 7-7

The force F does work on the block by accelerating it; f is the force of friction.

$$W = Fs$$

We must now relate this force to the velocity of the object. This can be done by recalling that a change in velocity per unit time is acceleration, and force and acceleration are related by Newton's second law of motion:

$$F = ma$$

We can substitute ma, which is equal to the force F, into the equation for work and obtain

$$W = mas$$

However, since kinetic energy is energy of motion, it is more usefully expressed in terms of the velocity rather than in terms of acceleration and distance covered. The easiest way of expressing the product of acceleration and distance in terms of velocity is to recall the relationship given on page 64.

$$v^2 = v_0^2 + 2as$$

The product of acceleration and distance appears directly in this equation. Solving this equation for as, we find

$$v^2 - v_0^2 = 2as$$

$$as = \tfrac{1}{2}(v^2 - v_0^2)$$

This can be substituted into the equation

$$W = mas$$

and we obtain

$$W = \tfrac{1}{2}m(v^2 - v_0^2)$$

This must therefore be the expression for a change in kinetic energy:

$$\Delta E_k = \tfrac{1}{2}m(v^2 - v_0^2) = \tfrac{1}{2}mv^2 - \tfrac{1}{2}mv_0^2$$

If v_0 is zero, that is, if the object is initially at rest with respect to some chosen frame of reference, the change in kinetic energy becomes simply the kinetic energy over and above what the object would have at rest in the frame of reference. (The Earth is often considered that frame of reference.) The expression for kinetic energy then becomes

$$E_k = \tfrac{1}{2}mv^2$$

Were we to make actual measurements of forces exerted on a block and the resulting increases in velocity, we would find that the work-energy principle for at least this part of our discussion is supported by observation: the work done does indeed equal the change in kinetic energy as we have expressed it.

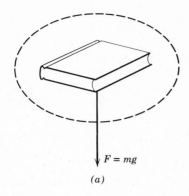

(a)

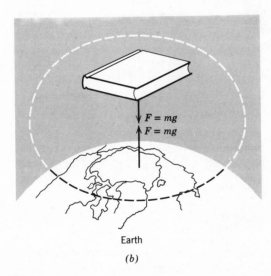

Earth

(b)

FIGURE 7-8

(a) The book alone forms the system; the force $F = mg$ is an external force. *(b)* In the Earth-book system the two forces, $F = mg$, are external forces.

Potential Energy

Suppose a book falls from a table top. The work-energy principle permits us to make the following analysis of the action. We can consider the book alone, call it our system, and recognize that the force of gravity acts on it (Figure 7-8*a*). The book accelerates as a result of this force, the velocity of the book increases and, consequently, so does its kinetic energy.

So far, our logic is successful, but let us describe another system and see whether our work-energy theorem still works. Rather than a system consisting of the book alone, let us consider the book and the Earth as a single system (Figure 7-8*b*). There are two forces to consider now: the force of gravity that the Earth exerts on the book and, by Newton's law of gravity, the force that the book exerts on the Earth. These two forces are equal in magnitude but opposite in direction. Each is exerted by one body of the system on another body in that same system. Neither is involved with anything outside the system. Therefore these two forces cannot do any work on the Earth-book system. How then can the change in kinetic energy be explained?

We must now rely on the principle of conservation of energy: if no work is done on the system by an external force, the energy of the system remains constant. This energy, however, can be transformed from one form to another. We have only to propose a new kind of energy which, during the book's fall, decreases at the same rate that the kinetic energy increases.

The new form of energy must be related to some aspect of the book that changes while it falls. This can only be the height of the book above the floor, or above the surface of the Earth, or above some convenient reference level. To make measurements so that we can test our ideas, we must express this new form of energy mathematically.

The change in kinetic energy ΔE_k must equal the final kinetic energy, the energy the book possessed an instant before it hit the floor, minus the initial kinetic energy, the energy the book possessed the instant it was released. The initial kinetic energy was zero, so the change in kinetic energy must be

$$\Delta E_k = \tfrac{1}{2}mv^2$$

By the principle of conservation of energy, the change in kinetic energy must equal the change in this new form of energy ΔE_n, where the subscript n refers to this new, as yet unnamed, form of energy. Therefore, since

$$\Delta E_n = \Delta E_k$$
$$\Delta E_n = \tfrac{1}{2}mv^2$$

We need now to find an expression relating the velocity to the distance through which the book has fallen, and we can again use the expression from page 64:

$$v^2 = v_0^2 + 2as$$

The initial velocity v_0 was zero, the acceleration a is the acceleration produced by gravity g, and the distance s through which it has fallen is the height h. Consequently,

$$v^2 = 2gh$$

This equation is very convenient, since we need an expression for the velocity squared to substitute into our expression for the change of energy:

$$\Delta E_n = \tfrac{1}{2}m(2gh)$$
$$\Delta E_n = mgh$$

The only thing left to do is to give this new form of energy a name and then assure ourselves that it is a valid description of falling objects.

This new form of energy is dependent on the height of the book above the floor, or above the Earth; in fact, it is actually related to the relative position of the book and the Earth in the Earth-book system. Energy dependent on relative position is called *potential energy*. If the system involves the Earth and some object, or the moon and some object, that is, if it involves a gravitational field, we call the energy *gravitational potential energy*. We can give this form of energy the symbol E_g, where g stands for gravitational potential energy. Therefore,

$$\Delta E_g = mgh$$

To assure ourselves that the use of potential energy is consistent with the work-energy principle, let us still consider the Earth-book system, but this time lift the book from the floor to the table top by some external force (Figure 7-9). You may exert this force, or it may be exerted by a motor such as the one James Joule used. The external force will do work on the Earth-book system, so the system's energy will change. The amount of change will depend on the work done, which in turn depends on the force exerted and the distance over which this force acts.

We would like to avoid any complications with kinetic energy in this example, so we will lift the book from the floor, where it has zero kinetic energy, to the table top, where it again has zero kinetic energy. The total change in kinetic energy is zero. We can, therefore, lift the book at a constant speed and neglect the kinetic energy so acquired because it eventually loses this kinetic energy.

To lift the book at a constant speed, we must exert a force F that is exactly equal to its weight mg. The amount of work W equals this force times the distance through which the force acts:

$$W = Fs$$

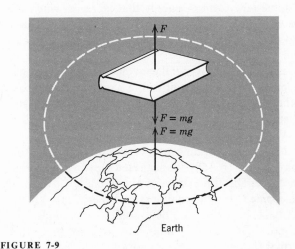

FIGURE 7-9

Energy is added to the Earth-book system by an external force F.

The force F equals mg, and the distance s must equal the height of the table above the floor h. Consequently, the work done on the system by the outside force is

$$W = mgh$$

This work has increased the energy of the system, and the energy increase is the change in the gravitational potential energy. Therefore,

$$\Delta E_g = mgh$$

This result should give us some confidence in the logic presented so far; the expression for potential energy is the same as the one we derived for the book falling freely. If the gravitational potential energy on the floor (or whatever reference level we choose) is set at zero, we can call the change in potential energy the potential energy itself:

$$E_g = mgh$$

Many measurements of lifting objects and falling objects have been made, from the time of James Joule on, and they support the results we have derived. There is, however, one factor we have neglected until now. What happens to the kinetic energy of the falling book when it hits the floor?

Thermal Energy

As the book hits the floor, the forces exerted by the floor on the book and the book on the floor are equal and opposite in direction, and both are internal forces. There being no external forces exerted on our Earth-book system, we must assume that the energy of the system does not change. But when the book hits the floor, both the potential and the kinetic energy become zero. Hence, if we are to maintain our principle of conservation of energy, we must invent some new form of energy that increases when the other two forms are reduced to zero. This still newer form of energy will have to be related to some aspect of the system that changes when the book collides with the floor.

One such collision does not reveal any significant change, but you might try repeated collisions such as flattening a nail with a hammer. Aside from the change in shape of the nail, has any other aspect of the nail changed? Yes, it is hot! We can, therefore, propose that the energy of our Earth-book system is transformed to thermal energy when the book hits the floor.

Thermal energy is, indeed, an important aspect of nature. In the example of the block sliding on the table top (p. 189), we neglected the force of friction between the block and the table. This force converts kinetic energy to thermal energy. James Joule transformed gravitational potential energy into kinetic energy and then into thermal energy by a number of different methods. The measurements he made indicated that no matter how he converted other forms of energy into thermal energy, 778 foot-pounds of mechanical energy (gravitational and kinetic energy) were converted into 1 Btu of thermal energy. In the mks system of units 4.18 joules of mechanical energy are equivalent to 1 calorie of thermal energy.

Thermal energy, however, is not a completely new form of energy. The modern atomic theory of matter includes a definition of thermal energy as the total kinetic energy of all the particles that form a substance, whether that substance is the air in a room, or water in a glass, the glass itself, or the sun. Consequently, our introduction of thermal energy should not alarm us by leading us to believe that for every action we will have to invent some new form of energy. If this were true, the principle of conservation of energy would become unwieldy. As it turns out, there are a fairly small number of different forms of energy, and so the process of accounting for energy changes in any one system is not burdensome.

Chemical energy, for example, is actually one form of potential energy, *electric potential energy*. Two objects with opposite electric charges have electric potential energy in the same sense that the Earth and the book have gravitational potential energy. Two objects of the same charge must be pushed together to increase their electric potential energy. When atoms combine to form molecules, and eventually the books and blocks we have been studying as examples, the forces that hold the atoms together are electric forces.

Joule's efforts to establish the principle of conservation of energy were considerable, and although his work made possible our current definition of thermal energy as the total kinetic energy of the molecules composing a substance, he was by no means the only one to develop a working hypothesis of energy and its conservation from a vague impression of an underlying similarity among various physical phenomena. At the same time that he made his investigations, at least a half a dozen men worked along similar lines, each from his own vantage point. They all recognized the fundamental importance of the principle of conservation of energy. This principle, when combined with Newtonian physics, formed an imposing framework upon which all of physics was based until the turn of the twentieth century.

Much of what is at present discussed in the other sciences, and a great deal of the material covered in engineering, is based upon this framework. Newtonian physics combined with the work-energy concept has proved eminently successful in desc.ibing the actions we observe in the world about us.

References

1. James B. Conant and L. K. Nash, (eds.), *Harvard Case Histories in Experimental Science*, Vol. I, Harvard University Press, Cambridge, Mass., 1957, p. 128.

2. *Ibid.*, p. 131.

3. James Joule, *Scientific Papers*, Vol. I, Taylor and Francis, 1884, p. 60.

4. *Ibid.*, pp. 123 f.

5. *Ibid.*, pp. 149 ff.

6. *Ibid.*, pp. 157 f.

Questions

1. Define (*a*) specific heat, (*b*) thermal equilibrium, (*c*) calorie, (*d*) Btu, (*e*) work.

2. Discuss Black's concept of thermal equilibrium according to both the caloric theory and the atomistic theory.

3. Count Rumford (1753-1814) learned by drilling cannon barrels that the drill becomes hot, the cannon barrel becomes hot, and the metal shavings are hot. Can this action be explained by the caloric theory of heat? If so, how? If not, why not?

4. If, in the generation of electricity, one voltaic battery gives off bubbles and the other does not, and if the two operate in exactly the same way in every other respect, which one will heat up more? Why?

5. Indicate clearly how the concept of *work* forms the bridge linking strict Newtonian physics with the concept of energy.

6. Discuss the question "How do we observe without disturbing what we observe?" As it applies to (*a*) a biologist, (*b*) an anthropologist, and (*c*) a sociologist.

Problems

1. Convert 40°F to (*a*) the Celsius scale, (*b*) the Kelvin scale.

2. Convert −40°C to the Fahrenheit scale.

3. How much heat is required to raise the temperature of 150 gm of water from 18° to 100°C?

4. How much heat is given off when 300 gm of aluminum "shot" (specific heat 0.22 cal/g °C) are cooled from 100° to 24°C?

5. If 250 g of steel pellets (specific heat 0.11 cal/g °C) at 100°C are mixed with 50 g of water at 10°C, what is the final temperature?

6. A 5-kg mass is lifted to a height of 10 m.
(*a*) How much work is done?
(*b*) What is the change in potential energy of the mass?
(*c*) If the 5-kg mass is released, what will be its velocity after falling through 10 meters?

7. A boy slides down a fairly smooth inclined plane, through a vertical distance of 8 m. His velocity upon reaching the bottom of the slide is 10 m/sec.
(*a*) How much energy (in joules) was lost to friction?
(*b*) How much heat (in calories) was gained by the slide (and the boy)?

Chapter Eight

Optics

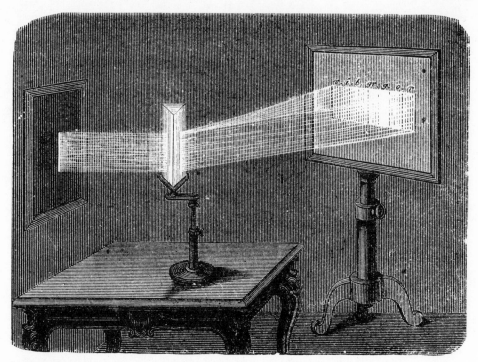

Bettman Archive

I f the nature of matter, motion, electricity, and magnetism has evoked considerable discussion over the centuries, so has the nature of light. Light is intangible; it travels with such velocity that at one time many considered its speed to be infinite. Light was thought to move from one place to another instantaneously, with no time elapsing during its passage. Aristotle, however, considered light to be a disturbance in the medium through which it travels. He felt that its velocity was finite, that it took a specific interval of time to travel, for example, from the sun to the Earth. What is more, Aristotle associated color with the object seen:

. . . [nothing] is visible except with the help of light; it is only in light that the color of a thing is seen. Hence our first task is to explain what light is (1).

REFLECTION AND REFRACTION

To understand light we must first study its behavior. The Greeks discovered that light reflects in a regular manner from a smooth surface. In Figure 8-1 let us call the angle between the incident ray and the normal (perpendicular) to the surface the *angle of incidence i* and the angle that the reflected ray makes with the normal to the surface the *angle of reflection r*. Ptolemy's *Optics* describes what Greek investigators had previously learned, namely, that the angle of reflection is equal to the angle of incidence.

Ptolemy and Refraction

When light strikes the surface of a transparent substance, such as water, some of the light is reflected from the surface, but some passes through the substance. If a light beam strikes the surface at right angles to that surface, it passes straight through the substance. If, however, it strikes the surface at some other angle, the light ray is bent on passing through the surface. This bending is called *refraction*. Ptolemy described the action of the transmitted light with a surprisingly objective approach to an observational problem. He measured the angle of incidence θ_1 and the corresponding angle of refraction θ_2, which is the angle between the normal to the surface and the ray of light after refraction.

The amount of refraction which takes place in water and which may be observed is determined by an experiment . . . [Figure 8-2].

On this disk draw a circle $ABGD$ with center at E and two diameters AEG and BED intersecting at right angles. Divide each quadrant into ninety equal parts and place over the center a very small colored marker. Then set the disk upright in a small basin and pour into the basin clear water in moderate amount so that the view is not obstructed. Let the surface of the disk, standing perpendicular to the surface of the water, be bisected by the latter, half the circle, and only half, that is, BGD, being entirely below the water. Let diameter AEG be perpendicular to the surface of the water.

Now take a measured arc, say AZ, from point A, in one of the two quadrants of the

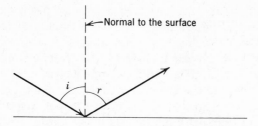

FIGURE 8-1

The angle of reflection equals the angle of incidence.

disk which are above the water level. Place over Z a small colored marker. With one eye take sightings until the markers at Z and at E both appear on a straight line proceeding from the eye. At the same time move a small, thin rod along the arc, GD, of the opposite quadrant, which is under the water, until the extremity of the rod appears at that point of the arc which is on a prolongation of the line joining the points Z and E.

Now if we measure the arc between point G and the point H, at which the rod appears on the aforesaid line, we shall find that this arc, GH, will always be smaller than arc AZ. Furthermore, when we draw ZE and EH, angle AEZ will always be greater than angle GEH. But this is possible only if there is a bending, that is, if ray ZE is bent toward H, according to the amount by which one of the opposite angles exceeds the other.

If, now, we place the eye along the perpendicular AE the visual ray will not be bent but will fall upon G, opposite A and in the same straight line as AE.

In all other positions, however, as arc AZ is increased, arc GH is also increased, but the amount of the bending of the ray will also be progressively greater (2).

Table 8-1 gives the experimental values obtained by Ptolemy and values obtained by measurements made today. From an analysis of Ptolemy's

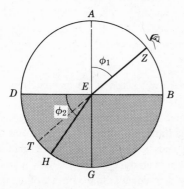

FIGURE 8-2

Ptolemy's method of measuring both the angle of incidence θ_1 in air and the angle of refraction θ_2 in water.

Table 8-1

Ptolemy	Today
When *AZ* is 10°, *GH* will be about 8°	$7\frac{1}{2}°$
When *AZ* is 20°, *GH* will be $15\frac{1}{2}°$	15°
When *AZ* is 30°, *GH* will be $22\frac{1}{2}°$	22°
When *AZ* is 40°, *GH* will be 29°	29°
When *AZ* is 50°, *GH* will be 35°	35°
When *AZ* is 60°, *GH* will be $40\frac{1}{2}°$	$40\frac{1}{2}°$
When *AZ* is 70°, *GH* will be $45\frac{1}{2}°$	45°
When *AZ* is 80°, *GH* will be 50°	48°

measurements* it becomes clear that he made some adjustments in the original measurements in an attempt to fit them into a regular pattern; but the pattern he chose is slightly different from the pattern we use today. Modern experimenters also make adjustments in their measurements, called "smoothing," but they report this adjustment along with an explanation of how the measurements were adjusted. (See question 2 at the end of this chapter.)

From Ptolemy's time until the seventeenth century, little work was done in the field of light and optics. The Arabs made some contributions, as did Robert Grosseteste and Roger Bacon of Merton College, Oxford. These two scholars, who were unaware of Ptolemy's work on refraction, expressed a strong interest in the study of light and optics. The rainbow was of special interest to them, although they considered the nature of light as well. Many philosophers felt that colors were mixtures of light and darkness; violet and blue supposedly had more darkness than yellow and red.

Galileo, Kepler, and Descartes each made contributions to the study of light and geometrical optics or the study of refraction through variously shaped surfaces. But the establishment of the study of physical optics (the study of the nature of light and color) as a mature and full fledged branch of physics was primarily the work of Isaac Newton.

Newton, the Prism, and 22 Feet

Newton's experiments in which he allowed a beam of light coming through a hole in his window blind to be dispersed into a spectrum (a little rainbow) by a prism are well known. The formation of spectra (plural of spectrum) by pieces of glass had been known since the time of the Romans or before. Newton, however, added a new element to those observations. All previous investigators had observed the spectrum by letting the light fall on some form of a screen held quite close to the prism; Newton placed the prism close to the blind of a window on one wall; the light then traveled 22 feet

* Morris R. Cohen and I. F. Drabkin, *A Source Book in Greek Science,* Harvard University Press, Cambridge, Mass., 1958, a footnote beginning on page 277.

across the room before it formed a spectrum on the opposite wall. The greater distance between the prism and the wall allowed the spectrum to be dispersed or spread out to a much greater extent. The colors formed could be studied in much greater detail than if the spectrum had been displayed on a screen close to the prism. As with Galileo, a simple change in the approach to a problem made all the difference in the solution.

Newton realized that the phenomenon of refraction as understood during his time could not account for the great length of the spectrum from the violet end through the blue, green, yellow, orange, to the red end of the spectrum. Studies had been made in the seventeenth century, similar to those done by Ptolemy, from which was derived a mathematical relationship between the angle of incidence and the angle of refraction; this relationship is called Snell's law. Before Newton, Snell's law was applied only to white light, not to individual colors, and hence the law could not be expected to predict or account for any differences of refraction among those colors. This failure of the then current theories about refraction stimulated Newton to experiment further in search of observations that would help him to understand better the formation of the spectral colors. He reported his initial results in 1672 in an article published in the *Philosophical Transactions of the Royal Society*. Parts of that article follow, but they have, to some extent, been rephrased in modern English.

Sir,

To perform my late promise to you, I shall without further ceremony acquaint you, that in the beginning of the year 1666 (at which time I applied myself to the grinding of optic glasses of other figures than spherical) I procured me a triangular glass-prism, to try therewith the celebrated phenomena of colors. And in order thereto having darkened my chamber, and made a small hole in my window-shuts, to let in a convenient quantity of the sun's light, I placed my prism at this entrance, that it might be thereby refracted to the opposite wall. It was at first a very pleasing divertissement, to view the vivid and intense colors produced thereby; but after a while applying myself to consider them more circumspectly, I became surprised to see them in an oblong form; which, according to the received laws of refraction, I expected should have been circular.

They were terminated at the sides with straight lines, but at the ends, the decay of light was so gradual, that it was difficult to determine justly, what was their figure; yet they seemed semicircular.

Comparing the length of this colored spectrum with its breadth, I found it about five times greater; a disproportion so extravagant, that it excited me to more than ordinary curiosity of examining from whence it might proceed. I could scarce think, that the various thicknesses of the glass, or the termination with shadow or darkness, could have any influence on light to produce such an effect; yet I thought it not amiss, first to examine those circumstances, and so tried, what would happen by transmitting light through parts of the glass of diverse thicknesses, or through holes in the window of diverse bignesses, or by setting the prism without so, that the light might pass through it, and be refracted before it was terminated by the hole. But I found none of those circumstances material. The fashion of the colors was in all these cases the same.

Then I suspected, whether by any unevenness in the glass, or other contingent irregularity, these colors might be thus dilated. And to try this, I took another prism like the former, and so placed it, that the light, passing through them both, might be refracted contrary ways [Figure 8-3], and so by the latter returned into that course, from which the former had diverted it. For, by this means I thought, the *regular* effects of the first prism would be destroyed by the second prism, but the *irregular* ones more augmented by the multiplicity of refractions. The event was, that the light, which by the first prism was diffused into an *oblong* form, was by the second reduced into an *orbicular* one with as much regularity, as when it did not at all pass through them [either prism]. So that, whatever was the cause of that length, 'twas not any contingent irregularity (3).

Determined to find out what property of sunlight causes the spectrum to take on such a great length (13½ inches) compared to its breadth (2⅝ inches), Newton first showed that the extra length could not, by the laws of refraction, result from the fact that the sun presents a disk in the sky of about one-half degree as seen from the Earth.

All the Colors of the Rainbow

Next, Newton had to convince himself (and others) that (1) white light is a mixture of all the colors of the spectrum, and (2) the extent of refraction does in fact depend on color.

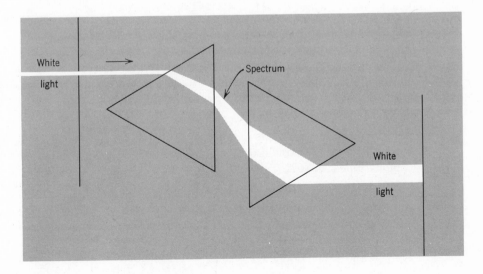

FIGURE 8-3

Newton showed that white light being dispersed into a spectrum by the first prism can be recombined to produce white light by a second prism placed in the contrary direction.

His experiment with the two prisms placed "contrary ways" clearly indicated that all the colors of the spectrum formed by the first prism are recombined by the second prism to form white light. Furthermore, he showed that if any color (or colors) is subtracted from the beam by placing a small object between the two prisms to block that color, the resulting mixture is no longer white. That is, white light is a mixture of *all* the colors of the spectrum. He also demonstrated, by inserting a piece of colored glass between the two prisms to reduce the intensity of one color (or colors), that the resulting mixture is again no longer white. That is, white light is a "proper mixture" of all the colors of the spectrum.*

The Experimentum Crucis

Equally important, Newton's experiments demonstrated that the prism itself does not add to or somehow transform white light into a spectrum, but that the refraction is different for each of the component colors that make up the light. To make this demonstration completely convincing, Newton first dispersed white light into a spectrum and then separated each color, one at a time, from the rest by allowing it to pass through a hole in a board. Then each particular color was made to fall separately on a second prism so situated that the refraction was in the same direction as that of the first prism (Figure 8-4). In this manner all the colors of the spectrum were examined. Furthermore, by keeping the angle of incidence of each of the colors falling on the second prism constant, any differences in the extent of refraction by the second prism could only be accounted for by differences in color. Newton called this experiment his

Experimentum Crucis, which was this: I took two boards, and placed one of them close behind the prism at the window, so that the light might pass through a small hole, made in it for the purpose, and fall on the other board, which I placed at about 12 feet distance, having first made a small hole in it also, for some of that incident light to pass through. Then I placed another prism behind this second board, so that the light, trajected through both boards, might pass through that also, and be again refracted before it arrived at the wall (4).

Reporting this experiment in his work *Optics*, which was not published until 1704, Newton explained the results in greater detail than in the original announcement in 1672.

By turning the prism *ABC* [Figure 8-4], slowly to and fro about its axis, this image [the spectrum] will be made to move up and down the board *de*, and by this means all its parts from one end to the other may be made to pass successively through the hole *g* which is made in the middle of that board. . . . I marked the places *M* and *N* of the opposite wall upon which the refracted light fell, and found that whilst the two boards

*The meaning of the term "proper mixture" depends entirely on the definition of white light. In this context white light is taken to be sunlight after it passes through a clear atmosphere while the sun is still high in the sky.

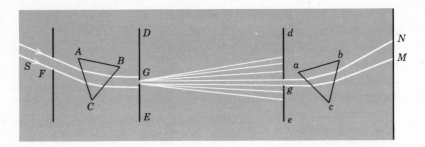

FIGURE 8-4

The *experimentum crucis* for Newton. The different colors of the spectrum striking the second prism were refracted differently even though the angle of incidence was kept constant.

and the second prism remained unmoved, those places, by turning the first prism about its axis, were changed perpetually. For when the lower part of the light [the red end of the spectrum] which fell upon the second board *de* was cast through the hole *g*, it went to the lower place *M* on the wall, and when the higher part of that light [the violent end of the spectrum] was cast through the same hole *g*, it went to a higher place *N* on the wall, and when any intermediate part of the light was cast through that hole, it went to some place on the wall between *M* and *N*. The unchanged position of the holes in the boards made the incidence of the rays upon the second prism to be the same in all cases. And yet in that common incidence some of the rays were more refracted, and others less (5).

With this crucial experiment Newton proved a number of things. First, he proved that the colors comprising the spectrum are basic. When blue light, for example, was passed through the hole in the second board and on through the second prism, the blue color was not altered by the second refraction, nor were any of the other colors altered by this second refraction. The spectral colors cannot be broken up into other colors.

Second, Newton showed that by maintaining the same angle of incidence on the second prism, different colors were refracted at different angles: violet light was refracted the most by both the first prism *and* the second, red light the least. Hence, white light incident upon a prism is broken up into its component colors because each color is refracted differently.

In the course of making these two major points about light, namely that white light is a proper mixture of all the colors of the spectrum, and that the refraction of light depends on its color as well as the angle of incidence, Newton established a new technique of scientific investigation. His writings on optics, both his early article in 1672 and his *Optics* of 1704, made it quite clear to other scientists that experimental evidence of all sorts must be obtained to support new ideas. One verifying experiment is not enough, nor are two experiments if three can be found. All experiments must be carried out with diligence and care, and with the utmost concern for every detail. Not only should the quantitative results of an experiment be reported, but details of the equipment and procedure should also be described so that other investigators can readily verify the work. As Newton's *Principia* is an unmatched theoretical and mathematical treatise, his *Optics* is a monument and served as a model for experimental research during the eighteenth century and even later.

WAVE MOTION

Newton and Particles

Newton has generally been considered a strong supporter of the particulate theory of light, the theory maintaining that light is composed of particles, or projectiles, emanating from the source and traveling to the observer. Newton certainly assumed light to be composed of particles, yet apparently he was convinced that these particles have some sort of wave nature about them.

Newton was compelled to impose a wave nature upon his light particles to explain the phenomenon of thin films, for example, a soap bubble, an oil film on water, or an air gap between two pieces of glass (see Figure 8-22).

White light incident upon such films reflects not as white light but as various colors of the spectrum. Newton showed conclusively that the thickness of the thin film determines what color will be reflected. If for a given thickness a color is reflected, this particular color is not transmitted, and some other color is transmitted. As the thickness of the film varies, so does the color of the light reflected, and hence the color transmitted. Under no circumstances is the same color both reflected and transmitted at places in the film where the thickness is the same.

To explain this startling observation, Newton used the expression "fits of easy reflection and easy transmission," which at first appears to be some sort of hocur-pocus explanation. Closer examination of Newton's statement, however, reveals his meaning:

light is in fits of easy reflection and easy transmission, before its incidence on transparent bodies. And probably it is put into such fits as its first emission from luminous bodies, and continues in them during all its progress. For these fits are of a lasting nature . . . (6).

In the paragraph just preceding this statement Newton referred to "alternate fits of easy reflection and easy transmission." These statements, understood in the context in which they were written, can only lead us to believe that Newton, who firmly established the particulate theory of light, felt strongly that there is a wave nature about these particles—a characteristic of light rays that varies in a regular alternating manner. Newton then could only describe this wave nature in terms of "alternate fits of easy reflection and easy transmission."

Our current explanation of thin-film phenomena does indeed ascribe wave characteristics to light, but this explanation was not forthcoming until the yearly years of the nineteenth century.

Wave Description

A Dutch mathematician and a contemporary of Newton, Christian Huygens (1629-1695), established a theoretical description of wave propagation that has survived all others and is still in use today. Before a wave can travel through a medium, whether it be a wave on the surface of water, a wave through a stretched spring, a sound wave through the air, or, as Huygens suggested, a light wave through an all pervading medium called *aether*, the medium must supply some resistance to the wave motion.

A wave motion is produced only if the medium, once distorted, supplies a restoring force to return that distorted portion to its equilibrium position. For example, if a taut spring, originally in an equilibrium position (Figure 8-5a), is pulled aside near one of its ends (Figure 8-5b) and then permitted to return to its initial position, a wave pulse is set up in the spring (Figure 8-5c, d). A restoring force F_r acts to return the spring to its equilibrium position. As a reaction to this restoring force, F' is exerted on either side of the pulse, but only on the right-hand side is the spring free to move, the left-hand side being held down. Consequently, the right-hand side of the spring is forced upward while the pulse itself returns to its equilibrium position. These motions amount to a transfer of energy along the spring which, having

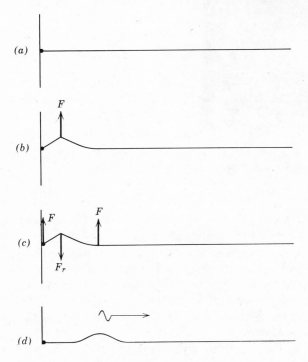

FIGURE 8-5

If a taut spring *(a)* is pulled aside *(b)*, the restoring force *(b)* will cause the spring to return to its equilibrium position and a wave will progress to the right *(d)*.

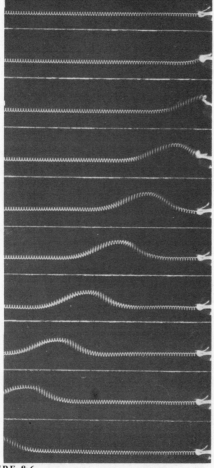

FIGURE 8-6

A transverse wave traveling from right to left along a taut spring. (From *Physics*, Physical Science Study Committee, D. C. Heath and Company, 1960.)

started to the right, continues in that direction. The actual propagation of such a pulse through a stretched spring is shown in Figure 8-6.

If the end of the spring where the wave pulse initiated continues to move up and down from the equilibrium position, a train of wave pulses is set in motion along the spring; wave pulses both above and below the equilibrium position follow each other alternately along the spring (Figure 8-7). Such a succession of wave pulses is called a *wavetrain*. This wavetrain is used to describe wave motion in general.

Various features of the wavetrain have special names that make it easy to discuss the action. Let us consider a wavetrain moving through a uniform medium to the right. Each portion of that wavetrain moves with the same velocity, called the *wave velocity v*. Should that wave pass into a different medium, however, the wave velocity might change; for example, sound

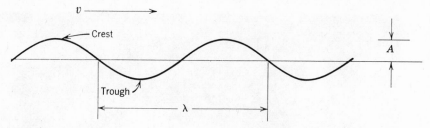

FIGURE 8-7

A portion of wavetrain with the wavelength λ and the amplitude *A* marked.

travels faster through steel than through air. Two adjacent pulses, one a *crest* and the other a *trough*, constitute one complete wave. The distance from one end of one complete wave to the other is called the *wavelength* λ (Greek lambda). The interval of time taken for one complete wave to replace the one in front of it is called the *period T*, which may also be thought of as the time needed for the wave to move a distance equal to one wavelength. The reciprocal of the period, $1/T$, is the *frequency f* of the wave and is a measure of the number of waves that cross a given point in a unit interval of time. The maximum displacement of the medium from the position of equilibrium is called the *amplitude A*.

If the wavelength is measured in meters, its velocity is measured in meters per second. Its period is then measured in seconds and its frequency in the reciprocal 1/sec, often called cycles per second (cps), or *hertz*, in honor of Heinrich Hertz (see Chapter 9).

How velocity, frequency, and wavelength relate to one another can be derived rather easily. The period *T* is the time it takes one wave traveling with a velocity *v* to traverse a distance equal to one wavelength λ. Therefore, since

$$v = \frac{s}{t}$$

$$v = \frac{\lambda}{T}$$

But the period *T* is equal to $1/f$, or $1/T$ is equal to *f*. Hence,

$$v = f\lambda$$

Any one particle in the spring oscillates in a direction perpendicular to the direction in which the wave travels (Figure 8-6). A wave in which the medium oscillates perpendicularly to the direction the wave is traveling is called *transverse*. A wave on the surface of a body of water is a transverse wave; a cork will bob up and down as the wave passes.

If the medium oscillates in a direction parallel to that in which the wave is traveling, the wave is called *longitudinal*. Sound travels through the air as longitudinal waves; the particles of air oscillate back and forth as the sound passes.

FIGURE 8-8

Ripples in a pool. (Francis Laping D. P. I.)

Water waves have breadth to them that waves in a single spring do not have. The water wave in the photograph shown in Figure 8-8 illustrates this breadth; it has a circular pattern. There are many particles of water all oscillating in unison or, as a physicist prefers to say, all oscillating *in phase*. All parts of one complete wave that are in phase with one another comprise the *wavefront*. The crest of an advancing water wave is such a wavefront, so is the trough. Many springs placed like spokes of a wheel could transmit a wavefront similar to the water wave in Figure 8-8.

Huygens' Wave Principle

Huygens contended that each particle in any wavefront behaves like a point source of a new wave. This has become known as Huygens' principle.

A point source of waves on the surface of water, as, for example, a small

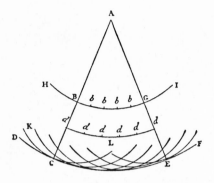

FIGURE 8-9

Huygens conceived the idea that secondary wavelets starting at $b, b, b, b,$ would result in a new wave front d-d; secondary wavelets starting at $d, d, d, d,$ would result in still another new wave front $DCEF$. (From *Treatise on Light* by Christian Huygens.)

pebble dropped in a pond (see Figure 8-8), sets up circular wavefronts; a point source emitting sound waves into the air sets up spherical wavefronts. Since it is easier to think of a circular wavefront, we shall restrict our discussion to surface water waves.

Huygens showed that if each point in a wavefront sets up small circular wavelets, the advancing wavefront is the summation of all these little waves, called secondary wavelets. The figure used by Huygens to illustrate this point is shown in Figure 8-9; each of the secondary wavelets is centered on the points $d, d, d,$ etc. Together they produce the effect of one advancing wavefront. The advancing wavefront is, according to Huygens, composed of the combined effect of many, many tiny secondary wavelets.

Obstacles and Diffraction

If any of those secondary wavelets are not permitted to advance, the wavefront will be changed. For example, if an advancing wavefront meets an obstacle, the secondary wavelets that strike the obstacle will not be permitted to make their contribution to the advancing wavefront, thus altering its wave structure. This alteration of a wavefront is called *diffraction*.

As an example of diffraction, imagine long straight ocean waves beating on the shores of a straight stretch of beach (Figure 8-10a). Jutting out into the ocean, however, is a wall to protect small boats from ocean waves. The secondary wavelet centered on point A in the advancing wavefront, which is immediately next to the edge of the wall, will expand without being interfered with by those secondary wavelets stopped by the wall. Thus the secondary wavelet centered on point A will expand as a regular circular wave and become the edge of the unobstructed advancing wavefront. The advancing wavefront therefore "bends around the corner."

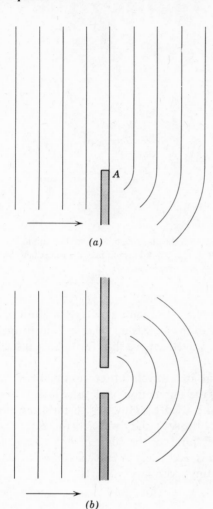

FIGURE 8-10

Wavefronts striking an obstacle will, by Huygens' principle, continue past that obstacle with an altered wavefront. Those secondary wavelets near the obstacle continue on as more or less circular wavefronts.

Had the sea wall been joined by another to form a narrow opening, both sides of the wavefront would have expanded out in a circular pattern (Figure 8-10*b*). The photographs in Figure 8-11 verify Huygens' principle—each point along the wavefront acts like a point source for new waves.

Water waves do in fact bend around a corner, as do sound waves. But the rather large-scale bending of these two kinds of waves is not at all evident with light. This point was crucial for Newton. Since light was not observed to bend around a corner in the manner of water waves and sound waves, Newton felt that light must be composed of particles. He maintained this

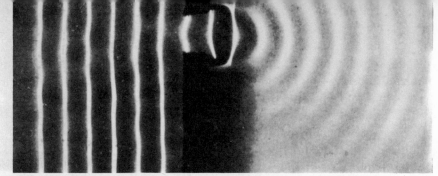

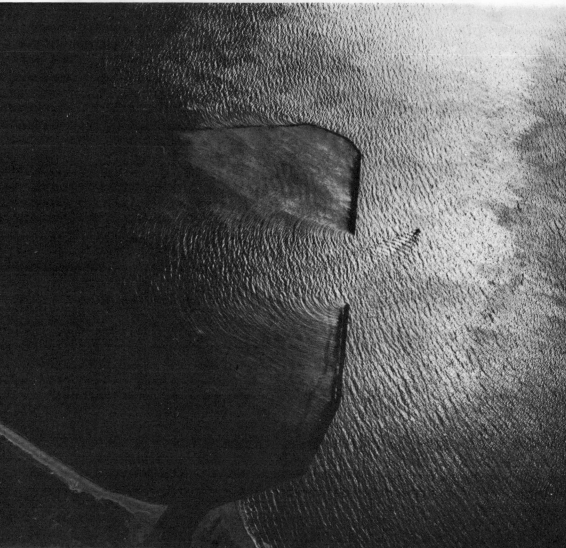

FIGURE 8-11

(a) Water waves (actually ripples in a tank) with straight-line wavefronts, spread out with circular wavefronts after passing through a narrow opening. (From *General Physics for Colleges* by Locke, Farwell, and Drew. Copyright 1923 by the Century Co. Reproduced by permission of Appleton-Century-Crofts, Division of Meredith Publishing Co.) *(b)* Ocean waves passing through an opening in a sea wall. (Photograph by John Shelton.)

view, even while ascribing a wavelike nature to these particles when dealing with thin-film phenomena.

Although Huygens established the technique by which any kind of wave motion can be analyzed, his wave theory of light did not immediately impress the scientific world. Everyone followed the lead set by Newton; at the same time they rejected his "fits of easy reflection and easy transmission." Newton became the authority for the particle theory of light, and that theory was accepted as indisputable during the eighteenth century. But in the year 1800 new observations were reported.

Young's Experiment and Interference

In the opening years of the nineteenth century, Thomas Young (1773-1829), an English physician, published a series of articles presenting observational evidence in support of the wave theory of light. The most famous and convincing of his arguments relied on evidence that two beams of light originating from the same source but traveling different distances can be recombined so that they apparently annihilate each other. However, the annihilation of the particles, supposedly composing the two beams of light, was in direct conflict with the conservation laws, which had even then made a deep impression on the scientific world.

The annihilation of one water wave by another was readily accepted, however, and is in fact easily observed. If small objects are dropped in a pond, the wavefronts radiating out from the points of impact do interfere with each other (Figure 8-12). At those points where the crests of one wave meet the crests of another, the amplitudes add together to produce a wave with an even greater amplitude, that is, an even higher crest. The intersection of two wave crests in a spring is shown in Figure 8-13a. Correspondingly, at those places where the troughs of one wave meet troughs of another, an even deeper trough appears. If a cork is placed where crests meet crests and, alternately, troughs meet troughs, the cork moves up and down more violently than if only one set of waves were passing.

The term *constructive interference* is applied to the interference of two waves meeting in such a way that the resultant amplitude is greater than that of either of the two interfering waves (Figure 8-13a).

At those points in the merging wavefronts where the crests of one wave meet the troughs of the other, the resulting amplitude is less than that of either the crest or the trough. If the height of the crest equals the depth of the trough, the two waves will annihilate or cancel each other.

The term *destructive interference* is applied to the interference of two waves meeting in such a way that the resulting amplitude is less than that of either of the two interfering waves (Figure 8-13b). A cork located in the water at a point where waves of equal amplitude interfere destructively does not bob up and down at all.

Young demonstrated both constructive and destructive interference of light by taking light from a single source, separating it into two beams, and then recombining those beams. The factor determining whether the interference is constructive or destructive is the difference in path length traveled by the two beams from the time they are separated until they are recombined.

Let S in Figure 8-14a represent a source of white light, whose shape, for convenience only, is a straight line; that is, the opening in the shield about the bulb is a narrow slit. Fairly close to this source Young placed two other slits, A and B, which permitted light to pass through them. Upon passing through the two slits, the light was diffracted and then permitted to fall on a screen (or into the eye). Young observed a series of small spectra.

To show that this is truly an interference phenomenon, let us consider Figure 8-14b. Light from the source S falls upon each of the two slits A and B through which it passes. According to Huygens' principle, each of these two slits becomes, in effect, a new source of light; circular wavefronts radiate

FIGURE 8-12

Waves set up by many tiny objects striking the water interfere with each other. (Jacques Wallach, Photo Researchers.)

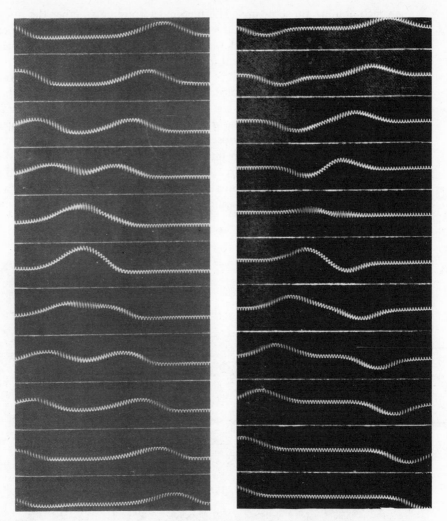

FIGURE 8-13

(a) Two crests of intersecting waves will add to yield one crest higher than either of the two original crests. *(b)* A crest and a trough of two intersecting waves will add algebraically to yield very little, if any, disturbance. (From *Physics,* Physical Science Study Committee, D. C. Heath, 1960.)

from each. These radiating circular waves interfere with each other in the same manner that waves produced on the surface of a pond by two pebbles interfere with each other. We are able, however, to observe the interference pattern of the light only on a screen (or with the eye); we cannot see the interference pattern between the slits and the screen as we can with water waves in Figure 8-12.

The light that arrives on the screen at point *O* (Figure 8-14)—one point on the line that is everywhere equidistant from the two slits—must arrive in

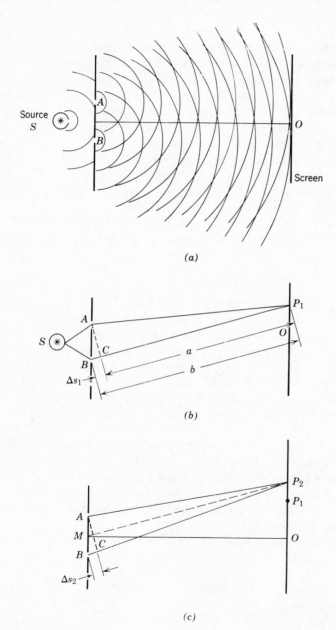

(a)

(b)

(c)

FIGURE 8-14

(a) Circular wavefronts originating at the source, continue as circular wave fronts after passing through each of the two slits. (b) Light arriving at point P_1 from slit B must travel farther by an amount Δs than light arriving from slit A. (c) If a point P_2 is farther from O than P_1, the distance Δs is increased, and constructive interference will occur for waves of longer wavelength.

such a way that constructive interference results. If two crests pass through two slits A and B simultaneously, these two crests will travel with the same velocity over the same distance and hence both arrive simultaneously at point O. At that point the light of all wavelengths, that is, of all colors, is more intense than it would be if only one of the slits permitted the light to pass. If the source S is white light, this central region is also white.

At some distance on either side of point O, the light from slit B must travel a different distance than that from slit A. How much different will determine whether they interfere constructively or destructively. (Since the diagram in Figure 8-14b is symmetrical above and below the point O, we shall, for simplicity, consider only those points above O).

The difference in the path length can be readily seen if we let P_1 be the center of a circle whose radius a is the distance from slit A to P_1. That circle will intersect line BP_1 at point C. Consequently, line segment CP_1 must also be equal to a. The distance b from slit B to the point P_1 is greater than a by an amount equal to $b - a$. This is the difference in the path length the two beams of light must travel before reaching the screen at P_1. Let us call that path difference Δs:

$$\Delta s = b - a$$

If Δs is equal to one wavelength of a particular color of light, that color will arrive at point P_1 in phase. The extra wavelength that appears in the light path BP_1 just exactly fits into the line segment BC; consequently, if a crest passes through A and B simultaneously, a crest also appears at C at that same instant of time. The crests appearing at slit A and at C simultaneously will travel the same distances, AP_1 and CP_1, at the same velocity, both arriving simultaneously at point P_1. They interfere constructively.

If, however, the difference in path length Δs is equal to only half a wavelength for a particular color of light, that color will arrive at point P_1 exactly out of phase. The one-half wavelength difference in the path length appears in the line segment BC; hence, if two crests pass simultaneously through slits A and B, a trough will appear at point C at that same instant of time. The trough at C and the crest at A will both travel the same distances, AP_1 and CP_1, at the same velocity, both arriving simultaneously at point P_1. They interfere destructively.

If the original source S is white light, the net result is that only one color arrives on the screen at point P_1; all other colors interfere destructively at that point. At some other point P_2 (Figure 8-14c) another color will arrive constructively, because this new Δs is longer than the one in Figure 8-14b. Should P_1 be that point on the screen where violet light appears and P_2 that point where red light appears, then as Young observed, the spectrum is spread out between these two points.

Using this analysis, we see that at the point on the screen where one color appears, the others must meet destructively. This constitutes proof that light is a wave phenomenon. Furthermore, since the path difference Δs is shorter for point P_1 than P_2, that part of any one spectrum appearing closer to the central region must be composed of light of the shortest wavelength. The color of each spectrum appearing closest to the central fringe is violet; red appears farthest from the central fringe. The wavelength of violet light is about 4×10^{-7} meter, that of red light about 7×10^{-7} meter.

Light also falls on the screen at points even farther from the central point O than P_2. At these more distant points, the path difference Δs is equal consecutively to two wavelengths, three wavelengths, four wavelengths, and so on, for constructive interference. Thus, Young observed not just one spectrum but a series of spectra on either side of the central position at O.

As we previously stated, all the light arriving at the central point O must arrive in phase since the path difference to that point is zero; hence the central fringe of light is the same color as the source, namely white.

It is very instructive to use a single-color source of light rather than white light. Such a source is called a *monochromatic* source, since it emits light of only one color. Sodium vapor under certain conditions emits light in a very small portion of the yellow region of the spectrum. Consequently, if a sodium arc lamp is used as the source, only yellow light will arrive on the screen and only the yellow portion of each of the spectra will appear. The remainder of the screen will be dark, and it will appear to be crossed by a series of yellow and dark bands, or *fringes* as they are called (Figure 8-15).

Fresnel and Diffraction

The work of Augustine Fresnel (1788-1827), a French scientist who communicated with Young and on one occasion journeyed to England to meet him, clarified Young's experiments and thus helped establish the wave theory of light. Fresnel was actually able to account for the diffraction of light observed in Young's double-slit experiment by employing Huygens' wave principle. His was a brilliant achievement, for the application of Huygens' principle to this problem, although it may appear fairly simple, is quite complex in detail.

Let us recall Huygens' principle by considering the plane waves in Figure 8-16a: each point on the wavefront acts like a point source of second-

FIGURE 8-15

The bright and dark fringes are the interference pattern of Young's double-slit experiment with a monochromatic source of light. (Reprinted with permission from *University Physics* by Sears and Zemansky, 3rd edition, Addison-Wesley Publishing Co. 1964.)

ary wavelets. The new wavefront consists of the many secondary wavelets added together. Consequently, if some of these wavelets are prohibited from contributing to the general interference pattern, the new wavefront will be altered.

An obstacle placed in the path of the oncoming plane waves will cast a shadow, and Fresnel showed that the edge of that shadow, along line *OB*, is not sharp (Figure 8-16*b*). Some of the secondary wavelets that are blocked would have made a contribution to the interference pattern at each of the

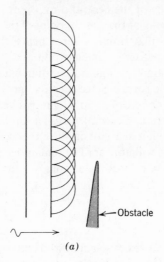

(a)

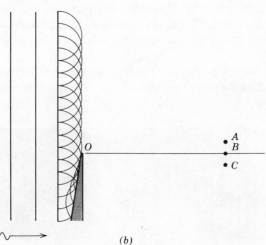

(b)

FIGURE 8-16

Fresnel showed by using Huygens' principle that light striking a sharp obstacle will not leave a sharp shadow. The light arriving at *A*, at *B*, and at *C*, is different from what it would have been had the obstacle not been placed in the path of oncoming light.

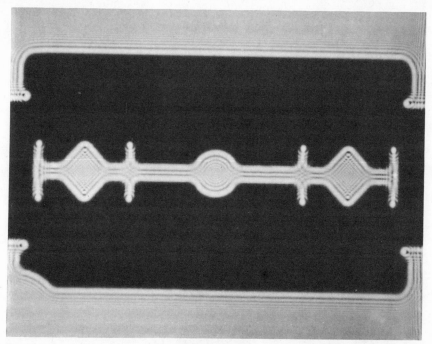

FIGURE 8-17

If the source of light is a point source, the edge of the shadow of a razor blade is not sharp, but composed of many bright and dark fringes that gradually fade out to the uniform brightness of the obstructed light. (Reprinted from *University Physics* by Sears and Zemansky, 3rd edition, Addison-Wesley Publishing Co. 1964.)

three points *A*, *B*, and *C*. Hence the resulting wavefronts arriving at points *A*, *B*, and *C* are different from what they would have been had these wavelets not been obstructed. To understand just how they are altered would require a more detailed study than is possible here, but the shadow formed by a sharp edge, such as a razor blade, is shown in Figure 8-17. The diffraction effects of light passing through a single slit are indicated in Figure 8-18. The bright and dark fringes are formed by a combination of diffraction and interference effects. In fact, diffraction effects are required to explain why Young saw so many spectra spread out on either side of the central fringe. Fresnel did more than just reinforce Young's conclusions; his work was the final verification that convinced the scientific world of the early nineteenth century that light is a wave phenomenon. The work of the two men demonstrated conclusively that the particle theory of light, so thoroughly established during the eighteenth century, had to be abandoned.

Actually the diffraction pattern produced by a straight edge was first observed by an Italian, Francesco Maria Grimaldi (1618-1663), and described in a publication of 1665. Newton, who maintained that light must be composed of particles because it does not diffract, had not only read Grimaldi's

FIGURE 8-18

The diffraction pattern of monochromatic light beam passing through a single slit. (Reprinted by permission from *Elementary Classical Physics* by R. T. Weidner and R. L. Sells, Allyn and Bacon, Inc. 1965.)

work but had, himself, made a careful study of what we now call diffraction effects. He did not, however, interpret the fringes as diffraction effects, for he had associated diffraction with sound and water waves, and hence was expecting large-scale diffraction.

Young thought that diffraction was dependent on wavelength, a fact that Fresnel later demonstrated. The longer the wavelength, the greater the angle γ (Greek gamma) through which the waves will diffract around an obstacle (Figures 8-19 and 8-20). Therefore, since the wavelength of light is extremely short, the angle through which the light diffracts is very small. Newton observed diffraction of light and yet, not recognizing it, used its supposed absence as an argument against the wave nature of light. It is not so much what we observe, but what we think we observe. Preconceptions have on many occasions influenced the course of science.

In the process of providing experimental evidence to support the wave theory of light, Young recalled the observations Newton had made on thin films and used his explanation of "alternating fits of easy reflection and easy transmission" as further proof of the wave theory. Young, however, developed the idea much further. He pointed out that if light reflects from

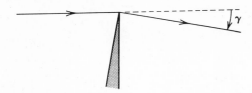

FIGURE 8-19

The angle γ through which a wave is diffracted decreases as the wavelength decreases.

two surfaces placed very close together, the reflection from one surface—for example, the top surface of the soap film in Figure 8-21—is joined by the reflection from the bottom surface. Upon joining, these two reflections interfere with each other. Whether those two reflections interfere constructively or destructively depends on the thickness of the thin film and on the wavelength of the light (Figure 8-22). Hence, as the thickness varies, the colors reflected will be different. Correspondingly, if the *reflected* light interferes destructively, the *transmitted* light will interfere constructively, and vice versa.

POLARIZED LIGHT

Although the wave nature of light had been firmly established, the *kind* of wave had not been determined. Is it a longitudinal or a transverse wave? The distinction between these two kinds of waves can be made on the basis

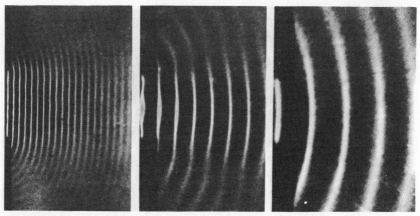

FIGURE 8-20

Waves of different wavelength passing through the same slit (opening in an obstacle) diffract differently. Those of shortest wavelength diffract the least, those whose wavelength is nearly equal to the width of the slit diffract the most. (From *Physics*, Physical Science Study Committee, D. C. Heath and Company, 1960.)

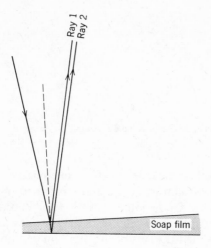

FIGURE 8-21

A ray of light incident on a thin film will reflect from both the top and the bottom surface of the film resulting in interference effects.

FIGURE 8-22

A piece of glass with one surface ground in the shape of a portion of a sphere is placed upon another piece of glass with one surface ground as a plane. The two surfaces meet only at the center, for the spherical surface is convex. Monochromatic light directed on the surfaces reflects an interference pattern that reveals the presences of the thin film of air between the two surfaces of glass. (Courtesy of Bausch and Lomb, Rochester, New York.)

of a phenomenon called polarization. The medium that transmits a *longitudinal* wave oscillates in a direction parallel to the direction in which the wave travels; there is only one such direction.

The medium that transmits a *transverse wave* oscillates in a direction perpendicular to the direction in which the wave travels. There are an infinite number of possible directions perpendicular to the direction of travel. As indicated by the double-headed arrows in Figure 8-23a, together they form a plane. But in which of the infinite number of possible directions does the medium oscillate? It may oscillate in only one direction, in which case we call the wave *polarized*; or it can oscillate in all possible directions, in which

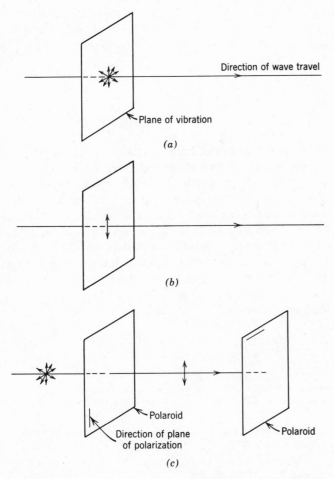

FIGURE 8-23

(a) The medium through which a transverse wave passes may vibrate in any direction at right angles to the direction of propagation. *(b)* If the medium vibrates in only one direction, the wave is said to be polarized. *(c)* A sheet of Polaroid will permit the passage of polarized light. A second sheet, turned at right angles to the first, will not permit the passages of that polarized light.

case we call the wave *unpolarized*. Thus a polarized wave is one in which the medium oscillates in only one of the many possible directions perpendicular to the direction of travel (Figure 8-23*b*). A transverse wave in a spring is polarized.

Even though sunlight and nearly all household lights are unpolarized, there are a number of different techniques for polarizing light. This fact convinced Fresnel and later Young that light is a transverse wave. If light is polarized by one piece of material, for example, a sheet of Polaroid* serves very well, another sheet of the same material turned at right angles to the first one will not permit any light to pass (Figure 8-23*c*). This is a test for polarization.

We see clearly that the term polarization has no meaning when applied to longitudinal waves, for in such waves the medium vibrates in only one direction, the direction of wave propagation. Sound waves cannot be polarized.

THE AETHER

The acceptance of the wave theory of light was not without its problems. What about the medium that was to transmit light? Surely light reached us from the sun and distant stars by traveling through "empty" space. Since there was no known medium in empty space, Huygens suggested one, the *aether*. The name, of course, was chosen from Aristotle's fifth element, the element of the heavens. The purpose of the aether was to transmit light; it had no other purpose, and in fact it proved troublesome with other theories and observations. Its properties had to be such that it could transmit light and yet not impede the motion of the planets about the sun. It was not easy for the nineteenth-century scientists and philosophers who followed Young and Fresnel to imagine a substance to meet these requirements.

References

1. Aristotle, *On the Soul*, Book II, Chap. 7, 418^b/1/, Harvard University Press, Cambridge, Mass.

2. M. R. Cohen, and I. E. Drabkin, *A Source Book in Greek Science*, Harvard University Press, Cambridge, Mass., 1958, p. 274.

3. I. B. Cohen, ed., *Isaac Newton's Papers & Letters on Natural Philosophy*, Harvard University Press, Cambridge, Mass., 1958, pp. 47 ff.

4. *Ibid.*, pp. 50 f.

5. Isaac Newton, *Opticks*, Dover Publications, New York (paperback), 1952, pp. 46 f.

6. *Ibid.*, p. 282.

* Polaroid is a twentieth-century invention, but naturally occurring crystals may also be used for the same purpose.

Questions

1. Define (*a*) reflection, (*b*) refraction, (*c*) wavepulse, (*d*) wavetrain, (*e*) wavelength, (*f*) wave velocity, (*g*) frequency, (*h*) period, (*i*) wavefront, (*j*) diffraction, (*k*) polarized wave.

2. Describe the *experimentum crucis* and indicate why it was so important to Newton.

3. To demonstrate how Ptolemy "smoothed" his measurements of the angle of refraction, set up a table so that you can find the differences between the successive values of the angle of refraction given in Table 8-1; that is, $15\frac{1}{2} - 8 = 7\frac{1}{2}$, $22\frac{1}{2} - 15\frac{1}{2} = 7$, etc. Now carry the table one more column by finding the second differences, the differences between the successive values of the first differences; that is, $7\frac{1}{2} - 7 = \frac{1}{2}$, etc. Make a similar difference table for the column giving measurements made today. Now comment on how the second differences have led historians of science to conclude that Ptolemy did indeed adjust his measurements.

4. By means of a drawing, demonstrate how Huygens' principle predicts that light will refract when it passes from one medium into another if its speed is different in the two media. You might draw a series of parallel wavefronts with the first striking the surface of the water at an oblique angle. During the same interval of time the secondary wavelets from this most forward wavefront will travel only three-fourths as far in the water as in the air.

5. Comment on one of the experiments used by Newton to demonstrate that white light is a proper mixture of all the colors of the rainbow.

Hitherto I have produced whiteness by mixing the colors of prisms. If now the colors of natural bodies are to be mingled, let water a little thickened with soap be agitated to raise a froth, and after that froth has stood a little, there will appear to one that shall view it intently various colors every where in the surfaces of the several bubbles; but to one that shall go so far off, that he cannot distinguish the colors from one another, the whole froth will grow white with a perfect whiteness.

6. Describe Young's double-slit experiment and indicate (*a*) why it could be used by Young to support the wave theory of light, and (*b*) how Young used it to obtain a value for the wavelength of red and blue light.

Problems

1. Find the period of oscillation of waves whose frequencies are (*a*) 25 cycles/sec; (*b*) 940 kilocycles/sec; (*c*) 80 megacycles/sec (mega is 10^6).

2. The velocity of light is 3×10^8 m/sec. Find the frequency of light at the (*a*) violet end of the spectrum if its wavelength is 4.0×10^{-7}m; (*b*) red end of the spectrum if its wavelength is 7.0×10^{-7}m.

3. In Figure 8-14*c*, assume that the triangles ABC and MOP_2 are similar. (This assumption is safe even if triangle MOP_2 is a right triangle and ABC is not, since the angles BAC and OMP_2 are so small.) The distances MO, OP_2, and AB can each be measured; therefore, the wavelength of light can be determined using the proportion

$$\frac{BC}{AB} = \frac{OP_2}{MO}$$

Calculate the wavelength of light observed if the following measurements are made of the second fringe, $BC = 2\lambda$: $OP_2 = 2.4 \times 10^{-4}$ m; $MO = 0.39$ m; $AB = 1.9 \times 10^{-3}$ m.

4. The velocity of sound is not far from 1100 ft/sec. How much later does the thunder arrive than the flash of light if a bolt of lightning strikes one mile away from an observer?

5. What is the wavelength of a musical note if its frequency is (*a*) 264 cps (middle C), (*b*) 528 cps (C above middle C), (*c*) 352 cps (F above middle C), (*d*) 440 cps (A above middle C).

Chapter Nine

Electromagnetic Waves

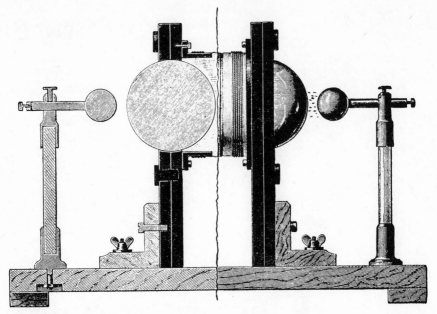

Culver Pictures, Inc.

W ith the wave theory of light well established and the aether conjec-
tured as the medium supplying passage for these waves as they travel
throughout the universe, the phenomenon of light seemed more or less
explained. Then an English physicist, James Clark Maxwell (1831-1879),
began to apply his profound understanding of mathematics to his keen
insight into physical ideas. Maxwell thought that it ought to be possible to
express the actions of Faraday's lines of force in mathematical form. He
knew that if this could be done, our understanding of electric and magnetic
phenomena would be greatly enhanced. But when he began his study, he
did not anticipate that the results would change the scientific concept of the
nature of light.

ELECTRIC AND MAGNETIC FIELDS

The Magnetic Field

The natural consequence of Maxwell's mathematical study of Faraday's
lines of force was the concept of the magnetic field. Maxwell knew that when
a small magnetic compass is used to investigate the magnetic "influence"
around a larger magnet, the detection of "influence" is not limited to certain
lines of force. No matter where the little magnetic compass is placed around
the bigger magnet, it will indicate the presence of the bigger magnet by
pointing in a certain direction. The region about a magnet is not broken up
into lines of force with no magnetic "influence" between them. Therefore,
Maxwell relied on the principle of continuity and used a form of mathematics
(calculus) to describe the magnetic field as a continuous thing.

The equations Maxwell developed describe changes in the magnetic
field as taking place in infinitely small steps rather than in a series of discontinuous jumps. By way of analogy, a ball rolling down a ramp will change
its height above the ground continuously. It will bounce down a flight of
stairs in a series of discontinuous jumps.

The magnetic field about a magnet can be investigated by using any small
magnet, for example, a small magnetic compass needle. Such a magnet is
called a *test magnet*. It indicates not only the direction of the magnetic field
but also the strength of that field. A magnetic compass needle used as a test
magnet will oscillate in a magnetic field; the more rapid the oscillations, the
stronger the magnetic field. In fact, the strength of the magnetic field can be
defined in terms of the frequency of oscillation of a small magnetic compass
needle.

Magnetic compass needles, however, are difficult to duplicate precisely,
and once duplicated they are subject to all sorts of influences. They cannot,
therefore, be used to define a precise concept such as the strength of a magnetic field. However, an electric current forms a magnetic field, and electric
currents can be controlled very precisely; the magnetic field strength is
therefore defined in terms of an electric current. The symbol for magnetic

FIGURE 9-1

J. Clerk Maxwell (1831–1879).

field strength is B, and although it is defined precisely in terms of an electric current, its meaning is made clear by the oscillation of a small test magnet.

The magnetic force F acting on the small test magnet is proportional to the square of the frequency of oscillation f of the test magnet,

$$F \propto f^2$$

The *magnetic field strength B* is by definition proportional to the force exerted:

$$B \propto F$$

Therefore,

$$B \propto f^2$$

Wherever we place a test magnet in a region about a larger magnet, the test magnet will point in a specific direction and will oscillate with a particular frequency. The direction of the magnetic field is defined as that direction indicated by the north end of a test compass (Figure 9-2a). The magnetic field strength B is proportional to the square of the frequency of oscillation of the test magnet. Indeed, we could make a plot of the entire region about a magnet by indicating the direction and strength of the magnetic forces at various points (Figure 9-2b). We know that quantities described by both direction and magnitude are called vectors, and it appears that the region about a magnet is best described by specifying all these vectors. A region so described is called a *vector field*, a concept that is one of Maxwell's many contributions to the study of physics.

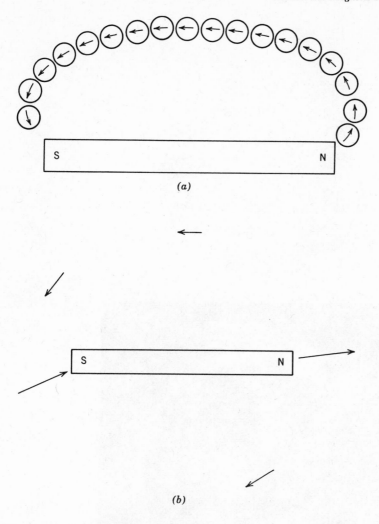

FIGURE 9-2

(a) Compasses placed about a bar magnet indicate the direction of the magnetic field. *(b)* The direction and magnitude of the magnetic field as shown at a few selected points in space.

The Electric Field

Just as a magnetic field can be described by vector concepts and vector mathematics, so can an electric field. To investigate and describe the electric field, a small positive charge is used as a *test charge* in the same sense that a small magnetic compass needle was used as a test magnet. The test charge, however, neither oscillates nor points in a particular direction like the test magnet. We are left, therefore, to imagine both the direction and magnitude of the forces exerted on it.

The magnitude of the vector field about an electric charge is given by

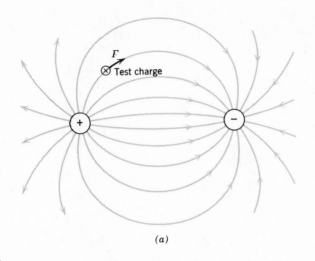

(a)

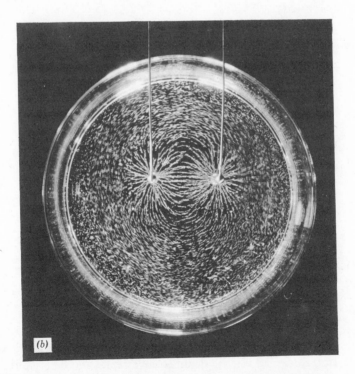

(b)

FIGURE 9-3

(a) A positive test charge, if initially at rest and free to move, will follow (trace out) a line of force in an electric field. *(b)* A photograph of small electrified particles orienting themselves in the electric field set up between two wires which contain opposite charges. *(c)* The electric field between two wires with the same electric charge. *(d)* The electric field between two metal plates with opposite electric charges. (From *Physics,* Physical Science Study Group, D. C. Heath and Company, 1960.)

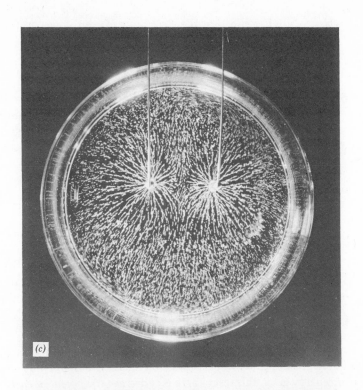

(c)

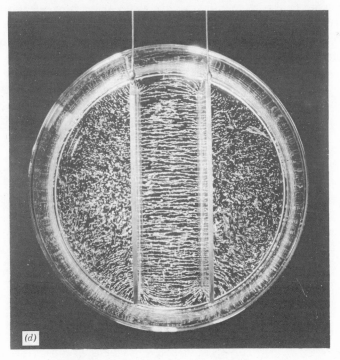

(d)

the *electric field strength* **E**, defined as the electric force exerted per unit test charge:

$$\mathbf{E} = \frac{F}{Q}$$

A force of 5×10^{-4} newton acting on a charge of 10^{-6} coulomb would indicate an electric field strength of 500 newtons per coulomb. This number describes the magnitude of the electric field at that point. If a charge of 2×10^{-6} coulomb is placed at that point in the vector field, the force acting on it will be

$$F = EQ$$
$$F = (500 \text{ nt/coul})(2 \times 10^{-6} \text{ coul})$$
$$F = 10^{-3} \text{ nt}$$

The direction of the vector field is the direction of the force acting on the positive test charge. Hence, the field is directed away from a positive charge and toward a negative charge (Figure 9-3). A positive test charge initially at rest in that field will be accelerated along the field line on which it is placed, traveling along that line from the positive charge toward the negative charge.

The Electric Field and Energy

In order to move the test charge from the negative charge toward the positive charge, work must be done. Work performed on an electric charge increases its energy. The energy given this test charge is called *electric potential energy*.

The concept of electric potential energy has become a convenient one to use; however, we never refer to absolute values of potential energy but rather to differences in potential energy. If an electric charge is moved through a potential difference, work must be done on that charge. The unit of potential difference is called the *volt* and *is defined as the potential difference through which one joule of work will move a charge of one coulomb*. That is, one volt is equal to one joule per coulomb,

$$V = \frac{W}{Q}$$

where V is the potential difference expressed in volts, W the work in joules, and Q the electric charge in coulombs.

The amount of energy gained by an electric charge falling through a potential difference can be found by using this equation. Suppose an electric charge of 0.01 coulomb falls through a potential difference of 120 volts; the amount of energy ΔE gained by that charge is equal to the work W done on the charge:

$$\Delta E = W = QV$$
$$\Delta E = (0.01)(120)$$
$$\Delta E = 1.2 \text{ joules}$$

If we know the mass of the body containing that electric charge, we can calculate its velocity, because energy acquired in falling through a potential difference is converted to kinetic energy, $\frac{1}{2}mv^2$.

The analogy between electric fields and gravitational fields is rather evident. As a stone falls to the Earth, gravitational potential energy is converted into kinetic energy; similarly, as a positive electric charge falls in an electric field, electrical potential energy is converted into kinetic energy. Gravitational forces are attractive and are described by an inverse square law; electrical forces are both attractive and repulsive, and for point charges are described by an inverse square law. It is possible that gravitational repulsive forces exist, but they have not yet been discovered.

We describe an electric field by specifying the electric field strength **E** at each point in space. The vector quantity **E** has both magnitude and direction. The electric field about a positive point charge is shown in Figure 9-4a. The field is directed outward because, as we have seen, the test charge used to investigate the field is, by convention, arbitrarily chosen as a positive charge. The field about a negative point charge is shown in Figure 9-4b. The gravitational field about mass m is shown in Figure 9-4c. In each of these three fields, the length of the arrows represents the magnitude of the field at the point at the rear end of the arrow.

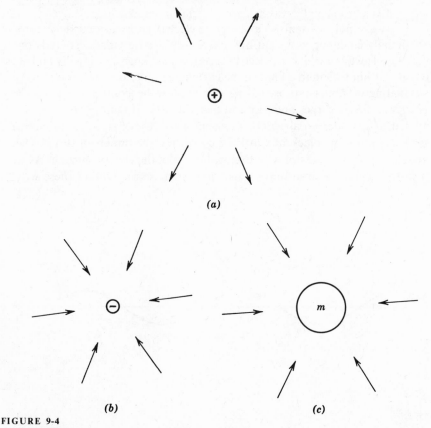

(a)

(b)   (c)

FIGURE 9-4

The magnitude and direction of the fields about *(a)* a positive electric charge, *(b)* a negative electric charge, and *(c)* a mass *m*,

Electromagnetic Waves

Oersted's experiment (see p. 151f) first suggested that the electric and magnetic fields are related, and Faraday's work confirmed the relationship. Just how they are related was made clear by Maxwell's investigations. Maxwell pointed out the fact that the electric vector and magnetic vector about a current-carrying wire are at right angles to each other, and he also described what happens to the electric and magnetic fields if the current in the wire changes. As the current changes, the magnetic field strength B changes. And as the magnetic field changes, an electric field is created. The electric field strength E and the magnetic field strength B are vectors that are at right angles to each other and that change by the same proportion. If one doubles, so does the other. Such a field is called an *electromagnetic field*.

Any change in an electromagnetic field takes time to propagate itself throughout the entire field. That is, an electromagnetic pulse travels at a specific velocity. From the mathematics Maxwell employed to describe these changing electromagnetic fields, he realized that the velocity with which an electromagnetic pulse travels is equal to the velocity of light! Maxwell therefore concluded that light is an electromagnetic wave. But how are a series of electromagnetic waves, such as light waves, produced?

If a wave pulse is sent by a changing electric current, a current that is continually changing will send out a series of electric pulses. There is one very simple way in which an electric current can change continually and yet remain within reasonable limits—the current can alternate its direction. An alternating current is an oscillating current, and hence its value is always changing; first it surges one way and then the other (Figure 9-5).

Let us consider an alternating-current generator of some sort or other represented by the black box in Figure 9-6a. To the outlets of this box we attach two short pieces of wire that will conduct the electric current. As we turn the generator (transmitter) on, the current starts. One of these wires

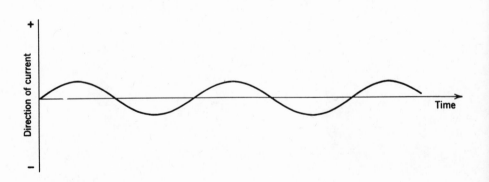

FIGURE 9-5

An alternating current reverses its direction twice every cycle, traveling in the positive and the negative direction during each cycle.

will become positively charged while the other becomes negatively charged (Figure 9-6*a*). An electromagnetic field will build up between these two wires. The direction of the electric and magnetic vectors is shown in Figure 9-6*a*, but for convenience the direction of only the electric vector is given in succeeding drawings.

As the current moves out to the end of the wire, the electromagnetic field continues to build up and moves out at the velocity of light. Just as the current reaches the end of the short wires, the direction of that current is reversed by the alternating-current generator (we adjust the length of the wires to correspond to the velocity and the frequency of the alternating current).

As the current reverses, it carries with it the ends of the electromagnetic field lines, but the field continues to travel out from the wires (Figure 9-6*c*). As the current continues to flow in the reversed direction, the charge on each of the two wires reverses, and with that reversal the electromagnetic field also reverses its direction. But the original electromagnetic disturbance continues to travel outward at the speed of light; it becomes effectively detached from the generator and travels out as a pulse of electromagnetic energy (Figures 9-6*d-h*).

As long as the generator is turned on, the current continues to reverse itself and the electromagnetic disturbances continue to travel out from the wires in alternating pulses. A succession of pulses of this nature is called a *wavetrain*. We are, after all, describing a wave phenomenon. A test charge placed at point *P* (Figure 9-6*i*) will, if it is free to move, oscillate back and forth as the alternating electric field passes it. Similarly, a small compass used as a test magnet would oscillate as the alternating magnetic field passes, were the compass needle able to oscillate that fast.

Maxwell developed his ideas on electromagnetic phenomena in the late 1850's and early 1860's, that is, some twenty-five to thirty years after Faraday began to publish his experiments with magnetically induced electric currents. From 1860 to 1865 Maxwell was Professor of Natural Philosophy at King's College in London. While at that post he was able to meet and enjoy discussions with Faraday whom he had long revered. In 1864 Maxwell published *A Dynamical Theory of the Electromagnetic Field* containing his mathematical extension of Faraday's imaginary lines of force. Maxwell's treatise marks one of the high points in the development of physics and reveals in a very striking way the power of mathematics to deal with physical concepts. A strictly mathematical treatment of electromagnetic fields is of little value if it does not guide theoretical and experimental research.

HERTZ AND RADIO WAVES

A number of scientists expressed their opinions of Maxwell's work, and not all of those opinions were favorable. One of the scientists most reluctant to accept Maxwell's conclusions was William Thompson (Lord Kelvin), who was working very closely with James Joule. But a young German

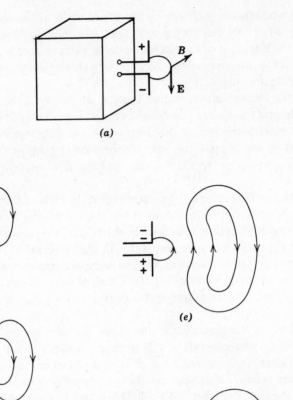

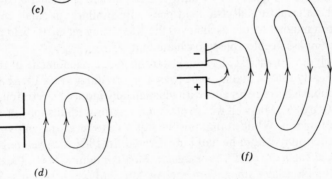

FIGURE 9-6
The process by which an alternating current generator (transmitter) sends out electromagnetic waves.

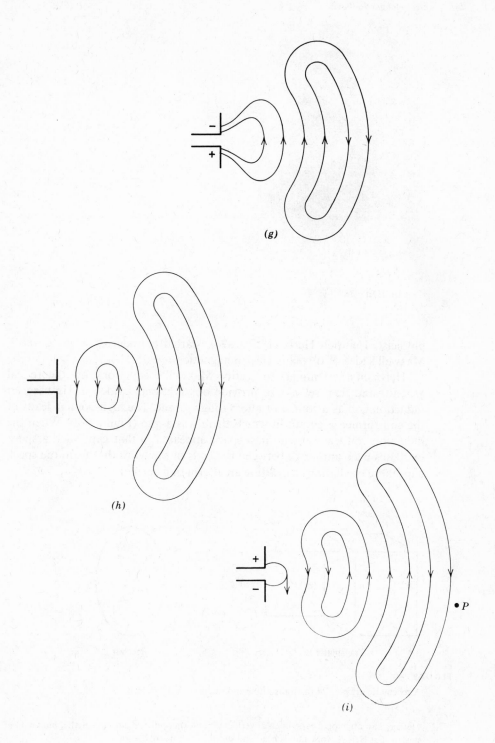

(g)

(h)

(i)

• P

FIGURE 9-7
Heinrich Hertz (1857–1894).

physicist, Heinrich Hertz (1857–1894), started considering ways to verify Maxwell's idea of traveling electromagnetic waves.*

Hertz first attempted to justify Maxwell's equations on theoretical grounds and then set out to provide observational evidence. He used an induction coil as a source of alternating current. The high-voltage leads of the coil formed a circuit in which there was a gap (Figure 9-8). When the induction coil was turned on, a spark appeared in that gap. Such a spark oscillates back and forth between the ends of the wires that form the spark gap; these oscillations constitute an alternating current.

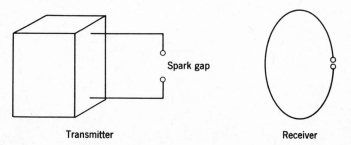

FIGURE 9-8
A schematic drawing of the transmitter and receiver used by Hertz.

* Others also attempted experimental verification of Maxwell's conclusions with a modicum of success, but Hertz's work stands head and shoulders above all others.

To determine whether or not electromagnetic waves were being emitted by the sparks, Hertz placed a wire ring with a similar gap some distance from the spark gap. He reasoned that an alternating electromagnetic field passing this ring should induce an alternating current in that ring. This current could then be detected by a little spark jumping the gap left in the ring.

The Transmitter and Receiver

In 1886 Hertz succeeded in showing that a current was indeed induced in the wire ring whenever the sparks jumped across the spark gap of what he referred to as the primary conductor and we now call the *transmitter*. Hertz called his ring the secondary conductor; today we refer to the *receiver*. The electromagnetic waves that Hertz succeeded in finding were, for many years, called Hertzian waves. They are now more generally called radio waves.

Having shown that electromagnetic waves did indeed propagate into space as predicted by Maxwell, Hertz then set about to demonstrate that these waves have the same properties as light. The properties that he concentrated on are (1) straight-line propagation, (2) reflection, (3) refraction, and (4) polarization.

The Radio Beam and a Prism

Utilizing metal sheets through which his waves could not pass, Hertz was able to define a reasonably narrow beam (Figure 9-9). With this narrow beam he was able to demonstrate that the waves he had detected do travel in straight lines, and that they can be reflected from a third metallic sheet as light reflects from a mirror; that is, the angle of incidence *i* equals the angle of reflection *r*.

It was a simple matter to prove that these rays reflect according to the observed laws of reflected light, but

in order to find out whether any refraction of the ray takes place in passing from air into another insulating medium, I had a large prism made of so-called hard pitch, a material like asphalt [Figure 9-10]. The base was an isosceles triangle 1.2 meters in the side, and with a refracting angle of nearly 30°. The refracting edge was placed vertical, and the height of the whole prism was 1.5 meters. But since the prism weighed about 12 cwt [1200 pounds], and would have been too heavy to move as a whole, it was built up of three pieces, each 0.5 meter high, placed one above the other. The material was cast in wooden boxes which were left around it, as they did not appear to interfere with its use. The prism was mounted on a support of such height that the middle of its refracting edge was at the same height as the primary and secondary spark gaps. When I was satisfied that refraction did take place, and had obtained some idea of its amount, I arranged the experiment in the following manner: The producing mirror [transmitter] was set up at a distance of 2.6 meters from the prism and facing one of the refracting

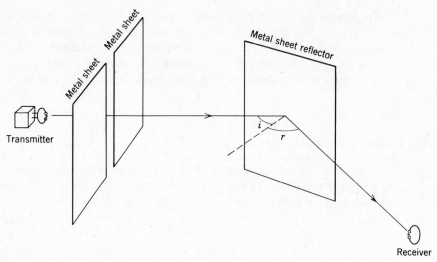

FIGURE 9-9

Hertz showed that he could define a straight beam of what we now call radio waves by inserting two metal sheets with a gap between them. This beam reflected from a third metal sheet as light reflects from a mirror.

surfaces, so that the axis of the beam was directed as nearly as possible towards the center of mass of the prism, and met the refracting surface at an angle of incidence of 25° (on the side of the normal towards the base). Near the refracting edge and also at the opposite side of the prism were placed two conducting screens which prevented the ray from passing by any other path than that through the prism. On the side of the emerging ray there was marked upon the floor a circle of 2.5 meters radius, having as

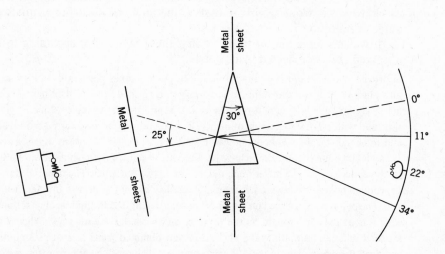

FIGURE 9-10

Hertz also showed that radio waves are refracted and dispersed into a spectrum in the manner of Newton's optical spectrum.

its center the center of mass of the lower end of the prism. Along this the receiving mirror [receiver] was now moved about, its aperture being always directed towards the center of the circle. No sparks were obtained when the mirror was placed in the direction of the incident ray produced; in this direction the prism threw a complete shadow. But sparks appeared when the mirror was moved towards the base of the prism, beginning when the angular deviation from the first position was about 11°. The sparking increased in intensity until the deviation amounted to about 22°, and then again decreased. The last sparks were observed with a deviation of about 34°. When the mirror was placed in a position ōf maximum effect, and then moved away from the prism along the radius of the circle, the sparks could be traced up to a distance of 5-6 meters. When an assistant stood either in front of the prism or behind it the sparking invariably ceased, which shows that the action reaches the secondary conductor through the prism and not in any other way (1).

Hertz refracted radio waves in the same manner that Newton had refracted light. Both men directed a mixture of wavelengths through a prism, observed the spectrum, and studied that spectrum. The radio wave spectrum of Hertz was spread out over an angle of nearly 23 degrees.

Polarized Radio Beam

To determine whether radio waves are polarized or unpolarized, Hertz rotated the receiving mirror about a line parallel to the direction in which the radio waves travel (see Figure 9-11). When the two mirrors were parallel to one another, a strong signal was received (Figure 9-11a); when the two mirrors were at right angles to each other, no signal whatsoever was detected (Figure 9-11b). Hertz therefore concluded that his radio waves were indeed polarized.

All these experiments proved to Hertz that the waves he was observing were the same as light waves, except for their wavelength. He determined that his radio waves had a wavelength close to 30 centimeters. Although he observed a slight diffraction of his radio waves, he did not detect any "bright" or "dark" fringes at the shadow's edge such as those seen in the shadow of the razor blade in Figure 8-17. Nor could he demonstrate the equivalent of Young's double-slit experiment. After all, his was the first radio transmitter and receiver ever built, and although ingenious, we could not expect it to be as refined as equipment built later. Today we can easily observe interference and diffraction effects with both radio waves and microwaves which have somewhat shorter wavelengths. Hertz, however, left little reason to doubt that what he observed, what we now call radio waves, are one variety of the phenomena predicted by Maxwell, traveling electromagnetic waves. He also established conclusively that the waves he had observed had many of the properties of light and that light must therefore be electromagnetic in nature. Strangely enough, however, Hertz was not the first to observe radio waves; Galvani had observed them in 1780.

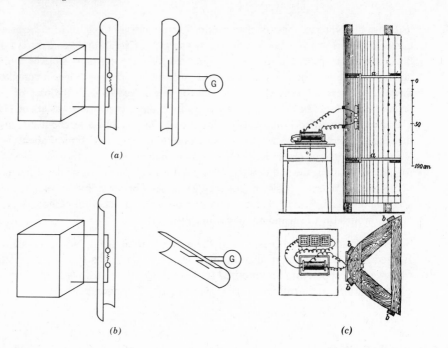

(a)

(b)

(c)

FIGURE 9-11

Hertz showed that with the transmitter and receiver parallel to one another *(a)* a signal was received. But with the receiver turned at right angles from the transmitter *(b)* no signal was received *(c)* from the transmitter used by Hertz. The spark-gap visible in the sideview would normally not be as visible. It is shown, for illustrative purposes, as if the reflecting surface were transparent.

Galvani and Frogs' Legs Continued

We learned earlier (see p. 149) about the chance observation that led to Galvani's discovery of the electric current. A frog's leg twitched when its nerve was touched by a scalpel at the same time that a spark was drawn from a nearby electrostatic generator. We now understand that the spark emitted electromagnetic waves, which in turn induced an electric current in the scalpel. The electric current transmitted to a nerve in the frog's leg caused the leg muscles to contract, producing the twitching. Galvani, however, did not pursue this aspect of his discovery, nor did anyone else at the time.

Galvani's descriptions of twitching frogs' legs, as we know, caught the attention of the scientific community and led to the almost immediate development of the voltaic cell, with its tremendous impact on physics. That the spark produced an effect some distance away, however, went unrecognized and uninvestigated. One wonders why no one recognized this effect. Certainly, invisible "rays" were not part of the thinking and imagination of the scientist in 1791, the year Galvani published his work. The existence of the invisible radio waves discovered by Hertz in 1886, however, had been predicted before he observed them. Maxwell had already indicated that they would have the same velocity as light waves, and Hertz went on to prove that they resembled light in every way except for their longer wavelength.

Today, a quick look at almost any neighborhood with television dipole antennae projecting above the housetops reveals the wide use of radio waves and microwaves. We are even transmitting radio waves from dipole antennae to help probe outer space (Figure 9-12).

THE ELECTROMAGNETIC SPECTRUM

The phenomena of light and radio waves comprise part of the electromagnetic spectrum, but only part. In 1800 (a mere nine years after Galvani's publication), Sir William Herschel (1738-1822), an oboe player but more famous as an astronomer, proved conclusively that light from both the sun and terrestrial sources carries what he called "rays of heat" with it.

FIGURE 9-12

Dipole antennae used by the M. I. T. Lincoln Laboratory to send radio signals into space and to receive their echo from such objects as the sun. (From the library of the American Institute of Physics).

Herschel and Infrared Light

Herschel demonstrated that if light is dispersed into a spectrum by a prism, the "rays of heat" are refracted even less than those of red light. If such a spectrum is directed onto a table and a thermometer placed in various parts of the spectrum, the thermometer records a temperature above that of the surrounding air in the room. If the thermometer is placed in the red-light portion, the temperature increase is greater than the temperature increase recorded in the blue-light portion. Finally, if the thermometer is placed beyond the red light, the temperature increase is still greater. In this region beyond the visible red portion of the spectrum, "rays of heat" apparently strike the thermometer.

Herschel carried out exhaustive studies comparing these "rays of heat" with light. The evidence he collected should have enabled him to conclude that he had found an extension of the visible spectrum. He was, however, predisposed to believe that the heat rays were different from light. It was not easy to accept the very first observations to indicate that the human senses might in some respect be missing part of what they presumably ought to perceive.

Hence it is established, by incontrovertible facts, that there are rays of heat, both solar and terrestrial, not endowed with a power of rendering objects visible.

It has also been proved, by the whole tenor of our prismatic experiments, that this invisible heat is continued, from the beginning of the least refrangible rays [those which are refracted the least] towards the most refrangible ones, in a series of uninterrupted gradation, from a gentle beginning to a certain maximum; and that it afterwards declines, as uniformly, to a vanishing state. These phenomena have been ascertained by an instrument, which figuratively speaking, we may call blind, and which, therefore, could give us no information about light; yet, by its faithful report, the thermometer, which is the instrument alluded to, can leave no doubt about the existence of the different degrees of heat in the prismatic spectrum.

This consideration, as has been observed, must alter the form of our proposed inquiry; for the question being thus at least partly decided, since it is ascertained that we have rays of heat which give no light, it can only become a subject of inquiry, whether some of these heat-making rays may not have a power of rendering objects visible, superadded to their now already established power of heating bodies.

This being the case, it is evident that the *onus probandi* [burden of proof] ought to lie with those who are willing to establish such an hypothesis; for it does not appear that nature is in the habit of using one and the same mechanism with any two of our senses; witness the vibrations of air that make sound; the effluvia that occasion smells; the particles that produce taste; the resistance or repulsive powers that affect touch: all these are evidently suited to their respective organs of sense. Are we then here, on the contrary, to suppose that the same mechanism should be the cause of such different sensations, as the delicate perceptions of vision, and the very grossest of all affections, which are common to the coarsest parts of our bodies, when exposed to heat ? (2)

In the succeeding decades other experiments proved that the invisible "rays of heat" observed beyond the red end of the visible spectrum have properties identical with those of light, and they became known as *infrared* rays. They differ from light waves only in having a longer wavelength. The infrared portion of the electromagnetic spectrum connects the visible part of the spectrum with that region first investigated by Hertz and now divided for convenience into the *microwave* and *radio wave* region.

Ritter and Ultraviolet Radiation

The year 1800 introduced us to the long wavelength side of the visible spectrum; the year 1801 introduced us to the short wavelength side. J. W. Ritter (1776-1810), a German physicist, announced the discovery of invisible rays beyond the violet, those rays now called *ultraviolet* radiation. Their existence was detected when they darkened a piece of paper dipped in a silver nitrate solution. This behavior formed the basis for the development of our extensive modern photographic techniques.

In 1804 Thomas Young, interested in extending his observations that proved light to be a wave phenomenon, demonstrated another way in which ultraviolet radiation could be considered similar to light. He discovered that ultraviolet radiation when reflected from thin films reveals interference effects. Young also suggested that Herschel's "heat rays" must similarly suffer interference effects and that the thermometer should register them, which it does.

Roentgen and X-Rays

The electromagnetic spectrum was not further extended until after the work of Hertz, and then within one year two new regions were found. In 1895 Wilhelm K. Roentgen (1845–1923), a German physicist and the first recipient of the Nobel prize in physics, discovered the x-ray region while he was working with cathode rays (see p. 313f). If the terminals of an induction coil are connected to two electrodes projecting into a glass cylinder from which air has been partially evacuated, a glow is seen to extend the length of the tube (Figure 9-13). This glow was noticed as early as the eighteenth century but was not studied in detail until better vacuum pumps were produced. Hertz made only a brief study of this glow, but after the discovery of radio waves interest in these cathode rays was renewed. They are called cathode rays because they were found to originate at the cathode (negative terminal) and travel the length of the tube to the anode (positive terminal).

While studying these cathode rays, Roentgen noticed that a cardboard coated with a fluorescent* material glowed in the darkened room whenever

---

* Fluoresence is the process by which a material will absorb light of one wavelength and emit light of a longer wavelength. For example, many rocks and minerals will absorb ultraviolet radiation and emit visible light.

FIGURE 9-13

A cathode ray beam. (Fundamental Photographs.)

the cathode ray was turned on. Influenced by Hertz's discovery of invisible radio waves, Roentgen suspected that an invisible ray was traveling from the cathode ray tube to the fluorescent material. Therefore, in a manner similar to that of Hertz, he tried to establish these rays as part of the electromagnetic spectrum. But his findings were much different.

1. If the discharge of a fairly large induction coil be made to pass through a [cathode ray tube] . . . which has been sufficiently exhausted, the tube being covered with thin, black cardboard which fits it with tolerable closeness, and if the whole apparatus be placed in a completely darkened room, there is observed at each discharge a bright illumination of a paper screen covered with barium platinocyanide, placed in the vicinity of the induction coil, the fluorescence thus produced being entirely independent of the fact whether the coated or the plain surface is turned towards the discharge tube. This fluorescence is visible even when the paper screen is at a distance of two meters from the apparatus.

It is easy to prove that the cause of the fluorescence proceeds from the discharge apparatus, and not from any other point in the conducting circuit.

2. The most striking feature of this phenomenon is the fact that an active agent here passes through a black cardboard envelope, which is opaque to the visible and the ultraviolet rays of the sun or of the electric arc; an agent, too, which has the power of producing active fluorescence. Hence we may first investigate the question whether other bodies also possess this property.

We soon discover that all bodies are transparent to this agent, though in very different degrees. I proceed to give a few examples: Paper is very transparent;[1] behind a bound book of about one thousand pages I saw the fluorescent screen light up brightly, the printers' ink offering scarcely a noticeable hindrance. In the same way the fluorescence appeared behind a double pack of cards; a single card held between the apparatus and the screen being almost unnoticeable to the eye. A single sheet of tin foil is also scarcely perceptible; it is only after several layers have been placed over one another that their shadow is distinctly seen on the screen. Thick blocks of wood are also transparent, pine boards two or three centimeters thick absorbing only slightly. A plate of aluminum about fifteen millimeters thick, though it enfeebled the action seriously, did not cause the fluorescence to disappear entirely. Sheets of hard rubber several centimeters thick still permit the rays to pass through them.[2] Glass plates of equal thickness behave quite differently, according as they contain lead (flint glass) or not; the former are much less transparent than the latter. If the hand be held between the discharge tube and the screen, the darker shadow of the bones is seen within the slightly dark shadow image of the hand itself (3).

---

[1] By "transparency" of a body I denote the relative brightness of a fluorescent screen placed close behind the body, referred to the brightness which the screen shows under the same circumstances, though without the interposition of the body.

[2] For brevity's sake I shall use the expression "rays"; and to distinguish them from others of this name I shall call them "x-rays."

The world eagerly received Roentgen's discovery. Within weeks of his announcement others were studying x-rays; the medical profession utilized the obvious benefits of the newly discovered phenomenon. Long after, however, investigators realized that x-rays are really dangerous and should be used only with great discretion and with every precaution.

Becquerel and Gamma Rays

In 1896, a year after Roentgen's discovery, Henri Becquerel (1852–1908), a French physicist and like Roentgen a Nobel prize recipient, was studying a phenomenon similar to fluorescence. Several substances absorb light of one wavelength, store that energy, and emit light of a longer wavelength at a later time, a phenomenon known as *phosphorescence*. Becquerel, having studied Roentgen's techniques, noticed that a certain phosphorescent substance, a uranium salt, emitted rays that traveled through thin sheets of metal. Becquerel used sunlight to stimulate his phosphorescent material.

It can be verified very easily that the radiation emitted by this substance when exposed to the sun or to diffuse daylight traverse, not only black paper sheets, but also different metals, for instance an aluminum plate and a thin sheet of copper. I have performed the following experiment, among others:

A *Lumière* plate with silver bromide emulsion was enclosed in an opaque housing of black cloth, closed at one side by an aluminum plate. When this arrangement was exposed to a bright sun, even for an entire day, the plate was not clouded. But when the uranium salt was placed on the outside, over the aluminum plate, and again exposed to the sun for many hours, it was observed in developing the plate by the usual procedure that the image of the crystalline material appeared black on the photographic plate. When the aluminum plate was a little thicker, the magnitude of the effect was less than that found after traversing two black paper sheets.

If, between the uranium salt and the aluminum plate or the black paper, a screen made of a copper sheet about 0.10 mm thick were interposed, for instance in the shape of a cross, it was observed on the negative that the image of the cross was lighter, but of a sort indicating that the radiation had traversed the copper plate. In another experiment a thinner copper plate (0.04 mm) weakened the active radiations very much less.

Phosphorescence produced, not by direct sunlight, but by the solar radiation reflected from the metallic mirror of a heliostat, and then refracted by a prism and a quartz lens, showed the same phenomena.

I insist particularly on the following fact, which seems to be very important and not in accordance with the phenomena that one would expect to observe: the same crystalline plates, placed with respect to the photographic plates under the same conditions and traversing the same screens, but sheltered from the incident radiations and kept in darkness, still produced the same photographic effect. I might point out how I was led to this observation:

Some of the preceding experiments were prepared during Wednesday the 26th and Thursday the 27th of February [1896], and since on those days the sun appeared only intermittently, I stopped all experiments and left them in readiness by placing the wrapped plates in the drawer of a cabinet, leaving in place the uranium salts. The sun did not appear on the following days and I developed the plates on March 1st, expecting to find only very faint images. The silhouettes appeared, on the contrary, with great intensity. I thought therefore that the action must continue in the dark and arranged the following experiment:

At the bottom of an opaque cardboard box I placed a photographic plate, then on the sensitive face I placed a crust of uranium salt which was convex, and touched the emulsion only at a few points; then, alongside, I placed another crust of the same salt separated from the emulsion by a thin glass plate; this operation was executed in a darkroom, the box was closed and then placed in another cardboard box, which I put into a drawer.

I did the same thing with a housing closed by an aluminum plate, inside which I placed a photographic plate, and then laid on it a piece of uranium salt. Everything was enclosed in an opaque box and placed in a drawer. After about five hours I developed the plates, and the images of the crystals appeared black, the same as in the preceding experiment, and as if they had been made phosphorescent by means of light. For the salt placed directly over the emulsion, there was scarcely any difference between the action at the points in contact and at other parts of the salt about one mm away from the emulsion; the difference may be attributed to the different distances of the sources of active radiation. The action of the salt placed over the glass plate was slightly weakened but the shape of the source was very well reproduced. In passing through the aluminum plate the effect was considerably weakened, but was nevertheless very clear.

It is important to note that this phenomenon does not seem to be due to luminous radiation emitted by phosphorescence, since the latter become very weak after about 1/100 of a second and are barely perceptible (4).

Between the years 1896 and 1903 many scientists worked with the new rays discovered by Becquerel. They learned that there are actually three distinct "rays" and named them alpha, beta, and gamma rays (see p. 323). In 1908 it was discovered that alpha rays are actually a stream of particles, the nuclei of helium atoms, and they henceforth became known as alpha particles. Beta rays became known as beta particles when in 1899 they were demonstrated to be a stream of electrons (discovered only a few years earlier in England—see p. 313ff). Gamma rays have retained their name, for the

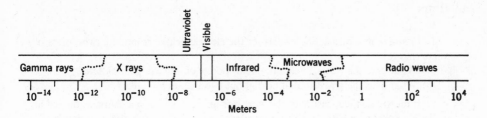

FIGURE 9-14

The electromagnetic spectrum showing that there is no sharp boundary dividing its various portions.

evidence is conclusive that they resemble other electromagnetic waves in every way except wavelength. The wavelength of gamma rays is even shorter than the wavelength of x-rays.

With the discovery of gamma rays, the electromagnetic spectrum was "complete." Figure 9-14 shows the range of the electromagnetic spectrum from the radio waves on the long-wavelength (low-frequency) end to the gamma rays on the short-wavelength (high-frequency) end. Maxwell's prediction of the general existence of such radiations (certainly he predicted none in particular) is proof of his genius. But it is significant that with the discovery of infrared radiation by Herschel and of ultraviolet radiation by Ritter, man began to realize that his senses are limited. From that time on, scientific investigation has had to rely more and more upon instruments. The senses that Nature has given us are not sufficiently sensitive for our study of her. The scientist found that he could indeed trust instruments to make both observations and measurements that he himself could not make. In effect, man extended his observational senses and his powers of measurement. As his reliance upon instruments became more complete, however, man found that he had more difficulty trying to "visualize" what his instruments observed for him. His mathematics was able to cope with the measurements, but his imagination began to exhibit limitations, just as his senses had previously proved themselves to be inadequate.

References

1. M. R. Shamos, ed., *Great Experiments of Physics*, Holt-Dryden, New York (paperback), 1959, p. 194.

2. William Herschel, *Philosophical Transactions*, 1800, pp. 507 f.

3. Shamos, *op. cit.,* pp. 201 f.

4. *Ibid.*, p. 213.

Questions

1. Figure 9-3*b* indicates the pattern of the electric field surrounding two points of unlike charge.

(*a*) What does the electric field about two points of like charge look like?

(*b*) What does the gravitational field about two masses look like?

2. If the volt is the measure of the work per unit charge when a charge is moved in an electric field, what is the concept equivalent to the volt in a gravitational field?

3. If an accelerating electric charge sets up electromagnetic waves, should an accelerating mass set up gravitational waves?

4. Describe the similarities and differences between an electric and a gravitational field.

5. Describe one method by which each of the various parts of the electromagnetic spectrum can be detected.

6. A small positive electric charge and a small magnetic compass are placed at *A*. A second positive charge is placed at *B*, not far from *A*. When all three (the two charges and the magnetic compass) are at rest, describe how the charge and the compass at *A* react to the charge at *B*.

7. If the charge at *B* in question 6 is moving in relation to the charge and compass at *A* (for simplicity, let us say that the charge at *B* moves at right angles to the line joining *A* and *B*), describe how the charge and compass at *A* react to the moving charge at *B*.

Problems

1. A charge of 4×10^{-6} coul has an electric force of 6 nt acting on it. What is the electric field strength at the point where the charge resides?

2. An electron ($q = -1.6 \times 10^{-19}$ coul) is in an electric field of strength $\mathbf{E} = 15$ nt/coul.

(*a*) What force is exerted on this electron?

(*b*) If the mass of an electron is 9.1×10^{-31} kg, what is this electron's acceleration?

(*c*) At this acceleration and starting from rest, how long would it take for this electron to acquire a speed of 3×10^6 m/sec, one-hundredth the speed of light?

3. An electron falls from rest through a potential difference of 24 volts. (*a*) What is its kinetic energy? (*b*) What is its velocity?

4. How fast must an electron be moving if it can "coast" up a potential difference of 200 volts before coming to rest?

5. The arrows in Figure 9-4 represent the magnitude of the electric field strength and gravitational field strength. How long would each of the arrows be if the distance of each from the center of charge or of the mass were (*a*) doubled, (*b*) reduced by a factor of one-half?

6. What is the gravitational field strength, $g = F/m$, of the Earth (a) at its surface, (b) at a point 8000 miles (two Earth radii) from its center, (c) at a point 240,000 miles (sixty Earth radii) from its center?

7. A compass needle at location P in a magnetic field oscillates with a frequency of 4 cps. It is then moved to location R where it oscillates with a frequency of 8 cps. What is the ratio of the magnetic field strength B at location R to that at location P?

8. Find the frequency of (a) x-rays with a wavelength of 10^{-10} m; (b) gamma rays with a wavelength of 10^{-15} m.

Chapter Ten

Four Failures of Classical Physics

Metropolitan Museum Of Art, Gift of Lyman G. Bloomingdale, 1901

During the eighteenth century Newton's laws were thoroughly studied by a number of mathematicians, notably the Frenchmen Laplace, Lagrange, and D'Alembert, the Bernoulli families in Switzerland, and Leonard Euler in Germany. By devising new forms of mathematics, these men vastly extended the use of Newton's laws. Although Newton himself could account not only for the behavior of an apple on the Earth but also for the motion of the moon about the Earth, the eighteenth-century mathematicians were able to show that in reality planetary orbits are *not* perfect ellipses.

Kepler's first law is strictly true only for a solar system consisting of the sun at its center with but one planet revolving about it. In reality, every planet attracts every other planet with the result that the elliptical orbit calculated for the rotation of any one planet about the sun is altered slightly. When Mars and Jupiter, for example, are on the same side of the sun, Jupiter, more distant from the sun and more massive than Mars, pulls Mars farther from the sun than Mars would travel were Jupiter not there. Jupiter and the sun have a tug of war over Mars, but the sun always wins. These refinements in the study of the interaction of planetary gravitational forces are part of what is called *perturbation theory* and are based on Newton's law of gravity and his second law of motion.

The accuracy of the more refined and exacting perturbation theory in predicting planetary motions was a great triumph of Newtonian physics. The laws of Newton became the true description, indeed the ultimate expression, of the behavior of the universe. With these laws the motion of any planet can be predicted for future centuries; its motions in past centuries can also be reconstructed.

By the mid-nineteenth century Maxwell added his powerful mathematics to the stream of physics, and light was shown to be but one of the many forms of electromagnetic waves. The world of physics began to look fairly complete. There were a number of loose ends, but those, it seemed, could be explained somehow in the framework made up of the theories of Newton, Joule, Maxwell, and others, that part of physics called *classical physics*.

By the turn of the century, however, the outlook was not so rosy. Observations indicated major failures of classical physics. We shall consider four of these failures.

THE FAILURE OF THE PLANET MERCURY

In the middle of the nineteenth century, belief in classical physics had been strengthened by U. J. J. Leverrier (1811-1877), a French astronomer. Yet, ironically, it was Leverrier who later revealed the unbelievable fact that Newton's laws may have limitations.

Leverrier and the Discovery of Neptune

Leverrier raised Newtonian mechanics to its zenith when he pursued the problem presented by the planet Uranus to its logical conclusion. In 1781 William Herschel (see p. 247), the great telescopic explorer of the universe, sighted Uranus, the first planet to be discovered by the telescope. Such a notable find of course prompted continued observations of the planet in order to determine its orbit.

The orbit of a planet cannot be well determined, however, until that planet travels through a large enough part of the orbit to define it uniquely. By Kepler's third law (see p. 110) a planet situated as far from the sun as Uranus (nineteen times as far as the Earth) travels quite slowly in its orbit. Uranus has an orbital speed of only 4 miles per second and thus takes 83 years to complete its journey about the sun. Consequently, a few decades elapsed before astronomers began to realize that there was some difficulty in accounting for the planet's motion. By the 1820's a discrepancy between the predictions of perturbation theory and the actual observations of Uranus became apparent. In 1845 Leverrier was encouraged to study the problem.

To begin, Leverrier determined for the first time the precise perturbing effects massive Jupiter and Saturn exerted on Uranus. Then he proved conclusively that those perturbing effects of the two giant planets could not be reconciled with all the observations of Uranus. His very thorough analysis revealed an unquestionable conflict between Newton's laws as then applied and the observations of the positions of Uranus. Such a conflict is not acceptable; a scientific law must be able to account for all observations within its realm.

There were three possible explanations. First, the observations might be incorrect. But it was not likely that the sightings of so many astronomers in different observatories over so many years could all be in error. Leverrier had no choice but to accept the observational evidence.

The second possible source of error noted by Leverrier was that Newton's law of gravity might not be quite an inverse square law; if the observations were to be trusted, perhaps the law was to be questioned. But that law had been questioned before and proved correct.

Only one other possibility remained. If the observations were correct, and the law was to be relied upon, the application of that law might be incomplete.

As early as 1829, a suggestion was advanced that there might be an unknown planet beyond Uranus whose perturbing influence, as yet unaccounted for by anybody, was the cause of the discrepancy. Leverrier took this approach, and in June 1846 he asked

Is it possible that Uranus' inequalities may be due to a planet located [in the plane of the solar system and beyond Uranus]? And were this so, where is the planet now? What is its mass? What are the elements of its orbit? (1)

These questions outlined a Herculean task. Both the unknown planet's mass and distance from Uranus were needed to calculate its effect on the orbit of Uranus. Moreover, since this unknown planet would actually move in an orbit about the sun, the distance between Uranus and the unknown planet would actually change over the rather long time period for which observations had been made of Uranus.* Leverrier was forced to work perturbation theory backward, that is, to determine the mass of the unseen planet, its orbit, and the position of the planet in that orbit, all by its perturbing influence on an observed planet.

By July 1846 he was able to publish the results of his analysis. He gave the predicted location and brightness of the planet and its size as seen from the Earth. But no observers checked his predictions by actually attempting to locate the planet. By September of 1846 the impatient Leverrier wrote to Dr. Galle, an astronomer at the Berlin Observatory, asking him to look with the observatory's large telescope at the region of the sky where the planet ought to be, in hopes that it might actually be seen. Since a planet appears as a disk, while a star is but a point of light, a large telescope makes the disk of a planet even easier to distinguish from a star.

In less than an hour after Dr. Galle began his search, he located the planet. It lay within only one degree of the position that Leverrier had predicted. The observation was checked the next night, and sure enough the planet had moved from its previous location. This motion further confirmed that

FIGURE 10-1

The planet Neptune with one of its two satellites showing. (Lick Observatory photograph.)

---

* Observations of Uranus had been made on a few occasions when it was considered a star, before it was identified as a planet. Leverrier used these sightings to extend the period of time over which the motion of Uranus could be studied.

it was not a star; stars do not measurably change their position in relation to other stars in one night.

The existence of the planet Neptune had been predicted by the theoretical application of Newton's laws and the prediction confirmed by observation. Newton's laws not only had permitted prediction of the planet's existence but had enabled Leverrier to estimate its size, its approximate orbit, and its location in that orbit. Never before had such a feat been accomplished. Rarely, if ever, is a scientific law so dramatically endorsed. Newtonian physics had never seemed in greater command of the universe!

The Failure to Discover Vulcan

During this period Leverrier was also analyzing the motion of the planet Mercury, a study that was to reveal a limitation of Newton's laws. He began this study in 1842, and by 1849 he was able to report that the observed motion of Mercury did not follow the motions predicted by perturbation theory.

Taking observations that ranged from 1753 to 1845, Leverrier found a discrepancy of such an extent that

a variation in 92 years must be seriously considered because of the exactitude of the observations from which they result. They cannot derive from unreliability in the observations; this would make it necessary to suppose that all astronomers have made great mistakes in measuring time. . . . These mistakes, moreover, would have to vary progressively in time, and differ by several minutes at the end of the period of 92 years. This is utterly absurd! (2)

Leverrier came to realize that the elliptical orbit of Mercury behaved in a manner that could not be accounted for by considering the perturbations of only the planets near Mercury, namely Venus and the Earth.

As we have stated, Kepler showed that the orbit of each planet is an ellipse (neglecting perturbations). The major axis of the elliptical orbit of each planet is aligned in space in relation to the stars, and this alignment can be determined by observations. The alignment is altered, however, by the perturbing effects of other planets; that is, the major axis of each elliptical orbit rotates (Figure 10-2). The rate at which the major axis rotates can be predicted by perturbation theory, but the observations of Mercury revealed that its axis, in fact, rotates too rapidly.

According to Leverrier's calculations, the major axis of Mercury's orbit should rotate 527 seconds of arc* per century; according to observation it does, in fact, rotate 565 seconds of arc per century. The rotation is thus 38 seconds of arc per century greater than predicted. This was not only greater than observational error could account for, but greater than theory could permit. Theory and observation clashed. Observations since 1845 indicate a discrepancy not of 38 seconds of arc, but of 43.

* A circle has 360 degrees, each degree is divided into 60 minutes of arc, and each minute is divided into 60 seconds of arc.

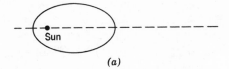

(a)

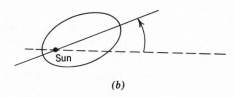

(b)

FIGURE 10-2

The major axis of each planetary orbit rotates relative to the distant stars. For Mercury, this rotation is too rapid to be accounted for by Newtonian mechanics.

Leverrier, who had succeeded so admirably in predicting the existence and location of Neptune, now applied that same technique to account for the unexplained motion of Mercury. Again he postulated an unknown planet, one revolving about the sun in an orbit even smaller than that of Mercury. The proposed planet, however, would have to be large and thus bright enough to see. Although a planet situated so near the sun would be visible only at sunrise, sunset, or during a total solar eclipse, no planet had ever been seen in this position and at these times.

Leverrier then suggested that theoretically the motion of Mercury could be accounted for by myriads of tiny planets revolving about the sun inside Mercury's orbit. But had anyone ever seen evidence that these minor planets exist? No.

Nevertheless, in 1859, as soon as he announced his hypothesis that minor planets exist inside Mercury's orbit, astronomers, professional and amateur, began to search. During the succeeding decades there were, of course, many reports of sightings—the eye often sees what the mind wants it to see—but not one sighting was verified! Vulcan, as one of the postulated minor planets was called, never materialized.

The continued failure to find any evidence whatsoever of a planet, or minor planets, or even clouds of dust particles to vindicate Leverrier's hypothesis proved embarrassing. The possibility that Newton's laws could not account for the motion of Mercury seemed ridiculous.

Finally, in the late nineteenth century, an American astronomer, Asaph Hall (1829-1907), suggested more seriously than anyone before him that Newton's law of gravity might not be precisely an inverse square law. He proposed that the law be written

$$F = G \frac{m_1 m_2}{d^{2+\delta}}$$

where the $2 + \delta$ means that the exponent of d may be 2.01 or 1.99 or some such value. (That is, $\delta = 0.01$ or $\delta = -0.01$.) Perhaps δ might be assigned a value that would make the revised gravitational law account for the unexplained motion of Mercury and yet not disturb Newton's admirable explanation of the motion of the other planets. Perhaps only a minor adjustment of Newton's law of gravity was necessary. Hall sought a compromise, thus following the tradition of earlier mediators, Simplicius (see p. 23) and Tycho Brahe (see p. 43f).

By 1916, however, Albert Einstein, (1879-1955) German-Swiss-American physicist and Nobel laureate, suggested that minor adjustments in Newton's law of gravity are not necessary, nor are they even desirable. He proposed his theory of general relativity which, he maintained, accounts for the motion of Mercury about the sun. And Einstein's theory appears to do just that. The motion of Mercury had been the principal observational support of the theory of general relativity, and in the 1960's even that observational support has been brought into question.

Robert Dicke of Princeton University believes that the apparent error in the motion of Mercury is not so large as had previously been calculated if we assume that the sun is neither a perfect sphere nor perfectly symmetrical to its core. He suggests that if the shape and structure of the sun are taken into account, neither Newtonian principles nor Einstein's relativistic principles explain the motion of Mercury. He proposes a new scalar-tensor theory, and in support of it he has made a series of observations demonstrating that the sun is not as spherical as had been thought.

On the other hand, John Wheeler, a colleague of Dicke's at the University of Princeton, maintains that the general theory of relativity is not only valid but should be extended to include the study of atomic particles.

Physics finds itself on a familiar shoreline; the tides of several theories ebb and flow. One of these two new ideas will become the most valued description of the current observations, and the other will retain only historical interest. Or perhaps another theory will be proposed to support still newer observations.

THE MICHELSON-MORLEY EXPERIMENT

Just as a detailed analysis of the orbit of Mercury brought Newton's laws into doubt, a thorough search for the postulated aether (see p. 226), also used to conduct Maxwell's electromagnetic waves, brought the aether theory under question. Since acceptance of the theory meant that the Earth as it revolves about the sun must travel through the aether, the motion of the Earth in relation to the aether must be observable. The most obvious way to observe this motion would be to measure the velocity of light in relation to the moving Earth.

If an oceanographer is to measure the velocity of ocean waves, either he must fix his observing platform (presumably a boat) at rest in relation to the Earth, or he must determine the velocity of his boat in relation to the Earth and subtract the boat's motion from the measured velocity of the ocean waves. For example, if he is traveling in the same direction as the waves, their velocity in relation to the boat will appear less than it would were he traveling in the opposite direction.

Two American physicists, A. Michelson (1852-1931), a Nobel prize-winner, and E. W. Morley (1838-1923), using arguments similar to our ocean wave example, attempted to detect the aether in 1887. They reasoned that the motion of the Earth in relation to the aether ought to be observable by measuring both the velocity of light in the direction in which the Earth moves and the velocity of light at right angles to the Earth's motion. If the aether in fact exists, two different velocities should be obtained.

Actually, the experiment can be performed without measuring the velocities directly. It is necessary only to determine whether there is a difference in the time light takes to travel two specified distances that are equal in length but at right angles to one another. If one of these distances is in the direction of the Earth's motion, the other is necessarily at right angles to this direction. Should the velocities of light be different, there would be a difference in the measured time intervals the light takes to travel these two distances.

The Boat-in-the-River Analogy

By analogy, let us consider a boat that has a certain velocity v in relation to the water. If the boat travels on still water, for example, on a lake, the boat then has a velocity v in relation to the Earth. If the same boat travels on a moving river, its velocity in relation to the Earth depends both on the boat's direction in the river and on the velocity u of the river in relation to the Earth.

Let us call the time it takes the boat to travel perpendicular to the flow of water a certain distance across stream and back again $t_\perp$; and the time it takes to travel parallel to the current downstream that same distance and back again $t_\parallel$. From an algebraic analysis of these motions, it can be shown that

$$t_\perp = t_\parallel \sqrt{1 - u^2/v^2}$$

If the boat is to arrive back upstream, u must be less than v. Therefore the term u^2/v^2 must be less than one. Consequently, $1 - u^2/v^2$, and finally $\sqrt{1 - u^2/v^2}$, is also always less than one. The time required to make the trip across stream and back is therefore less than the time it takes to make the round trip upstream and down. This difference in time of travel is the most important part of our analogy.

The Michelson Interferometer

Our transition from the boat in the river to light traveling through the aether is aided by examining an instrument designed by Michelson, called the Michelson interferometer because it measures the amount of interference of light. The stream of water is replaced with the stream of aether that should flow past the Earth as it moves in its orbit about the sun.* The boat in the river is replaced by a beam of light that is broken into two beams, one traveling perpendicular to the flowing aether and the other parallel to it (Figure 10-3).

A beam of monochromatic light leaves source S and travels to a beam splitter BS, a mirror that reflects half of the light to a mirror M_1 and refracts the other half of the light to mirror M_2. These two beams of light return to the beam splitter, which again splits each of the beams into two parts. The only parts with which we are concerned, however, are the two beams that travel to the telescope and are there seen by the observer.

If the distance L_1 (from the beam splitter to mirror M_1) is made equal to the distance L_2 (from the beam splitter to M_2), and if the instrument is at rest in the aether, the light that travels from the beam splitter to M_1 will take the same time as the light that travels from the beam splitter to M_2. Therefore, the beams returning to the beam splitter will meet with constructive interference. If, however, the light going to M_2 and back should take a little longer to travel than the light going to M_1 and back, the two beams might meet with destructive interference. This difference in time of travel is the crux of the Michelson-Morley experiment.

The Interferometer and The Aether

If we place the interferometer on the Earth so that the aether supposedly flows past our instrument in a direction parallel to L_2, the setup of the apparatus simulates a boat in a river. The light that travels the round trip to M_2 and back to the beam splitter travels upstream and downstream through the aether. The light that travels the round trip to M_1 and back to BS travels across the stream and back.

By applying the boat-in-the-river analogy, the time the beam of light takes to travel across the aether stream and back, $t_\perp$, is related to the time the other beam takes to travel downstream the same distance and back again, $t_\parallel$, by the equation

$$t_\perp = t_\parallel \sqrt{1 - v^2/c^2}$$

The velocity of light c replaces the velocity of the boat in the river, and the velocity of the Earth v in relation to the aether replaces the velocity of the river water in relation to the Earth.

This equation permits us to predict whether the light traveling upstream and downstream through the aether takes a longer time than that traveling

---

* We are assuming in this argument that the aether is at rest in relation to the sun. The results of the experiment, however do not depend on this assumption.

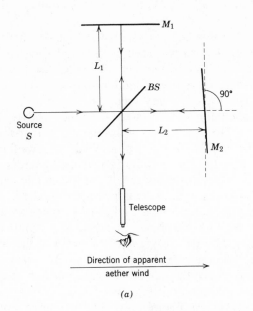

FIGURE 10-3

    (a) Michelson interferometer. Light from the source *S* travels to the beam splitter *BS*. From the beam splitter it travels to both mirror M_1 and M_2, returns, and then travels to the telescope and the eye. *(b)* A Michelson interferometer with the beam splitter in the foreground. The extra plate of glass between the beam splitter and one of the mirrors serves the purpose of making the optical paths equal, for without it one of the beams would travel through the glass of the beam splitter three times and the other beam only once. (Courtesy of Central Scientific Company.)

across stream and back, that is, whether $t_\perp$ is greater than or smaller than $t_\parallel$. Since the aether wind is supposedly caused by the Earth's motion about the sun, the speed of the aether wind must be 18.6 miles per second (the orbital velocity of the Earth). The speed of light, however, is 186,000 miles per second. The ratio v/c is therefore very small, and the ratio v^2/c^2 even smaller. This extremely small number subtracted from one yields a number only slightly less than one. The square root of this number is therefore also only slightly less than one, so that when it is multiplied by $t_\parallel$, we find $t_\perp$ to be slightly less than $t_\parallel$. Light therefore travels the distance L_1 across stream and back in less time than it travels the distance L_2 upstream and back.

In actual practice, however, it is virtually impossible to make L_1 and L_2 exactly equal. Therefore, the two mirrors are adjusted so that they are not quite perpendicular (see M_2 in Figure 10-3a). The distances from the beam splitter to the two mirrors will then change over the width of the beam of light. In some regions the two returning beams will interfere constructively and in others destructively. Since the mirrors are made flat, these regions will alternate. The eye will see a very definite interference pattern, namely a series of bright and dark fringes (Figure 10-4). The two mirrors produce the equivalent of the thin-film phenomenon, with the film in the form of a wedge.

We cannot determine, however, which single bright fringe results from the constructive interference of two light beams traveling the same distances. Therefore, we cannot really tell whether light takes the same time to make the two round trips or not by just looking at the fringes. But, Michelson reasoned, if the Earth does move through the postulated aether, rotating

FIGURE 10-4

The interference fringes seen with a Michelson interferometer. (Reprinted with permission from *Introduction to Atomic and Nuclear Physics* by Harvey E. White, D. Van Nostrand Company Inc. 1964.)

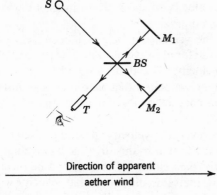

FIGURE 10-5

If the interferometer is oriented so that the paths of light form an angle of 45 degrees with the motion of the Earth (and with respect to the supposed aether wind), then by rotating the interferometer one way or the other, the fringes should shift one way or the other.

the interferometer through an angle of 90 degrees while the observer watches the fringes ought to shift the fringe pattern.

To explain the fringe shift, we shall consider the interferometer when it is located so that the two beams of light going to the two mirrors each make an angle of 45 degrees with the direction of flow of the aether (Figure 10-5). Since in this position both beams travel the same path through the aether, it may be used as a reference position. For the one fringe for which $L_1 = L_2$, the intervals of time the two light beams travel are equal.

If the interferometer is now rotated clockwise, the light going to M_2 travels more and more perpendicular to the flow of aether; hence this light beam will take less time to make the round trip than the light traveling to M_1. The fringes should shift as a result of this rotation. Let us say that they shift to the right in the field viewed by the telescope.

If, however, the interferometer is rotated in a counterclockwise direction, the light going to M_2 travels more and more parallel to the flow of the aether. It consequently takes more time to make the round trip than the light traveling to M_1. As a result, the fringes should again shift, but in the opposite direction, to the left.

Michelson and Morley mounted their interferometer in a tub of mercury which enabled one to rotate it through the angle of 90 degrees while the other looked for a shift of fringes during the rotation. The result of many such trials was that they found no perceptible fringe shift!

This result is as startling as saying that two boats which travel with the same speed in still water take the same time interval to travel equal distances in a river, one upstream and back, and the other across stream and back. The result clearly indicates that whatever the aether had been imagined to be, and whatever duties it had been assigned to perform, it does not in fact exist. Maxwell's electromagnetic waves travel through empty space, an aetherless space. Moreover, the velocity of these electromagnetic waves is independent of the motion of the observer.

We can easily imagine that the velocity of light is independent of the motion of the source, for this is true for all wave motion. The sound emitted by a jet plane traveling with a velocity twice that of sound (Mach 2) does not itself travel through air with any velocity other than that of sound. The velocity of sound is dictated by the characteristics of the atmosphere through which it travels.

If the observer is moving, however, the velocity of sound in relation to him depends on his own motion in relation to the air. If he is traveling with a velocity of Mach $\frac{1}{4}$ (one-fourth the velocity of sound) and he measures the velocity of sound traveling in the same direction, that velocity will be Mach $\frac{3}{4}$. If, on the other hand, he measures the velocity of sound traveling in the opposite direction, that velocity will be Mach $1\frac{1}{4}$ in relation to him.

What is true for sound, however, is not true for light. *The Michelson-Morley experiment proved that the velocity of light is independent of the motion of the observer.* Light traveling in the two arms of the interferometer moves with the same velocity, no matter which way the arms point or which way the Earth is moving.

The results of this famous experiment clearly indicate that were a physicist to make a large number of very accurate measurements of the velocity of light received from stars, whatever their positions in relation to the Earth's motion, these velocities would all be the same. Light from a star toward which the Earth is moving (star *a* in Figure 10-6) would have the same velocity in relation to the Earth as light received from a star from which the Earth is receding (star *b*).

Although the results of the Michelson-Morley experiment were clear, they were not immediately accepted because they were so much at variance with classical theory. It was not until 1905 and Einstein's initial publication on the theory of special relativity that the experiment could be reconciled with theory—a new theory.

Light from star *a*

Direction of motion
of earth

Light from star *b*

FIGURE 10-6

The Michelson-Morley experiment indicates that if we on a moving Earth were to measure the speed of light from star *a* and from star *b*, very accurately, these measured velocities would be equal.

LIGHT AND THE RADIATION CURVE

The other two failures of classical physics were not so closely involved with space and time measurements as were the problems presented by Mercury and by the Michelson-Morley experiment. The first of these, presented by what is called the radiation curve, became apparent in the 1890's, although the concept of the radiation curve had been developed much earlier.

The Three Kinds of Spectra

Newton had discovered the techniques by which sunlight can be spread out into a spectrum allowing detailed study of the various colors. To Newton the spectrum was a continuous array of colors from the violet to the red, an unbroken sequence that we now call the *continuous spectrum* (the rainbow).

The sun's light is white, but the light of a flame is not. Salts placed in a flame color that flame, each salt adding its own distinctive color. In 1752 Thomas Melvill (1726-1753), a Scottish physicist, was the first to pass the light of a flame through a small hole and then through a prism. He noted that only certain colors came through. If "sea salt" (sodium chloride) was used, the only color he could see through the prism was yellow, and since the image of the circular opening through which the light passed was uniformly bright, he rightly supposed that the break between the yellow color and the darkness on either side was quite sudden.

Later discoveries revealed that if the light of a flame is sent first through a narrow slit rather than a hole, the images become straight, bright lines. These are now called *spectral lines*. Most salts placed in a flame give off spectral lines of many colors; between these brightly colored lines is darkness. A spectrum composed only of the bright spectral lines has become known simply as a *bright-line spectrum* (Figure 10-7a).

More than half a century passed before Joseph Fraunhofer (1787-1826), a Bavarian instrument maker, noticed and studied the opposite effect of the bright-line spectrum. In 1814 he observed that the sun's spectrum is not really a continuous array of colors. If sunlight is allowed to pass through a narrow slit and then through a prism, the spectrum so formed is a continuous spectrum crossed by many dark lines. Later it was realized that these dark lines are produced when a relatively cool gas absorbs very selectively only certain discrete regions of the continuous spectrum. Such a spectrum is called an *absorption spectrum* (or a dark-line spectrum) (Figure 10-7b). If sodium gas is used to absorb light from a continuous spectrum,

(a)

(b)

FIGURE 10-7

(a) A bright line spectrum of iron. *(b)* A portion of the solar spectrum—an absorption spectrum. (Lick Observatory photograph.)

the dark lines so formed appear in the yellow region of the spectrum, in fact, in exactly the same region in which the bright spectral lines of sodium appear. The bright lines of sodium have the same wavelength as the absorption lines of sodium.

In 1823 John Herschel (1792-1871), son of the famous astronomer William Herschel, correctly suggested that each chemical element when vaporized, for example, by a flame or spark, emits its own unique and thus identifying sequence of spectral lines. Consequently, the spectrum, either the bright-line or the absorption spectrum, can be used for chemical analysis. A study of the spectral lines of each of the chemical elements formed the basis for the development of the theory of atomic structure in the early part of the twentieth century (see Chapter 12). Until that time there was no satisfactory explanation for the origin of the spectral lines.

Cavity Radiation

The continuous spectrum presented problems that could not be solved with classical theory. Previously, investigators had learned that any solid heated to incandescence will emit a continuous spectrum. The spectrum is the same no matter what material composes the solid. In 1859, the German physicist G. R. Kirchhoff (1824-1887) observed that the emitting power of a surface is directly proportional to its absorbing power. That is, a black object, such as cast iron, will emit more light than a bright object, for example, a shiny metal, even if both are heated to the same temperature. After making this observation, Kirchhoff began a search for the perfect absorber, an object that would absorb *all* the light that falls upon it; that object would then be a perfect radiator. We now recognize, however, that it is perfect only in the Platonistic sense, for it can be achieved only in theory, not in practice.

The closest approximation to a perfect radiator is a cavity hollowed out of a chunk of material, usually metal, with a very small hole as the only opening through which light can enter or leave (Figure 10-8a). Certainly, any light entering the small hole from the outside will be trapped inside the cavity. Correspondingly, when the piece of metal is heated to incandescence, the hole glows brighter than the metal surrounding the hole (Figure 10-8b). Radiation so formed is called *cavity radiation*. A perfect radiator is often called a blackbody, and the term blackbody radiation describes this type of emission.

When a blackbody is heated to incandescence, it first glows a deep red color, then a bright red, then an orange, and finally almost a yellow or white color. The term "white heat" is justifiably a stronger term than "red-hot." If the temperature of a body is raised still higher, its color becomes blue. Heating a substance to such a high temperature that it becomes "blue-hot" is difficult on Earth. Astronomers had long been aware, however, that some stars are red, others white, and still others blue. Now they know that these colors denote differences in temperature and that blue stars are even hotter than white stars.

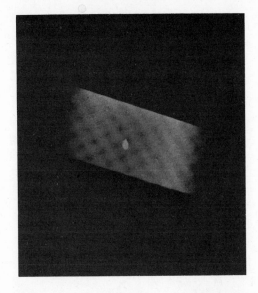

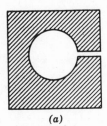

(a)

FIGURE 10-8

(a) A schematic drawing of a cavity resonator. *(b)* A hole drilled in a tungsten filament constitutes a cavity resonator; it glows more brightly than the filament itself. (Reprinted with permission from *Physics*, Halliday and Resnick, John Wiley & Sons, Inc., 1966.)

To determine the intensity of each color at various temperatures, it is
necessary to investigate each color of the light emitted by cavity radiation.
In order to attain the necessary degree of precision, however, rather than
measure the amount of radiation in each color, we measure the amount of
radiant energy in each of a series of small subdivisions of the spectrum. That
is, the spectrum is divided into many small regions and the energy in each
region is measured. When the measured energy in each of these small regions
is plotted against the wavelength of light in the middle of each region, a
steplike curve is obtained (Figure 10-9a). A smooth curve is then drawn
through the midpoints (or the approximate midpoints) of each of the treads
of the steplike graph (Figure 10-9b). When this process is repeated for radiation emitted by cavities at various temperatures, the smooth curves of
Figure 10-9c are obtained.

Comparing the three curves in Figure 10-9c, we note that the amount of
energy emitted at each wavelength increases as the temperature of the cavity
is raised. Because the energy emitted at each wavelength (more specifically,
each of the many subdivisions of the spectrum as shown in Figure 10-9a)
is greater for higher temperatures, the total amount of energy emitted increases as the temperature of the cavity increases.

The energy in each subdivision of the spectrum can be represented by
the area of each little rectangle in Figure 10-9a; therefore, the total energy
emitted by the cavity is represented by the total area under the curve. Thus,
if we add up the areas of all the rectangles in Figure 10-9a, we will obtain

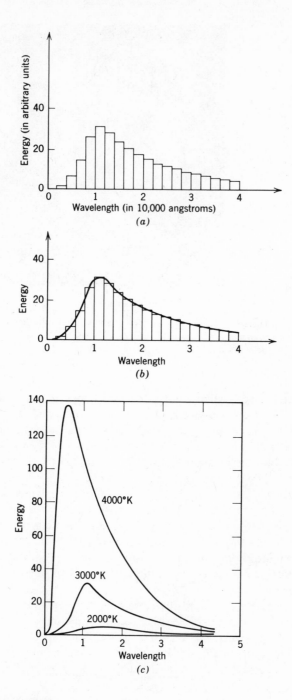

FIGURE 10-9

(a) Measurements of energy in successive small regions of the spectrum plotted against wavelength. (b) The steplike curve can be smoothed out. (c) Radiation curves for cavity resonators at different temperatures.

the total area under the curve for radiation from a cavity with a temperature of 3000°K in Figure 10-9c.

Actually, the area under the radiation curve does not simply represent the energy emitted, for we have not specified the size of the source or the time interval involved. Obviously, more energy will be emitted by a large incandescent body than by a small one at the same temperature. Equally obvious is the fact that more energy will be emitted by a given body in two seconds than in one second. Thus, the area under the radiation curve represents the amount of energy emitted per unit area of the source per unit of time.

Figure 10-9c also indicates clearly that the peak of the curve shifts to the shorter wavelengths as cavity sources of higher temperatures are considered. This shift corresponds to our previous observation that blue stars are hotter than white stars, which in turn are hotter than red stars.

Early Descriptions of the Radiation Curve

In the 1890's a great deal of work was done on the radiation curve. The combined efforts of the Austrians' Josef Stefan (1835-1893) in 1879 and Louis Boltzmann (1844-1906) revealed that the energy emitted per area of the source per unit time is proportional to the fourth power of the temperature:

$$E \propto T^4$$
$$E = \sigma T^4$$

This is now known as the Stefan-Boltzmann law, where σ is the Stefan-Boltzmann constant. Although useful for certain aspects of physics and astrophysics, the law gives us very little information about the radiation curve, for it does no more than relate the area under the curve to the fourth power of the temperature.

In 1893 Wilhelm Wien (1864-1928), a German physicist, showed that the peak of the radiation curve is related in a simple manner to the temperature. If the wavelength of the peak of the curve is called λ_{peak}, then

$$\lambda_{peak} \propto \frac{1}{T}$$

$$\lambda_{peak} = \frac{A}{T}$$

The constant of proportionality A is called Wien's displacement constant. Again, although this may be a very simple and useful relationship, it really tells us very little about the radiation curve.

In 1896, Wien extended his work and derived a mathematical expression that was intended to represent the *shape* of the radiation curve. The shape, after all, is the most important characteristic of the curve, for by knowing it we can locate the peak of that curve as well as the area underneath it. Wien's work was based to some extent on classical theory and to some extent on empirical data. He was intrigued by the similar shapes of the observed radiation curve and the curve (first arrived at theoretically) that indicates how many molecules in a gas travel at each particular velocity* (Figure 10-10). Wien therefore surmised that the wavelengths of the radiation should bear some relationship to the velocities of the molecules which themselves must emit the light. Presumably, the acceleration is proportional to the velocity, for according to Maxwell's equations an accelerating electric charge radiates electromagnetic waves. The equation for a curve derived by Wien successfully "fit" or matched the observed radiation curve at the short-wavelength end of the spectrum, but it failed at the long-wavelength end (Figure 10-11).

In the year 1900, Lord Rayleigh (1842-1919), an English physicist, published the results of his study of the radiation curve. Unlike Wien, who relied on the similarity between two curves—one empirical and the other theoretical—Rayleigh relied solely on classical theory to derive a mathematical expression for the shape of the radiation curve. Rayleigh assumed that the aether postulated to transmit electromagnetic waves was free to vibrate inside of an enclosure, such as a cavity, much as air vibrates in a closed pipe.

The mathematical expression derived by Rayleigh defined a curve that matched the observed radiation curve at the long-wavelength end but failed rather completely at the short-wavelength end. Rayleigh's expression predicted that the amount of energy emitted would *increase to infinity* as the wavelength *decreased to zero*—no matter what the temperature of the radiating cavity. Because this prediction obviously did not agree with experimental results, why did Rayleigh bother to publish it?

Rayleigh's theoretical curve was important for the very fact that it *did not* match the observed or experimental curve. After all, Rayleigh's curve was based on classical physics, and if classical physics had been so successful

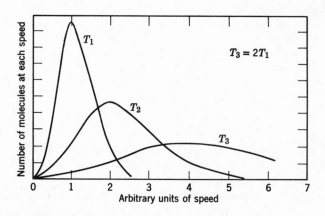

FIGURE 10-10

Maxwellian velocity distributions for gases of different temperatures.

---

* The work on this curve, the velocity distribution curve, was done by Maxwell, and is called the Maxwellian distribution of velocities.

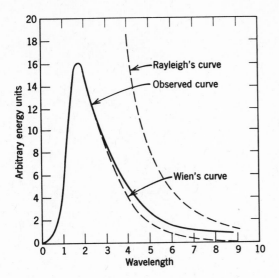

FIGURE 10-11

Neither the mathematical expression of Wein, nor that of Rayleigh was able to account for the observed radiation curve.

in other fields, why could it not predict the shape of the radiation curve? What was wrong? Rayleigh's work was of such great significance that the failure of his curve at the short-wavelength end became known as the "ultraviolet catastrophe."

Planck's Law

In a matter of months Max Planck (1858-1947), a German physicist and Nobel prizewinner, answered Rayleigh's question. Planck observed that Wien's expression was accurate for the short wavelengths and Rayleigh's was accurate for the long wavelengths. Perhaps the two expressions could be combined somehow to yield an expression that would represent the observed radiation curve over the entire spectrum.

Planck found a way to combine Wien's and Rayleigh's expressions and derived a new equation; in October 1900 he presented his work to the German Physical Society. At that time, he could say only that his equation* was based partly on theory (the laws of thermodynamics) and partly on observation (the work of Wien and Rayleigh); he could give no theoretical support for it, but it did correspond to the shape of the radiation curve. By December 1900 Planck offered justification for the equation, but the arguments he used contradicted the ideas of classical physics!

Classical physics maintains that an oscillator (anything that oscillates

---

*For our purposes the complexities of this equation preclude all but a casual glance:

$$\Delta E = \frac{8\pi ch\lambda^{-5}\, \Delta\lambda}{e^{ch/k\lambda T} - 1}$$

FIGURE 10-12

According to the laws of Newton and the law of continuity, the amplitude of the swinging pendulum can assume *any* value in the continuum between 0 degrees and that angle where the pendulum ceases to swing regularly.

regularly) can have any amount of energy within certain limits. The lower limit is zero and the upper limit is the amount of energy that prevents it from functioning as an oscillator. For example, a simple pendulum can have any amount of energy as long as it does not swing higher than some angle at which it will cease to swing regularly.

The amount of energy possessed by a simple pendulum can be measured by the angle from the vertical through which it swings (Figure 10-12). Newton's laws of motion, which describe the motion of a simple pendulum, do not in any way limit that angle to certain discrete values. The angle could, for example, be 5 degrees or 5.000001 degrees or any value between these two. Newton's laws are tied to the law of continuity (see p. 130f).

The Quantum Postulates

Planck realized that to explain his mathematical expression defining the radiation curve, he must think of and describe electric oscillators in a different way; they could not have just any amount of energy. The difference in behavior is usually stated as *Planck's two quantum postulates*.

The term *quantum* designates the discreteness of any system. It means that some aspect or characteristic property of the system comes in only certain sizes. For example, when we place a plate in an empty cupboard, we must place it on one of the shelves; it cannot be successfully placed between two shelves. The gravitational potential energy of that plate in the cupboard is quantized. It can have only certain discrete values; no other values of potential energy are permitted, since the plate cannot float between two shelves.

The term *postulate* was chosen because Planck's two statements are given only in defense of his equation. These two statements are true only if the equation is true; or more to the point, the equation is correct if these two statements can be shown to be valid.

The *first postulate* is as follows: *Each electric oscillator is quantized.* That is, each of Planck's electric oscillators is permitted only certain discrete values of energy; the oscillator *cannot* have an amount of energy between the permitted levels of energy. And indeed the term *energy level* has come to describe a particular amount of energy permitted to an oscillator.

Planck stipulated that the amount of energy possessed by each oscillator is proportional to its frequency of vibration f:

$$E \propto f$$

Again by inserting a constant of proportionality, called *Planck's constant*, we obtain an equation:

$$E = hf$$

The *second postulate* relates the *change* in the amount of energy of the oscillator with any energy the oscillator might emit. More important, this postulate stipulates that the radiation is not emitted continuously as sound is emitted by a vibrating string. The second postulate is stated as follows: *An oscillator emits or absorbs radiation only when it changes from one energy level to another.* The oscillator will emit energy when it changes from one energy level to a lower energy level. Similarly, the oscillator will absorb energy by changing from one energy level to a higher energy. The amount of energy emitted or absorbed must equal the change in energy of the oscillator; energy is conserved. While the oscillator remains at one level of energy, it will not radiate energy. Consequently, Planck's description of an oscillator requires that it emit radiation in packets or bunches; and it can emit packets of energy of only certain discrete amounts. These amounts depend on the energy levels of the particular oscillator.

Planck knew that these postulates are obviously contrary to the classical laws of physics, and this knowledge disturbed his basically conservative nature. His genius, however, permitted him, first, to recognize their importance and, second, to incorporate them into his mathematical expression. This was the first step in the development of what has become a fundamental branch of physics—*quantum physics*.

Physicists were reluctant to adopt Planck's two quantum postulates to explain the observed radiation curve until they could be verified by some other means, that is, by means independent of blackbody radiation. The verification was supplied less than five years later by Albert Einstein in his description of the perplexing photoelectric effect.

THE PHOTOELECTRIC EFFECT

The observations that began the study of the photoelectric effect are curiously similar to and connected with the chance observation of radio waves by Galvani and their later identification by Hertz. In 1887, during the course of his work on radio waves, Hertz observed that the spark in the secondary (or receiving) circuit jumps across the terminals more readily if two conditions were fulfilled: (1) the terminals (metal spheres) are polished, and (2) a second spark throws light on the polished terminals. Hertz did not, however, pursue these initial observations with further investigations, for he was too busy with the identification of radio waves. He simply reported his discovery and left its investigation to others.

Ultraviolet Radiation and Electrons

In the next year (1888), the discovery was made that a plate of negatively charged zinc metal will lose its charge if ultraviolet radiation is directed on it. Furthermore, it was found that the same is not true of a positively charged plate. Ultraviolet radiation does not cause the plate to lose a positive charge.

After the discovery of the negatively charged electron in 1897 (see pp. 313ff), it became clear that whenever ultraviolet radiation is directed onto a metallic surface, that surface ejects electrons. Ultraviolet radiation causes the electrons to escape from the metal; hence the name of the phenomenon, the photoelectric effect (*photo* is the Greek word for light). In the next few years the velocity of the ejected electrons was measured, and with this measurement the fourth discrepancy between classical physics and observations became apparent.

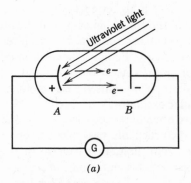
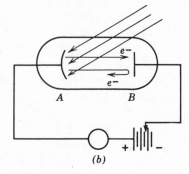

FIGURE 10-13

    (a) Ultraviolet light will cause the ejection of electrons from the photosensitive surface *A*. Those electrons, with kinetic energy, will travel to the plate *B* giving that plate a negative electric charge. They can return to plate *A* by way of the galvanometer. *(b)* If a variable potential difference is set up between plates *A* and *B*, only those electrons with enough kinetic energy will make it up the potential gradient to plate *B*.

If ultraviolet radiation is incident on (falls on) a photosensitive plate (plate *A* of Figure 10-13*a*), electrons are ejected, giving that plate a net positive charge. Since the ejected electrons have kinetic energy, they are able to travel through the evacuated tube to plate *B*, and by collecting there they give that plate a net negative charge. If the two plates are now connected through a galvanometer, an electric current is established. This is the basic circuit of the light meter used by photographers.

If now a variable direct-current source is added to the circuit (Figure 10-13*b*) so as to place a potential difference between the two plates, with *A* positive and *B* negative, the electrons are repelled by *B*. They can climb the electrical potential hill from *A* to *B* only because they were ejected with an initial kinetic energy. By gradually increasing the potential difference across *AB*, a value for the potential difference will eventually be reached at which no current passes through the galvanometer. At that value of potential difference the most energetic of the ejected electrons will be brought to rest just before reaching plate *B*. We can express this mathematically.

Since the electrons are brought to rest, their initial kinetic energy $\frac{1}{2}mv^2$ is equal to their loss of kinetic energy in coasting up the electrical hill from *A* to *B* (but never quite making it to *B*). We can relate this kinetic energy to the electrical potential difference *V* if we recall the equation (see p. 236)

$$\Delta E_k = QV$$

where ΔE_k is the chance in kinetic energy of a particle with a charge *Q* as it moves through an electrical potential difference *V*. The change in kinetic energy of the ejected electrons is their initial kinetic energy, $\frac{1}{2}mv^2$; their charge is usually designated as *e*. Therefore, the equation becomes

$$\frac{1}{2}mv^2 = eV$$

The charge of the electron *e* was measured repeatedly with increasing accuracy during the first decade of the nineteenth century (see chapter 13); the kinetic energy of the ejected electrons could therefore be calculated.

Predictions of the Wave Theory

From the study of wave motion in general the energy of the incident radiation was known to be proportional to the square of the amplitude of the waves. For example, water waves with an amplitude of 10 feet carry a great deal more energy (100 times as much) than water waves with an amplitude of 1 foot, if their wavelengths are the same.

The intensity (brightness) of light is also proportional to the square of the amplitude of the light waves. Hence, the measurement of the intensity is, according to wave theory, a measurement of the energy contained in a wave. By doubling the intensity of the light, the energy is doubled.

Correspondingly, if the amount of energy contained in the light waves is doubled and that energy is imparted to electrons, causing them to be ejected, the kinetic energy of the ejected electrons should also be doubled. This ought to be true, even if the electrons cannot absorb all the energy of any one given wave. Presumably, as a wave strikes an electron the wave gives some energy to that electron, but the wave itself continues on to strike other particles of matter until all its energy is spent. For example, a ten-foot wave striking a boat will impart much more kinetic energy to the boat than a one-foot wave. Both waves will continue on, however, until all their energy is transformed into heat by some means or other.

Classical physics, the wave theory of light in this instance, makes a rather straightforward and easily verifiable prediction: doubling the intensity of incident ultraviolet radiation should double the kinetic energy of the ejected electrons.

The Observations

As was soon discovered, not all the ejected electrons have the same initial kinetic energy. Consequently, let us concern ourselves with the electrons having the maximum kinetic energy, since this maximum kinetic energy is characteristic of the particular metallic surface used. To test our prediction, we must be able to project ultraviolet radiation upon a piece of metal and to measure the kinetic energy of the ejected electrons, and we must also be able to vary the intensity of the projected ultraviolet radiation. If we increase the intensity of the ultraviolet radiation, classical theory predicts that the kinetic energy of the ejected electrons will be increased.

Observation reveals, however, that if the intensity of the incident ultraviolet light is changed, the maximum kinetic energy of the ejected electrons is *not altered!* The prediction of classical physics is not borne out by experiment. To be sure, *more* electrons are ejected as the intensity is increased, but the maximum kinetic energy remains constant.

Further experiments reveal, however, that the maximum kinetic energy of the electrons can be altered, but by a quite different and unsuspected technique. If the ultraviolet radiation is first directed through a prism before it is made to fall on the piece of metal, the wavelength, or frequency, of the incident ultraviolet radiation can be varied by directing various parts of the ultraviolet spectrum on the metal. Observations reveal that the maximum kinetic energy of the ejected electrons increases as the frequency of the ultraviolet radiation increases.

Even stranger is the fact that if the frequency of the incident radiation is decreased, that is, as the wavelength is lengthened, there is a particular frequency below which no electrons whatsoever are ejected. This particular frequency, called the *threshold frequency*, is observed to depend on the metal. For most metals it occurs in the ultraviolet part of the spectrum, and from these metals no electrons at all are ejected by visible light, no matter how intense the beam that strikes the surface. On the other hand, a very weak source of ultraviolet radiation will cause the ejection of some electrons. These few electrons will have a maximum kinetic energy that depends on the metallic surface from which they come and on the frequency of the incident light, *but not on the intensity of that light*.

About all that can be done with classical physics to rescue it from this dilemma is to suppose that there is a time lag between the incidence of the light and the ejection of the electrons. Supposedly a given electron will absorb some energy with the incidence of each wavefront (electrons are small compared to a wavefront), but it will take many such wavefronts before that electron will have stored up enough energy to be ejected. It would, according to wave theory, require an individual electron something like a day to store up the required amount of energy.

Again, observation was not in agreement with classical physics. Electrons were observed almost immediately after the incidence of the ultraviolet radiation. Later and more refined experiments have indicated that an electron takes only about 10^{-9} second to acquire the amount of energy needed to be ejected from the metallic surface.

Einstein's Photoelectric Equation

Albert Einstein first saw the connection between Planck's work and the photoelectric effect. Planck proposed that light is emitted in packets, not in a continuous wavetrain. The amount of energy of each packet is equal to the change in energy of the electrical oscillator emitting that packet. Planck felt, however, that the packet of light was an electromagnetic wave as described by Maxwell, that it traveled as an electromagnetic wave, and that its energy would be absorbed as an electromagnetic wave.

Einstein reasoned that since there is almost no time lag between the incidence of light and the ejection of the electron, the energy of light cannot be spread uniformly over the wavefront; the energy must be concentrated in individual packets with no energy between them. One packet is completely absorbed by each electron ejected from the metal. The electron absorbs all the energy concentrated in the packet rather than a portion of the energy spread uniformly over the whole wavefront. These packets or quanta of light energy have become known as *photons*.

Einstein contended that each photon has an amount of energy,

$$E = hf$$

where h is Planck's constant and f is the frequency of the light of which the photon is part. We recall Planck's contention that the amount of energy E is equal to the energy lost by the oscillator which emitted the photon in the first place. Consequently, the light is emitted as a unit, a packet, a particle, a quantum, a photon. Einstein proposed that the photon is absorbed as a unit, that is, as a particle. He made it quite clear that to account for the photoelectric effect, light must be concentrated in particles rather than spread uniformly over waves.

Einstein maintained that all the energy of the photon is converted to kinetic energy of the electron. But when the energy is converted, the electron is still under the surface of the metal. Before it can be observed in our laboratory, it must escape from the metallic surface. Solids, in our example a metal, are held together in their rigid shapes by certain forces, called intermolecular forces. These intermolecular forces bond the material together, and in order to escape from the surface of a metal, the electron must lose some of the initial kinetic energy given it by the photon to overcome these forces. The amount of energy lost is called the *work function W*. Consequently, the amount of kinetic energy measured by our laboratory equipment is less than the amount given the electron by the photon.

The incident photon has an energy hf; all this energy is given the bound electron. But the electron loses an amount of energy W in escaping from the metal, so it escapes with a kinetic energy E_k:

$$E_k = hf - W$$

This is Einstein's photoelectric equation, presented by him in a paper published in 1905.

Verification and Support of Planck's Quantum Postulates

The photoelectric equation was experimentally verified with considerable accuracy in 1916 by R. A. Millikan (1868–1953), an American physicist. As part of the verification, Millikan made an independent calculation of Planck's constant which agreed closely with the value Planck himself obtained by quite a different method. The numerical value of Planck's constant is

$$h = 6.6 \times 10^{-34} \text{ joule-sec}$$

In Millikan's verification, the kinetic energy of the ejected electrons was determined as the frequency of the incident light is changed. After repeated measurements of this sort, we can draw a graph plotting the kinetic energy of the ejected electrons against the frequency of the incident light. Figure 10-14a indicates the results of such an investigation. At the threshold frequency f_0 the electrons are ejected, but supposedly with zero kinetic energy. Einstein explained that the ejected electron has no kinetic energy when the amount of energy contained by an ejecting photon (whose frequency is equal to the particular threshold frequency) is just equal to the energy required by the electron to escape from the metal. That is, the work function W equals the energy of the incident photon:

$$W = hf_0$$

For frequencies greater than f_0 the electrons will be ejected. When plots of kinetic energy versus frequency are made for a number of different metals, we see that the straight-line curves are all parallel (Figure 10-14b). Each curve intersects the $E_k = 0$ axis in a different place; each metal has a different threshold frequency. But with frequencies higher than the threshold frequency, the kinetic energy of the ejected electrons increases uniformly

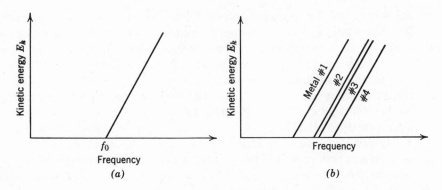

FIGURE 10-14

(a) The observed relationship between the frequency of incident ultraviolet light and kinetic energy of the ejected electrons. No electrons are ejected for frequencies less than the threshold frequency f_0. *(b)* The photoelectric relationship for a number of metals. The slope of each curve represents Planck's constant and so these curves are all parallel.

with increasing frequency of the incident light. This uniform increase, called the slope of the curve, is the graphic expression of Planck's constant. The value of Planck's constant was determined by Millikan by making accurate measurements of the slopes of many such curves.

Millikan's work verified Einstein's photoelectric equation completely. There is now no doubt that Einstein's interpretation of the photoelectric effect corresponds precisely to contemporary observation. The general acceptance of the photoelectric equation resulted in the general acceptance of Planck's radiation law. Planck's two quantum postulates are considered a valid description of the manner in which electrical oscillators work.

Planck's oscillators are now recognized as the atoms of which a substance is composed. These atoms are quantized; they can have only certain discrete amounts of energy. They emit light only when they change from one level of energy to another, lower level of energy. The light so emitted is a quantum of light, a photon. When this photon, traveling with the velocity of light, strikes an electron in the surface of a piece of metal, the entire photon—all its energy—is absorbed by the electron. The electron converts the energy it has received from the photon into kinetic energy. If the kinetic energy is great enough, the electron will escape from the surface of the metal, losing some energy in the process. The electron, with its remaining amount of kinetic energy, then travels free of the metal until it is captured by another piece of matter.

The net result of Planck's radiation law and Einstein's photoelectric equation was the founding of a revolutionary branch of physics called quantum physics, a very important part of modern physics. But why had not the quantum nature of matter and light been discovered earlier? Certainly the particulate theory of matter has long held the attention of many physicists, but the quantum theory delves deeper. The quantum theory describes more than just the atomistic nature of matter. It indicates a great deal about the atoms themselves and specifies the internal structure of matter.

The particulate or quantum nature of light was not observed earlier because the experiments performed by Young and Fresnel were based on the interference of light. And it is just this interference phenomenon that established the wave theory of light. Maxwell extended the wave theory of light by mathematical analysis, but his mathematics was the mathematics of continuity. Only after research in physics had reached the necessary stage of sophistication could the interaction of light and electricity be observed, and only after electricity was shown to be composed of tiny particles called electrons could the photoelectric effect be explained.

And so science progresses. Explanations and theories can extend only as far as the observations upon which they are based permit. The predictions that are in turn formulated on the basis of these theories may require that more precise measurements be taken or may indicate that entirely new observations are there to be made, observations undreamed of before the theory indicated a new direction.

A new direction was taken at the turn of the twentieth century, a direction forced upon us by observation and illuminated by Einstein's principles of relativity and by the principles of quantum physics.

References

1. Norwood Russell Hanson, *Isis*, Vol. 53, part 3, no. 173, September 1962, p. 362.
2. *Ibid.*, p. 366.

Questions

1. Leverrier was faced with three choices when he realized that the motion of Uranus did not conform to Newtonian physics as it was then applied. Comment on these three choices and how progress in science is influenced by such incidents.

2. Describe how the Michelson interferometer is optically similar to a thin film.

3. The spectrum of nearly every star is an absorption spectrum. What does this observation indicate about the region above the surface of a star? (The "surface" is the outermost region of a star that emits a continuous spectrum.)

4. Comment on the contradiction between the quantized states postulated by Planck and the law of continuity upon which Boscovich relied.

5. The potential energy of a plate in an empty cupboard is quantized. What other aspects of the world about us represent a quantization of nature? For example, is a guitar string quantized?

6. Why were the inception and development of the quantum theory dependent on increased precision of observation?

7. According to Einstein's explanation of the photoelectric effect, it is possible for a photon to disappear completely as an entity. Why is this not a contradiction to the principle of conservation of energy?

8. Comment on the value of compromise in scientific theories, recalling Asaph Hall (see p. 263), Simplicius (see p. 23), and Tycho Brahe (see p. 43).

Problems

1. When Mars is directly between the sun and Jupiter, it is nearly 2.5 times as far from Jupiter as it is from the sun. The mass of the sun is about 1000 times that of Jupiter. Find the ratio of the force F_s exerted on Mars by the sun to the force F_j exerted on Mars by Jupiter. This ratio and the fact that the distance between planets varies are the basis of perturbation theory.

2. A boat travels 10 mph in still water; a river flows at 8 mph.

(*a*) How long will it take the boat to make a round trip 1 mile and back again in still water?

(*b*) How long will it take the boat to make a round trip 1 mile up the river and back again?

(*c*) How long will it take the boat to make a round trip 1 mile directly across the river and back again?

3. What is the energy of a photon whose frequency is (*a*) 7.5×10^{14} cps (violet light); (*b*) 4.0×10^{14} cps (ultraviolet); (*c*) 3.0×10^{18} cps (x-rays); (*d*) 3.0×10^{23} cps (gamma rays).

4. Electrons with a maximum kinetic energy of 6.0×10^{-20} joule escape from a metallic surface as ultraviolet radiation of frequency 6.0×10^{14} cps is directed on that surface. What is the work function of that surface?

5. Using only the Stefan-Boltzmann law and Wien's law, compare the radiation of three stars whose surface temperatures are respectively 3000, 6000, and 12,000°K. (The values of the constants are given in Appendix C.)

6. What is the wavelength of the peak of the radiation curve for an object whose temperature is 310°K, the temperature of a healthy human body?

7. How much energy does a healthy body ($T = 310°K$) radiate each second? Assume that the area of the body is the same as a rectangular solid, $30 \times 60 \times 200$ cm.

Chapter Eleven

Einstein and Special Relativity

Bettmann Archive

In the closing decade of the nineteenth century, classical physics came upon a number of stumbling blocks. The previous chapter discussed four failures, but there were other difficulties as well. Many theorists attempted to explain these difficulties by making relatively minor changes and additions to classical physics. Some of the men who contributed important observations and ideas during these very critical years were H. A. Lorentz (1853–1928), a Dutch physicist and Nobel prizewinner; G. F. FitzGerald (1851–1901), an Irish physicist; and J. H. Poincaré (1854–1912), a French mathematician. The individual contributions of each of these men (and of others as well) will not be discussed in full, for that is not the purpose of this chapter. The development of new ideas in physics is a complex process involving the contributions of many men and the interplay of numerous observations and ideas. A critical study of the early history of the principle of relativity must consider all these complex details.

Certainly the Michelson-Morley experiment was one of the more important observations to stimulate new thinking in the 1890's. That experiment indicated that the measured velocity of light is independent of the motion of the observer. Why is it that velocities of ordinary things, such as automobiles, add together very nicely and yet cannot be added to the velocity of light? Velocity is a basic concept; it deals directly with length and time, two of the fundamental quantities of physics. Let us now reexamine those fundamental quantities, length, mass, time, and electric charge.

CLASSICAL SPACE, TIME, AND MASS

Newton considered space, the three-dimensional counterpart of length, to be absolute. There is in this universe an absolute space against which every motion can be judged. "Absolute space, in its own nature, without relation to anything external, remains always similar and immovable" (1).

He recognized that there is also a relative space, a space carried by some body such as the Earth as it moves through absolute space.

For if the earth, for instance, moves, a space of our air, which relatively and in respect of the earth remains always the same, will at one time be one part of the absolute space into which the air passes; at another time it will be another part of the same, and so, absolutely understood, it will be continually changed (1).

Newton's absolute space was the stationary universe. The system of "fixed" stars was at rest in that absolute space; the sun, according to Newton, was at rest in absolute space.

If defining space is difficult, defining time is even more difficult. Newton wrote that

absolute, true, and mathematical time, of itself, and from its own nature, flows equably without relation to anything external, and by another name is called duration: relative, apparent, and common time, is some sensible and external (whether accurate or unequable) measure of duration by the means of motion, which is commonly used instead of true time; such as an hour, a day, a month, a year (1).

As with the definition of space, Newton felt that an absolute time existed independent of life and matter. For Newton there was also a relative time measured by means of some moving object, preferably one that oscillates.

Others, such as Leibnitz, disagreed with Newton and felt that "there can be no time independent of events: for time is formed by events and relations among them, and constitutes the universal order of succession." The existence of absolute versus relative space and time was discussed and debated by both physicists and philosophers until Einstein's theory of special relativity was accepted. The definition of mass, however, was not questioned.

The mass of a body became associated with the amount of matter in that body. This association, when considered with the principle of conservation of mass, leads to the conclusion that the mass of a given body is constant, that is, absolute. Even defining mass as a measure of the resistance of a body to acceleration leads to the same conclusion. Newton's second law of motion makes it quite clear; the same force always produces the same acceleration in bodies of equal mass:

$$a = \frac{F}{m}$$

Classical physics held space, time, and mass as invariants; that is, each of these three fundamental units was the same for all who used them, no matter under what circumstances they were used. A meter is always a meter long; an hour is always an hour; a kilogram is always a kilogram.

Then Michelson and Morley showed that the velocity of light is the same for all observers whatever their motion. How is it possible for two observers, let us say A and B, traveling in opposite directions (Figure 11-1), to measure the velocity of a beam of light traveling in the same direction as A (and thus in the direction opposite to B) and record the same value? This observation conflicts with our basic conception of space and time, for velocity is a derived unit, based on space and time.

FIGURE 11-1

Two observers traveling in opposite directions A and B will both measure the same speed of light for a beam traveling in the direction c.

LORENTZ-FITZGERALD CONTRACTION

In an attempt to explain the results of the Michelson-Morley experiment, FitzGerald and Lorentz suggested that the length of the measuring instrument depends on the motion of that instrument. Their hypothesis, the so-called Lorentz-FitzGerald contraction, absurd as it may seem, is as follows:

Every body which has the velocity v with respect to the aether contracts in the direction of motion by the fraction (2)

$$\sqrt{1 - v^2/c^2}$$

In this fraction, c is the velocity of light.

Needless to say, this suggestion did not receive immediate acceptance by all physicists; however, it appeared to "explain" the results of the Michelson-Morley experiment.

Michelson, and almost everyone else, assumed that the two arms of the interferometer were equal in length, $L_2 = L_1$. The Lorentz-FitzGerald contraction hypothesis states that the one arm that is parallel to the direction of the Earth's motion, L_2 in Figure 10-3a, contracts so that

$$L_2 = L_1\sqrt{1 - v^2/c^2}$$

According to classical physics, light requires two different time intervals to travel the two arms of the interferometer; the first interval $t_\parallel$ is for traveling parallel to the direction of the Earth's motion, and the second interval $t_\perp$ for traveling perpendicular to it,

$$t_\parallel = \frac{2L_2}{c(1 - v^2/c^2)}$$

$$t_\perp = \frac{2L_1}{c\sqrt{1 - v^2/c^2}}$$

where $L_1 = L_2$.

Lorentz and FitzGerald suggested that the arm parallel to the direction of motion shrinks slightly, so that L_2 no longer equals L_1. In place of L_2 we should substitute its contracted length $L_1\sqrt{1 - v^2/c^2}$. Hence, the time for light to travel the arm becomes

$$t_\parallel = \frac{2L_1\sqrt{1 - v^2/c^2}}{c(1 - v^2/c^2)}$$

which reduces to

$$t_\parallel = \frac{2L_1}{c\sqrt{1 - v^2/c^2}}$$

This is the same interval of time required for light to travel the arm that is perpendicular to the motion of the Earth, and consequently the light returns from the two arms always in phase; under such circumstances no fringe shift would be expected, even if the interferometer were rotated through 360 degrees.

Lorentz went even further; he came to the conclusion that the rate of time for two individuals moving relative to each other should be different.

FIGURE 11-2

Albert Einstein (1879–1955). (New York Times photograph.)

He called these "local times." A few years later, in 1905, Albert Einstein combined the ideas of Lorentz, FitzGerald, and others with his own to develop a new logical foundation for physics, a foundation that goes beyond Newton's laws.

THE PRINCIPLE OF SPECIAL RELATIVITY

The *principle of special relativity* is "special" because it considers only inertial systems (see p. 100), that is, systems in which an object free to move will not accelerate in any direction. Any relative motion between inertial systems must therefore be uniform motion, motion of a constant velocity. Two space vehicles drifting in free space, and therefore free of all gravitational fields, constitute two such systems. For our purposes, however, we can consider the Earth as one inertial system by neglecting the very small accelerations resulting from its daily rotation and yearly revolution, and by not considering falling objects which, of course, accelerate. For the other inertial system we shall use Einstein's example of a train moving at a constant speed along a straight stretch of track. Each of these two inertial systems will constitute a frame of reference.

Difficulties with Simultaneity

Einstein saw clearly one fundamental difficulty in comparing two systems: people on the train and people on an embankment near the track may not be able mutually to establish two events as occurring simultaneously.

The train, in fact, may now be a poor example for such a description, but in 1905 it was the fastest vehicle around. Today we would probably choose a space vehicle traveling at extremely high velocities. However, we shall continue with Einstein's example, bearing in mind that the rather low velocity of the train is really not adequate for the accuracy of our timekeepers. Nevertheless, Einstein made the point: it is impossible to establish on different inertial systems that any two events are simultaneous. Einstein, in 1916 (eleven years after the original announcement of the principle of special relativity), explained this point by means of a thought experiment (see Figure 11-3).

Lightning has struck the rails on our railway embankment at two places *A* and *B* far distant from each other. I make the additional assertion that these two lightning flashes occurred simultaneously. If I ask you whether there is sense in this statement, you will answer my question with a decided "Yes." But if I now approach you with the request to explain to me the sense of the statement more precisely, you find after some consideration that the answer to this question is not so easy as it appears at first sight. . . .

After thinking the matter over for some time you then offer the following suggestion with which to test simultaneity. By measuring along the rails, the connecting line *AB* should be measured up and an observer placed at the mid-point *M* of the distance *AB*. This observer should be supplied with an arrangement (e.g., two mirrors inclined at 90°) which allows him visually to observe both places *A* and *B* at the same time. If the observer perceives the two flashes of lightning at the same time, then they are simultaneous (3).

Einstein then makes it clear that with this perfectly adequate definition of simultaneity "good" clocks could be synchronized, and that thereafter each of these "good" clocks would always have identical settings. But

up to now our considerations have been referred to a particular body of reference, which we have styled a "railway embankment." We suppose a very long train traveling along

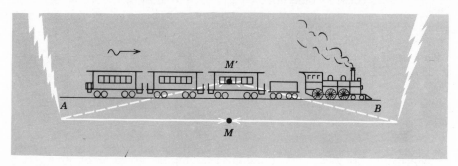

FIGURE 11-3

If two bolts of lightning strike simultaneously for an observer at *M* on the railway embankment, those same bolts will not strike simultaneously for an observer at *M'* on a moving train.

the rails with the constant velocity v and in the direction indicated in [Figure 11-3]. People traveling in this train will with advantage use the train as a rigid reference-body (co-ordinate system); they regard all events in reference to the train. Then every event which takes place along the line [AB] also takes place at a particular point of the train. Also the definition of simultaneity can be given relative to the train in exactly the same way as with respect to the embankment. As a natural consequence, however, the following question arises:

Are two events (e.g., the two strokes of lightning A and B) which are simultaneous *with reference to the railway embankment* also simultaneous *relative to the train*? We shall show directly that the answer must be in the negative.

When we say that the lightning strokes A and B are simultaneous with respect to the embankment, we mean: the rays of light emitted at the places A and B, where the lightning occurs, meet each other at the mid-point M of the length A-B of the embankment. But the events A and B also correspond to position A and B on the train. Let M' be the mid-point of the distance A-B on the traveling train. Just when the flashes* of lightning occur, this point M' naturally coincides with the point M, but it moves towards the right in the diagram with the velocity v of the train. If an observer sitting in the position M' in the train did not possess this velocity, then he would remain permanently at M, and the light rays emitted by the flashes of lightning A and B would reach him simultaneously, i.e., they would meet just where he is situated. Now in reality (considered with reference to the railway embankment) he is hastening towards the beam of light coming from B, while he is riding on ahead of the beam of light coming from A. Hence the observer will see the beam of light emitted from B earlier than he will see that emitted from A. Observers who take the railway train as their reference-body must therefore come to the conclusion that the lightning flash B took place earlier than the lightning flash A. We thus arrive at the important result:

Events which are simultaneous with reference to the embankment are not simultaneous with respect to the train, and *vice versa* (relativity of simultaneity). Every reference-body (co-ordinate system) has its own particular time; unless we are told the reference-body to which the statement of time refers, there is no meaning in a statement of the time of an event (3).

Had the preceding discussion referred to the thunder rather than the lightning flash, we would not have hesitated to draw the conclusion that the observer on the train heard the thunder from B before he heard the thunder from A. Sound travels very slowly compared to light; anyone who has observed a thunderstorm will vouch for that. Had sound rather than light been used to regulate our clocks, we would have encountered difficulties with simultaneity long ago. But we are accustomed to *looking* at clocks, and we tacitly assume that, for all practical purposes, light travels with essentially an infinite velocity. We simply do not take the speed of light into consideration. Einstein did! He realized that since light travels faster than anything else in this universe, not only must we recognize the difficulties with simultaneity, but we must incorporate this recognition in theories of physics that consider transformations from one coordinate system to another.

*As judged from the embankment.

Questions about Time

If two events, such as the striking of two clocks, can be considered simultaneous only in those systems at rest with respect to each other, there is no such thing as absolute time. There is no universal clock against which everyone can set their own clocks. This conclusion directly contradicts Newton's concept of absolute time. Einstein recognized only relative time and then clarified its meaning.

How, he asked, can we be sure that a second of time ticked away on the embankment is of the same duration as the second of time ticked on the train? We cannot be sure, and in fact the following arguments should convince us that these two time intervals are of different duration as seen by one observer.

Let us assume that there are three men on the train; two of them, G and J, are sitting across the aisle from each other, and a third, the trainman T, is standing in front of these two men (Figure 11-4a). Man G throws a ball to man J and the trainman measures the velocity with which the ball traveled. (He has a "good" measuring device by which this can be done.) He claims that the velocity is u. That is, he measures the velocity of the ball u with respect to the train.

Over three centuries ago Galileo pointed out (see p. 66ff) that an observer on board a ship sees a stone dropped from the mast fall straight down and strike the deck at the foot of the mast. To an observer located on the shore, however, the stone's path is a parabola; the stone moves forward with the ship as it falls. Hence the description of the stone's path depends on the motion of the observer.

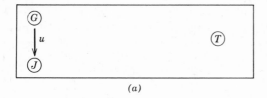

(a)

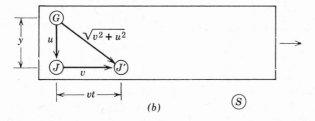

(b)

FIGURE 11-4

As passenger G throws passenger J a ball *(a)*, Galileo's concepts of transforming a velocity from one frame of reference to another (moving relative to the first) work very adequately *(b)*.

The same analysis applies to our example of the path and thus the velocity of a ball thrown by man G on the train. A stationman S, standing on the railway embankment, claims that the ball went forward with the train while it traveled from G to J's new position J' (Figure 11-4b). Consequently, he measures its velocity to be the vector sum of the ball's velocity with respect to the train, u, plus the train's velocity with respect to the ground, v. That vector sum is

$$\sqrt{u^2 + v^2}$$

The trainman T and the stationman S measure different velocities for that one toss of the ball. No one has disputed this point, which was first made more than three hundred years ago.

But supposing man G, instead of throwing a ball, "throws" a flash of light? Is the situation the same? Will the velocities measured by trainman T and stationman S again be different? The Michelson-Morley experiment demands that the answer to this question be no. *These two measured velocities must be the same!*

How, then, is it possible for one flash of light to leave man G and travel these two different distances as seen by the two observers, and yet arrive at man J with the same velocity? An object cannot, it would seem, travel two different distances with the same velocity in the same interval of time! The distances *are* different, but Michelson and Morley claim that the velocities *are* the same. Consequently, the only alternative left is that the *times* recorded *are different*! The watches of the stationman and trainman, which keep the same time while both are in the station, must keep different times when one is moving relative to the other. Let us see if this astonishing conclusion can be expressed more precisely.

The time interval Δt indicated by each watch for the transit of the light beam from man G to J must equal the distance traveled s divided by the velocity v:

$$\Delta t = \frac{s}{v}$$

Let us call the distance separating the two men y (Figure 11-5a); this distance is perpendicular to the direction of travel of the train and consequently is not altered by the Lorentz-FitzGerald contraction. Therefore, the time interval Δt_T measured by the trainman would be

$$\Delta t_T = \frac{y}{c}$$

where c is the velocity of light.

The stationman S sees the beam of light travel from G to J', and according to the Michelson-Morley experiment this velocity must also be c (Figure 11-5b). Consequently, the vector component of the velocity that is parallel to the direction y, the distance we are considering, must (by the

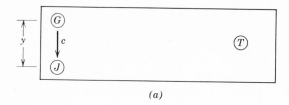

(a)

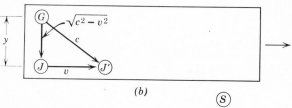

(b)

FIGURE 11-5

As passenger *G* throws passenger *J* a beam of light *(a)*, the Michelson-Morley experiment indicates that Galileo's transformation of velocities no longer applies *(b)*.

Pythagorean theorem) be $\sqrt{c^2 - v^2}$. Therefore, the time interval for the beam of light to travel a distance y as seen by the stationman would be

$$\Delta t_S = \frac{y}{\sqrt{c^2 - v^2}}$$

The distance y is the same for both the trainman and the stationman, for the length of that line is at right angles to its motion. We can therefore compare these two time intervals by writing the expressions in terms of y. For this comparison let us consider the train as our frame of reference, since the action takes place on the train. The three men *T*, *G*, and *J* are each at rest on this frame of reference; the stationman *S* moves relative to it. For the trainman *T*

$$y = c\,\Delta t_T$$

For stationman *S*

$$y = \sqrt{c^2 - v^2}\,\Delta t_S$$

Consequently,

$$c\,\Delta t_T = \sqrt{c^2 - v^2}\,\Delta t_S$$

The time interval as measured by the stationman is different from the time interval measured by the trainman:

$$\Delta t_S = \Delta t_T \frac{c}{\sqrt{c^2 - v^2}}$$

This expression may be written in more familiar form by multiplying the numerator and denominator of the fraction by the quantity $1/c$, which for the denominator may be written as $\sqrt{1/c^2}$:

$$\Delta t_S = \Delta t_T \frac{c(1/c)}{\sqrt{c^2 - v^2}\sqrt{1/c^2}}$$

$$\Delta t_S = \Delta t_T \frac{1}{\sqrt{1 - v^2/c^2}}$$

Let us analyze the expression $\sqrt{1 - v^2/c^2}$; the reader may realize by now that it is an important expression. For ever-increasing values of v, the quotient v^2/c^2 becomes larger, but as long as v is less than c, this quotient never exceeds one. Consequently, the denominator $1 - v^2/c^2$ must also be less than one, for we see that its value is one minus some fraction less than one. Hence, the fraction

$$\frac{1}{\sqrt{1 - v^2/c^2}}$$

must be larger than one, for it is equal to one divided by some number less than one. Therefore, the time interval Δt_S exceeds the time interval Δt_T. If this time interval Δt_T is assumed to be one second, trainman T then sees that the stationman's watch takes longer to tick this second than his own watch. Consequently, the three men on the train claim that the stationman's watch runs slow compared to theirs; the stationman is moving relative to their frame of reference, the train.

We have considered the situation as seen by the three men on the train. How does the stationman see the same situation? If the two men G and J are on the railway embankment instead of the train, a similar derivation shows that the trainman's watch runs slow compared to the watch of the stationman on the ground. It works both ways. If two frames of reference are moving with a constant velocity relative to each other, the people in either frame of reference claim that their own time is "normal" but that the clocks in the other frame of reference run more slowly than their own. It appears that not only is absolute time nonexistent, but the rate of relative time depends on the observer's motion relative to the clock. This consideration of time is basic to the principle of special relativity.

In the nineteenth century the aether had supplied a universal frame of reference; it was supposed that all motion could be referred to it. The Michelson-Morley experiment, however, demonstrated that the speed of light is the same in all inertial systems; consequently, there is no absolute frame of reference in this universe. In a thought experiment we considered two different inertial systems, the train and the Earth, and recognized the universal value for the speed of light. Furthermore, by assuming that no signal can travel faster than light, we were forced to conclude that the rate of a clock is not the same in all inertial systems. The speed of light therefore becomes a fundamental universal constant and is represented by the letter c.

Questions about Mass

As Lorentz had begun to question the Newtonian concept of space (length) and time, Henri Poincaré began to question the Newtonian concept of mass. He pointed out that in a free aether electromagnetic energy might possess mass. According to this view, energy was directly proportional to mass, that is,

$$E \propto m$$

Therefore,

$$E = Km$$

where K is the constant of proportionality. There was some question about the value of K, but one fact was obvious: in order to make the units of the equation come out correctly, K had to have units of velocity squared. Kinetic energy, it will be recalled, is equal to $\frac{1}{2}mv^2$. In fact, Poincaré suggested that the constant might be the velocity of light squared:

$$E = mc^2$$

The Postulates of Special Relativity

At this point it is appropriate to ask why Einstein is given so much credit for the development of the principle of special relativity when Lorentz, FitzGerald, Poincaré, and others had previously derived some of the equations that are a fundamental part of the principle. The answer is that Einstein was the only one who derived these expressions and still others from one set of basic principles. Lorentz, FitzGerald, and Poincaré were each working with separate problems and came up with only limited solutions. Einstein combined all these problems into one; he considered the total picture.

The principle of special relativity is based on two postulates.

1. The laws of physics that are valid in one inertial frame of reference are valid in any inertial frame of reference.
2. The speed of light is always the same, independent of both the motion of the source of light and the motion of the observer relative to that source.

The first postulate is an extension of Galileo's comments on the motion of objects in moving frames of reference, such as ships moving uniformly in tranquil waters. An inertial frame of reference, it will be recalled, is any reference frame in which an object free to move, moves uniformly and does not accelerate.

The second postulate stems from the Michelson-Morley experiment and asserts the fundamental importance of the speed of light. A consequence of this postulate is that there is no fundamental frame of reference in this universe. The speed of light is the same for *all* inertial frames of reference.

Relativistic Transformations

If the two postulates (premises) of special relativity are applied consistently and logically, the natural consequence is that whenever the laws of physics are transformed from one inertial frame of reference to another, the three quantities mass, length, and time are transformed according to these three equations:

$$\Delta t = \frac{\Delta t_0}{\sqrt{1 - v^2/c^2}}$$

$$L = L_0\sqrt{1 - v^2/c^2}$$

$$m = \frac{m_0}{\sqrt{1 - v^2/c^2}}$$

The first two equations refer to intervals, not points in coordinate systems. The interval of time Δt_0 is the interval of time witnessed in the observer's own frame of reference, the frame of reference in which he is at rest; Δt is the interval of time he observes in a frame of reference moving relative to him. The interval of space L_0 is a distance in the observer's own frame of reference; L is a distance observed in a frame of reference moving relative to him. In the last equation the mass m_0 is that of an object in the observer's own frame of reference and is called *rest mass*. The mass m is called *relativistic mass* and is the mass observed in an inertial frame of reference moving relative to the observer.

Using the postulates of special relativity, Einstein showed that Lorentz and FitzGerald had been correct in their suggestion that space is affected by relative motion. A meterstick in the physics laboratory is at rest in our frame of reference. A meterstick moving relative to us with a speed approaching that of light will appear shorter to us as long as its length is parallel to the direction of motion. If that meterstick were in a space vehicle, we on Earth would claim the meterstick in the vehicle to be shorter than ours by the amount

$$L = L_0\sqrt{1 - v^2/c^2}$$

where L_0, the *rest length*, is the length of an object at rest in the observer's frame of reference (in this case, on Earth). Correspondingly, those people in the space vehicle who are traveling with a velocity v relative to us would claim that our meterstick is shorter than theirs. Who is right?

Einstein claimed that both are right in the same sense that both Galileo's sailor and his landlubber were correct when they each described different paths for the stone falling from the top of the mast of a moving ship. The sailor saw it fall straight down; the landlubber saw it move forward as it fell down. The description depends on one's frame of reference. Now

Einstein claimed that not only does the path of an object depend on relative motion, but the object's length, the object's mass and, should the object be a clock of some kind, the intervals of time it keeps also depend on relative motion.

This is a rather difficult concept for us to accept, we who travel at such low velocities as 65 miles per hour, or in a jet airplane at 650 miles per hour. The contraction of length and dilation of time are completely negligible at these velocities; our everyday experiences and thus our "common sense" are based on this negligibility. These effects become evident only when velocities approaching that of light are reached. Consequently, relativistic effects seem strange to us; relativistic concepts appear contrary to common sense. Which is to be trusted? Which is the more reliable?

Commonsense notions have previously led to erroneous descriptions of the physical universe. It was common sense that led Aristotle to maintain that the Earth is the very center of the universe. Common sense long dictated that the Earth is immobile. Now we follow Einstein's lead and maintain that nothing is immobile; all things are moving. There is nothing in this universe that is absolutely at rest.

For the sake of clarity, let us perform some calculations indicating the effects of motion at relativistic speeds. We shall assume that a space vehicle passes the Earth traveling with a velocity of eight-tenths that of light ($v = 0.8c$). Further, let us assume that we on Earth can make measurements of mass, length, and time aboard that vehicle and that the space vehicle's passengers can make the corresponding measurements of equipment in our physics laboratory. The calculations are not difficult; the velocity $v = 0.8c$ was chosen so that the numbers work out nicely.

$$\frac{v}{c} = \frac{0.8c}{c} = 0.8$$

$$\frac{v^2}{c^2} = 0.64$$

$$1 - \frac{v^2}{c^2} = 0.36$$

$$\sqrt{1 - v^2/c^2} = 0.6$$

The equation for transformation of length becomes

$$L = L_0 \times 0.6$$

Suppose that L_0 is the length of one meterstick. The length of that stick in the space vehicle whizzing past at $0.8c$ *as measured by us on Earth* would be only 60 centimeters ($L = 100$ cm $\times 0.6$). The length of that same stick as measured by those aboard the vehicle would be L_0 or 100 centimeters, its rest length. The passengers in turn would measure a meterstick in one of our terrestrial physics laboratories to be 60 centimeters, if that meterstick were also parallel to the direction of relative motion.

A chunk of material in the space vehicle, which the travelers would consider to have a mass of one kilogram, would for us on Earth be

$$m = \frac{m_0}{\sqrt{1 - v^2/c^2}}$$

$$m = \frac{1.0 \text{ kg}}{0.6}$$

$$m = 1.7 \text{ kg}$$

We must now return to our definition of mass to help us understand this statement. We have declared mass to be the measure of resistance to acceleration. Consequently, special relativity theory simply says that as the velocity of an object relative to the observer approaches the speed of light, the object becomes more and more difficult to accelerate. If a force F causes an acceleration a when the mass m_0 is at rest or traveling at "everyday" velocities, that same force F will cause an acceleration less than a when the object is traveling so fast that its relativistic mass becomes significant. As velocities ever closer to that of light are considered, the relativistic mass approaches infinity, that is, the acceleration approaches zero. Apparently the velocity of light is the limiting velocity.

If we were to compare our clocks on Earth with those of the passengers in the same space vehicle, we would find their second to be 1.7 times longer than our second. While our clocks ticked away 1.7 seconds, theirs would tick away only one second. As seen by us, their clocks would be slower than ours. But if they were to measure the rate of our clocks, they would claim that our clocks are slow compared to theirs, and by the same amount.

The relationships involving length and time lead us to the same conclusion about the limiting value of the speed of light as does the transformation equation for mass. The time relationship indicates that if a space vehicle were to reach the velocity of light relative to us, we would say that its time has stopped, that its second of time is of an infinite duration. Furthermore, we would claim that the dimensions in the direction of relative motion of all objects on the spaceship (including the spaceship itself) become zero.

If the velocity of an object is increased, it acquires kinetic energy. As velocities approaching that of light are reached, special relativity indicates that the kinetic energy becomes

$$E_k = (m - m_0)c^2$$

The relativistic kinetic energy is equal to the difference between relativistic mass and the rest mass multiplied by the square of the velocity of light.

Pursuing this argument still further, Einstein showed that not only kinetic energy but all forms of energy are related to mass; he proposed the equivalence of mass and energy:

$$E = \Delta mc^2$$

This equation means that any mass Δm may be converted from one form of energy—mass-energy—to another form of energy, such as electromagnetic energy or kinetic energy. Associated with every particle that has mass is a quantity called *rest energy*, $\Delta m_0 c^2$.

It is interesting to note that the quantity of electric charge Q is invariant when transformed from one frame of reference to another. That is, if Q_0 is the charge of an object at rest in relation to the observer, and Q the charge of the same object as measured by an observer moving in relation to that charge, then $Q = Q_0$.

Correspondence Principle

Before the principle of special relativity can be accepted—and it is difficult enough for us to accept—we must show that it conforms to or obeys the *correspondence principle*. This principle specifies that any new theory or any new description of nature, formulated to explain new observations or new conditions, must, when applied to the older conditions, revert to the principles that successfully described nature before the advent of the new conditions. This means that the mechanics of special relativity must reduce to Newtonian mechanics when velocities much less than that of light are considered.

The relativistic transformation equations do in fact conform to the correspondence principle. If v is much less than c, the ratio v^2/c^2 may be taken to be zero, and the relativistic transformation equations become

$$\Delta t = \Delta t_0 \frac{1}{\sqrt{1-0}} = \Delta t_0$$

$$L = L_0 \sqrt{1-0} = L_0$$

$$m = m_0 \frac{1}{\sqrt{1-0}} = m_0$$

The principle of special relativity does indeed specify that mass, length, and time are invariant when they are transformed from one system to another if those systems move relative to each other with a velocity comparable to our "everyday" velocities. This invariance is a basic assumption in Newtonian physics.

Relativistic kinetic energy can also be shown to reduce to Newtonian kinetic energy when velocities much less than that of light are considered; that is, when v is much less than c,

$$E_k = (m - m_0)c^2$$

becomes

$$E_k = \tfrac{1}{2} m_0 v^2$$

In Newtonian mechanics, however, there is no need for the special term rest mass.

Observational Confirmation of Special Relativity

It is essential that the principle of special relativity conform to the correspondence principle, but before the ideas of relativity can be completely accepted, we must also have observational proof to support its claims and predictions.

Since special relativity deals with objects traveling at velocities approachthat of light, our experimental evidence will not include observations of large objects (with the exception of some astronomical objects). For a space vehicle to achieve a velocity approaching that of light would require energies far beyond our technical capabilities today. An electron, however, has an extremely small mass, 9×10^{-31} kilogram, and electrons can be accelerated by the electric fields in machines called particle accelerators. As velocities approaching that of light are reached, electrons do indeed become more difficult to accelerate. Their mass does increase, and the increase in mass is precisely that predicted by Einstein. Figure 11-6 gives both the theoretical and observational values for the relativistic mass of an electron as its velocity is increased.

Relativistic mass becomes a practical concern in the design and operation of modern particle accelerators which accelerate electrons and other charged particles to speeds of $0.9999c$ and faster. These huge accelerators, including the two-mile linear accelerator at Stanford University in California and the synchrotron at the Brookhaven National Laboratory on Long Island, New York, are designed according to relativistic mechanics.

We have further observational support for the principle of special relativity in the conversion of mass to energy, most dramatically achieved in

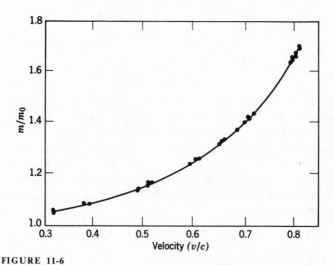

FIGURE 11-6

The smooth curve is the theoretical relationship m/m_0; the dots are values of m/m_0 obtained from high-speed electrons. (Reprinted by permission from *Introduction to Special Relativity*, by Robert Resnick, John Wiley & Sons, Inc., 1968.)

nuclear bombs. The conversion of mass to energy for peaceful purposes, however, is increasing. A number of power plants throughout the world utilize nuclear reactors to generate electric energy. Ships and submarines are being propelled for thousands upon thousands of miles without refueling because huge amounts of energy can be liberated by the conversion of only a small amount of mass.

Relativistic time dilation has also been observed. In studying the world of sub-atomic particles, physicists have learned that not all particles are stable; some have a very short lifetime before they decay into other particles. One particle, the pi meson or pion (see p. 388ff), has an average lifetime of about 2.5×10^{-8} second when it is essentially at rest in the physics laboratory. But particle accelerators can produce pions with a velocity of $0.9999c$. At this very high velocity scientists have observed the lifetime of pions to increase by a factor of about 70, in complete agreement with the principle of special relativity. Pions are actually clocks; their rate of decay is a direct and accurate indication of the passage of time. Their rate of decay slows down as their relative velocity approaches that of light.

The Twin Problem

We on Earth claim that our clocks are "normal." Those in the space vehicle whizzing past the Earth claim that their clocks are "normal." Clocks are used to measure the rate of change of any event; therefore, if clocks on the space vehicle appear to us to run slow, so must all actions measured by those clocks in that space vehicle. The interesting aspect of this statement is that it applies to the human body. The people in the space vehicle would appear to age less rapidly than we here on the Earth. But those people in the space vehicle would claim that we age less rapidly than they. And here we are led, unwittingly, into the "twin problem."

Let us suppose that one of two identical twins is placed aboard a space vehicle and sent away for some years on a voyage into space with a velocity of $0.8c$; the other twin remains here on Earth. We on Earth claim that the traveling twin ages less rapidly than the stay-at-home twin. But the twin in the space vehicle would look at the Earth moving relative to him with a velocity of $0.8c$ and claim that the stay-at-home twin ages more slowly than he. What would happen when the twin in the space vehicle returns? Who would have aged less?

We are led into this predicament because we have not considered the problem completely. If the traveling twin is to travel at a *constant velocity* (uniform speed in a straight line) of $0.8c$ relative to the Earth, he will never return! Under these conditions the ages of the twins could never again be compared.

Before that twin can return, he must somehow reverse the direction of his travel. To do this, he must either bring his vehicle to a stop (relative to the Earth) and start it going in the opposite direction, or take the space vehicle in a large curved path. In performing either of these maneuvers, however, the traveling twin would experience accelerations that the stay-at-home twin does not feel. The experience of these two twins would not be the same.

In addition to the sensations the traveling twin feels in his stomach, he could, without question, devise experiments to determine that it is he who is accelerating. A glass tube with a ball held in the middle by two springs will indicate any acceleration in the direction parallel to the axis of the tube (Figure 11-7). Three such tubes, all at right angles to each other, will indicate accelerations in any direction. Consequently, the traveling twin would know that it is he who has traveled, and that his brother has stayed home. Therefore, since their experiences have been different, they should accordingly expect some difference upon reuniting; the principle of special relativity maintains that this difference would be in the ages of the twins. The traveling twin would not have aged as much as the stay-at-home twin.

The accelerations are not considered in establishing the age difference, for they were involved only in starting and stopping the vehicle. Once the vehicle has achieved a constant velocity, there are no accelerations. But the longer the constant velocity is maintained, the greater the age difference between the two twins, no matter what the accelerations. Special relativity is concerned only with frames of reference moving with a constant relative velocity.

It is important to recognize that the principle of special relativity does not ascribe any "cause" for the apparent changes. Quite the contrary. Special relativity maintains that the difference in length, mass, and the rate of a clock is only *apparent*; the length of a meterstick, the mass of a one-kilogram chunk of material, the rate of a clock depend entirely on the relative motion between the observer and the observed. The apparent difference is the natural consequence of uniform relative motion in a universe in which the speed of light is finite and has a constant value for all frames of reference.

Once the traveling twin has returned to the Earth, neither the meterstick nor the one-kilogram chunk of material will bear any record of the journey. The *rate* of the clock will again match ours which never left the physics laboratory. But the *setting* of the clock (clock and calendar, if the trip was long and fast enough) will be different. The clock (and calendar) that traveled would be behind the stay-at-home clock (and calendar).

Comparing the clocks indicates that a trip has been made. Had the ac-

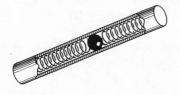

FIGURE 11-7

Two springs holding a ball centered inside a tube act as an accelerometer. If the tube accelerates in the direction of its axis the ball will compress one of the springs.

celerometers aboard the space vehicle recorded their activities, that record together with the clock would tell us a great deal about the trip. Certainly there would be no question that the trip had been made, nor would there be any question about who made it.

The statement that the traveling twin would age less rapidly than the stay-at-home twin is a prediction; the experiment itself has not yet been performed. But pions age less rapidly when they travel at speeds approaching that of light, and we are all made of pions and other such particles.

References

1. Isaac Newton, *Principia*, University of California Press, Berkeley (paperback), 1962, p. 6.
2. Max Born, *Einstein's Theory of Relativity*, Dover Publications, New York (paperback), 1962, p. 220.
3. Albert Einstein, *Relativity, the Special and General Theory*, © 1961 by the Estate of Albert Einstein. Used by permission of Crown Publishers, Inc.

Questions

1. Explain why simultaneous events cannot be established on systems moving relative to each other. What fundamental aspect of nature prevents this?

2. What fundamental aspect of nature dictates that the thought experiment with two men on a train throwing a beam of light is different from an experiment in which they throw a ball? Explain.

3. Is there *a priori* any reason why space, time, and mass should be invariants? Can we realistically extend our commonsense observations of everyday life to describe observations made between two frames of reference moving relative to each other at speeds close to that of light?

4. Discuss the significant differences between the experiences of the stay-at-home twin and his traveling brother. How many of these differences are measurable? How many may be used after the trip to obtain information about that trip?

5. Indicate why the speed of light is the limiting velocity.

6. Would the principle of special relativity prove inadequate should the speed of light be shown to change slightly over the years?

7. Discuss the importance of the correspondence principle to science.

8. (*a*) A flat car moving along a track with uniform motion contains two insulated

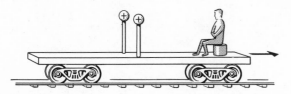

FIGURE 11-8

spheres, each with a positive electric charge (Fig. 11-8). A man rides on the flat car and observes the effect the two spheres have on each other. Describe what he observes.

(b) A second man stands on the railway embankment, and he also observes the effect the two spheres have on each other. Are the observations of the man on the railway embankment different from those of the man on the flat car? How? Explain.

9. How will the observations of the two men in question 8 differ if the flat car is accelerating?

Problems

1. Calculate the relativistic mass of an electron ($m_0 = 9.1 \times 10^{-31}$ kg) at speeds of $0.80c$, $0.90c$, and $0.95c$. (*Note.* When v equals $0.80c$, $0.90c$, and $0.95c$, $\sqrt{1 - v^2/c^2}$ equals 0.60, 0.44, and 0.32 respectively.)

2. Using the information of problem 1 and Newton's second law of motion, calculate the instantaneous acceleration of the electron at speeds of $0.80c$, $0.90c$, and $0.95c$ when that electron is placed in a uniform electric field of strength 100 nt/coul.

3. Again using the information from problem 1, calculate the relativistic kinetic energy of the electron at speeds of $0.80c$, $0.90c$, and $0.95c$.

4. Calculate the values of kinetic energy of an electron at speeds of $0.80c$, $0.90c$, and $0.95c$, using only the Newtonian relationship, $E_k = \frac{1}{2}mv^2$. Compare these values of kinetic energy with those obtained in problem 3.

5. How much energy would be derived from the conversion of the mass of one electron into energy when that electron is (a) at rest, and (b) traveling with a speed of $0.95c$ (see problem 3)? (c) What is the meaning of the difference between the answers in parts a and b?

Chapter Twelve

The Atom and Quantum Mechanics

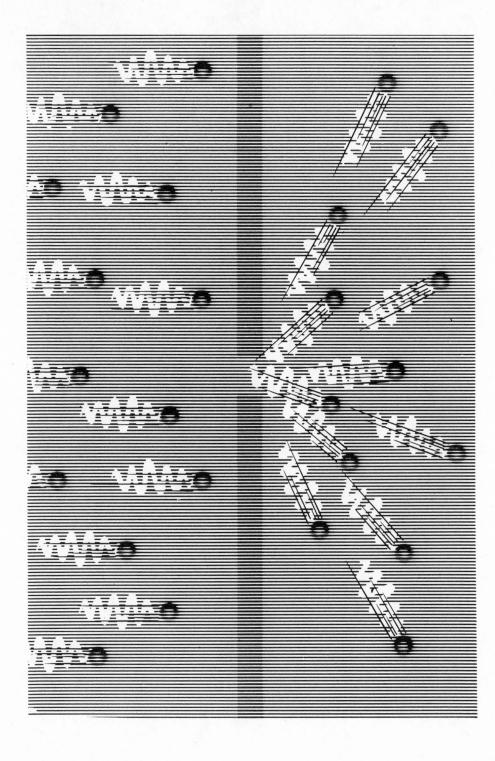

M odern atomic theory is the foundation upon which the present descriptions of the physical world are based. This theory stems not so much from the Greeks as from John Dalton (1766–1844), a Quaker teacher in Manchester, England. Atomism, as proposed by Democritus, was almost completely philosophical in nature; he did not pretend to account for the many actions, reactions, and changes in the world about him. Today we try to explain all the actions in our physical universe in terms of the atomic theory. Its successes are both numerous and encouraging.

The nineteenth century also witnessed the acceptance of Maxwell's electromagnetic wave theory which united the wave theory of light with the theories of electricity and magnetism. The atomic nature—or, better, the particulate nature—of electricity was not finally identified until the turn of the century. With the development of the concept of a minimum electric charge, a particle called the *electron,* it became necessary to consider how a *particle* of electricity could produce a *wave* of electromagnetic energy.

The particulate nature of electricity had been hinted at by Faraday, although he himself did not believe that electricity was composed of particles. Nonetheless, in the 1870's and 1880's several physicists pointed out that Faraday's experiments on electrolysis (see p. 166f) clearly indicated that electricity is composed of small particles, each with the same electric charge. Most physicists did not believe this view until experimental evidence began to mount in favor of the particulate nature of electricity. Of the many experiments indicating the existence of the electron, two have been more frequently cited in the literature. Because of their excellence and intrinsic interest, and because each demonstrates a unity to physics, we continue to give them credit for the general acceptance of the idea of the electron and the establishment of its electric charge and mass.

J. J. THOMSON AND THE CHARGE-TO-MASS RATIO

J. J. Thomson (1856–1940), English physicist and Nobel prizewinner, completed experiments in 1897 providing strong circumstantial evidence in favor of the electron theory. That evidence, however, was more readily accepted by those who already were favorably inclined toward the particulate nature of electricity. For the others it did not *prove* the existence of the electron. In fact, in the 1890's many of Thomson's colleagues thought he was less than serious. In 1897 Lord Kelvin still wrote of electricity as an electric fluid. Nevertheless, Thomson's experiments are historically significant and mark a turning point in the development of physics.

Cathode Rays and Negative Electricity

After considerable work with cathode rays, such as those used by Roentgen when he discovered x rays, Thomson reported in 1897:

The experiments discussed in this paper were undertaken in the hope of gaining some information as to the nature of the cathode rays. The most diverse opinions are held as to these rays; according to the almost unanimous opinion of German physicists they are due to some process in the aether to which—inasmuch as in a uniform magnetic field their course is circular and not rectilinear—no phenomenon hitherto observed is analogous: another view of these rays is that, so far from being wholly aethereal, they are in fact wholly material, and that they mark the paths of particles of matter charged with negative electricity. It would seem at first sight that it ought not to be difficult to discriminate between views so different, yet experience shows that this is not the case, as amongst the physicists who have most deeply studied the subject can be found supporters of either theory (1).

Thomson continued with a report on improvements of two experiments first performed by the physicists Jean Baptiste Perrin (1870–1942) and Heinrich Hertz (whose work with electromagnetic waves we discussed in Chapter 9). Perrin tried to show that cathode rays carried a negative charge with them, but his experiment was open to criticism; Thomson's work removed all earlier objections.

The arrangement used was as follows: Two coaxial cylinders [Figure 12-1] with slits in them are placed in a bulb connected with the discharge tube; the cathode rays from the cathode A pass into the bulb through a slit in a metal plug fitted into the neck of the tube; this plug is connected with the anode and is put to earth [ground]. The cathode rays thus do not fall upon the cylinders unless they are deflected by a magnet. The outer cylinder is connected with the earth, the inner with the electrometer. When the cathode rays (whose path was traced by the phosphorescence on the glass) did not fall

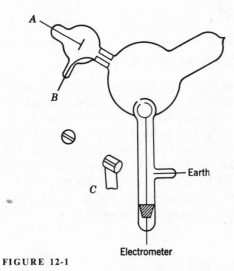

FIGURE 12-1

The Perrin tube. If the cathode ray originating at A is deflected by a magnet so as to enter the opening in the shield, the electroscope will register a negative electric charge. (J. J. Thomson, *Philosophical Magazine,* Vol. 44, Series 5, 1897, page 293.)

on the slit, the electric charge sent to the electrometer . . . was small and irregular; when, however, the rays were bent by a magnet so as to fall on the slit there was a large charge of negative electricity sent to the electrometer. . . . Thus this experiment shows that however we twist and deflect the cathode rays by magnetic forces, the negative electrification follows the same path as the rays, and that this negative electrification is indissolubly connected with the cathode rays (2).

Cathode Rays in an Electric Field

Having established that cathode rays carry a negative electric charge, Thomson found it necessary to demonstrate that they can be deflected by electric forces—a demonstration that even so capable an experimentalist as Heinrich Hertz had not been able to perform.

An objection very generally urged against the view that the cathode rays are negatively electrified particles, is that hitherto no deflection of the rays has been observed under a small electrostatic force, and though the rays are deflected when they pass near electrodes connected with sources of large differences of potential, such as induction coils or electrical machines, the deflection in this case is regarded by the supporters of the aethereal theory as due to the discharge passing between the electrodes, and not primarily to the electrostatic field. Hertz made the rays travel between two parallel plates of metal placed inside the discharge tube, but found that they were not deflected when the plates were connected with a battery of storage cells; on repeating this experiment I at first got the same result, but subsequent experiments showed that the absence of deflection is due to the conductivity conferred on the rarefied gas by the cathode rays. On measuring this conductivity it was found that it diminished very rapidly as the exhaustion increased; it seemed then that on trying Hertz's experiment at very high exhaustions [vacuum] there might be a chance of detecting the deflection of the cathode rays by an electrostatic force. The apparatus used is represented in [Figure 12-2].

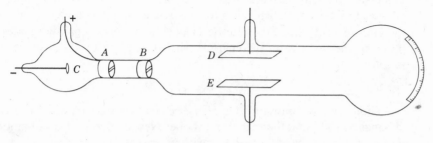

FIGURE 12-2

By projecting a cathode ray through slits *A* and *B*, Thomson produced a narrow beam which was then deflected by the electric field between plate *D* and *E*. The amount of deflection was measured by the scale on the end of the evacuated tube. (J. J. Thomson, *Philosophical Magazine*, Vol. 44, Series 5, 1897.)

The rays from the cathode C pass through a slit in the anode A, which is a metal plug fitting tightly into the tube and connected with the earth; after passing through a second slit [to narrow the beam still more] in another earth-connected metal plug B, they travel between two parallel aluminum plates about 5 cm long by 2 broad and at a distance of 1.5 cm apart; they then fall on the end of the tube and produce a narrow well-defined phosphorescent patch. A scale pasted on the outside of the tube serves to measure the deflection of this patch. At high exhaustions the rays were deflected when the two aluminum plates were connected with the terminals of a battery of small storage cells; the rays were depressed when the upper plate was connected with the negative pole of the battery, the lower with the positive, and raised when the upper plate was connected with the positive, the lower with the negative pole. The deflection was proportional to the difference of potential between the plates, and I could detect the deflection when the potential-difference was as small as two volts. . . .

As the cathode rays carry a charge of negative electricity, are deflected by an electro-static force as if they were negatively electrified, and are acted on by a magnetic force in just the way in which this force would act on a negatively electrified body moving along the path of these rays, I can see no escape from the conclusion that they are charges of negative electricity carried by particles of matter. The question next arises, what are these particles? are they atoms, or molecules, or matter in a still finer state of subdivision? To throw some light on this point, I have made a series of measurements of the ratio of the mass of these particles to the charge carried by it (3).

Cathode Rays in Both an Electric and a Magnetic Field

Although Thomson measured the value of this ratio, now called the charge-to-mass ratio e/m, by several different methods, one of his methods was much more accurate than the others. Thomson had previously discovered that a cathode ray is deflected when it is directed between two parallel plates if one of these plates has a positive charge and the other a negative charge. Adopting the premise that the cathode ray and hence an electric current is composed of tiny particles, each with charge e and mass m, Thomson reasoned that each particle has a force exerted upon it as it travels through the electric field. Since the electric field $\mathbf{E}$ between two parallel plates is uniform, the force F on the charged particle is given by the relation (see p. 236)

$$F = e\mathbf{E}$$

Thomson's assumption that each electron is a particle of matter (i.e., has mass) meant that the particle must obey Newton's second law of motion, $F = ma$. The forces in these two equations are the same, and so

$$ma = e\mathbf{E}$$

or

$$\frac{e}{m} = \frac{a}{\mathbf{E}}$$

Since Thomson could calculate the electric field strength **E** from the charge on the plates and from their size and separation, he had only to determine the acceleration of the electrons to be able to calculate the charge-to-mass ratio e/m.

As each electron enters the uniform electric field, a force directed toward the positively charged plate is exerted on it. Consequently, it accelerates in that direction and its direction of travel is altered (Figure 12-3a). While its velocity in the direction of the applied electric force increases, the velocity in the original direction of travel remains the same. The motion of the electron is equivalent to the motion of Galileo's stone dropped from the mast of a moving ship. The horizontal velocity of the stone remains constant, but the downward velocity increases at a uniform rate. The problem of determining the acceleration of the electron is therefore not a difficult one.

The acceleration can be measured by the angle through which the beam is deflected, but that angle also depends on the initial velocity of the electrons. To determine the initial velocity of the charged particles, Thomson made use of a magnetic field and the fact that the magnitude of the magnetic force acting on charged particles moving in a magnetic field depends on the velocity of these charges. If the charges travel at right angles to the magnetic field, the magnetic force acting on the particles is experimentally found to be

$$F = qvB$$

where q is the charge, v the velocity, and B the magnetic field strength. He reasoned that if he applied a magnetic force just equal in magnitude but opposite in direction to the electric force, the beam of particles would pass

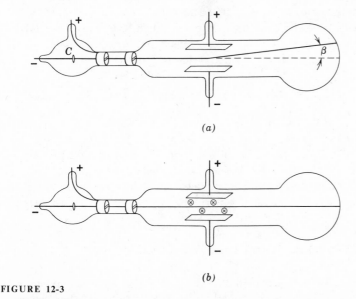

(a)

(b)

FIGURE 12-3

(a) Cathode rays are deflected in an electric field as if they carry a negative charge. (b) A magnetic field can be used to deflect the cathode rays. If the magnetic and electric fields are properly adjusted, electrons of a particular velocity will travel in a straight line.

through the tube undeflected, that is, the acceleration of the particles would be zero. Newton's second law of motion ($F = ma$) would then permit him to determine the initial velocity of the charged particles. The two forces acting on the electrons (for which $q = e$) are $e\mathbf{E}$ acting upward and $-ev\mathbf{B}$ acting downward (Figure 12-3*b*). The sum of these forces is zero:

$$F = ma$$

$$e\mathbf{E} - ev\mathbf{B} = 0$$

Therefore,

$$-ev\mathbf{B} = -e\mathbf{E}$$

and

$$v = \frac{\mathbf{E}}{\mathbf{B}}$$

This equation permitted Thomson to determine the velocity of the electrons by adjusting the electric and magnetic fields. At particular values of $\mathbf{E}$ and B, charged particles of only one velocity passed undeflected through the tube. A modification of Thomson's instrument is now called a velocity selector.

Knowing both the initial velocity of the particles and the angle through which the beam was deflected when the electric field alone was applied, Thomson was able to calculate the acceleration of the charged particles and consequently their charge-to-mass ratio. Thomson's experimental results yielded a value of e/m of 0.77×10^{11} coulombs per kilogram. The accepted value today is 1.76×10^{11} coulombs per kilogram.

After the efforts of Thomson and others to determine the exact value of e/m, the separate values for this charge and the mass of the electron were still not known. Since the electron's charge is much more easily detected, this was determined first. Once the charge was determined, the mass of the electron was easily found from the charge-to-mass ratio.

MILLIKAN AND THE OIL DROP EXPERIMENT

In the early part of the twentieth century, a number of attempts were made to determine the charge on the electron. Many of these methods were developed around the idea that a small water or oil droplet might pick up a free electron and thus become electrically charged; once charged, the oil drops could be influenced by an electric field. Since the electron's mass was believed to be extremely small, it was reasoned that the oil drop's weight would not be affected by the addition of even a few electrons.

A Charged Drop in Two Fields, Electric and Gravitational

Between 1908 and 1917, Robert Millikan (1868–1953), an American physicist and subsequently a Nobel prizewinner, accomplished the feat of measuring the charge on the electron with increased accuracy. His technique was to permit small oil droplets to fall between two plates across which he applied a high voltage. If an oil drop was electrically neutral, or the electric field was turned off, the drop fell under the influence of the gravitational field of the Earth (Figure 12-4a).

If, however, the drop acquired an electric charge and the electric field was turned on, the drop was no longer subject to the force of gravity alone (Figure 12-14b). Two forces act upon a charged oil drop in an electric field, the force of gravity mg, and the force exerted by the electric field Eq. If the upper plate is positive and the lower negative, and if the drop acquires a negative charge, the drop will rise when the high voltage is turned on and fall when the high voltage is turned off.

The motion of the drop is, however, impeded by the air. The drops are so small that the resistance of the air prevents an acceleration beyond a terminal

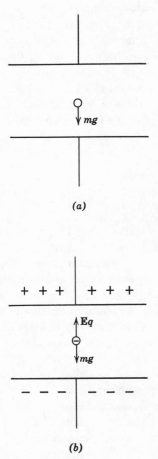

(a)

(b)

FIGURE 12-4

(a) An oil drop in a gravitation field has only the gravitational force mg acting on it. (b) A negatively charged oil drop placed in an electric field directed downward will have two forces exerted on it: a gravitational force mg downward, and an electrical force Eq upward.

velocity. Similarly, a leaf, or a man hanging from a parachute, falls with a constant terminal velocity (see p. 71).

The downward terminal velocity v_1 is proportional to the weight of the drop mg. The upward velocity v_2 is proportional to the resultant force on the drop, $\mathbf{E}q - mg$. Hence

$$\frac{v_1}{v_2} = \frac{mg}{\mathbf{E}q - mg}$$

By solving this equation for q we find

$$q = \frac{mg(v_1 + v_2)}{\mathbf{E}v_2}$$

The quantity g (the acceleration produced by gravity) is known, the electric field strength $\mathbf{E}$ was easily calculated, and the velocities v_1 and v_2 were measured. This leaves only the mass of the oil drop unknown.

Since the drops were so small, Millikan could measure neither their mass nor their size directly. He therefore had to infer the mass of each drop from an analysis of the drop's motion through the air. The resistance of the air to the motion of each drop, resulting in the drop attaining terminal velocity, depends on the mass and the size of each drop (i.e., the density of the oil used), and on the density and the pressure of the air. By measuring these quantities, Millikan was able to obtain a good value for the mass of each drop, and thus to calculate the charge of each of the thousands of drops he observed.

The Charge Changes—The Motion Changes

In the actual performance of the experiment, Millikan took all precautions necessary. For example, he controlled the temperature between the two plates across which the voltage was applied so that no air currents would develop in the region. The oil drops were formed by an atomizer and permitted to fall through a small hole in the upper plate (Figure 12-5). He projected the light of an arc lamp into the region between the plates to illuminate the oil drops and then observed these drops with a small telescope. In his field of view he placed two horizontal lines whose distance apart he knew quite accurately. He could then determine the velocity of the drops by timing their passage between these lines. On many occasions he was able to observe a single drop for several hours as it alternately fell with the electric field turned off and rose with the electric field turned on. In one instance he observed one drop for as long as four and a half hours. The measurements made during the first few minutes of observing this particular drop are given in Table 12-1. [The unit of electric charge used by Millikan, the electrostatic unit (esu), has been converted to coulombs: 1 esu = 3.33×10^{-10} coul.] The voltage across the plates for this series of observations was 7950 volts.

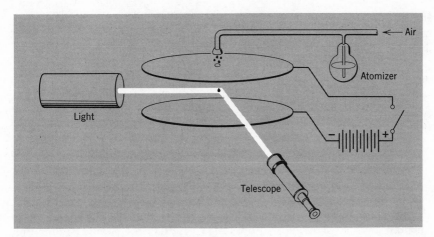

FIGURE 12-5

A schematic drawing of the Millikan oil drop apparatus. By looking through the telescope, the light-reflecting oil drops can be seen.

We see from Table 12-1 that the upward velocity changed considerably; the time interval required for the oil drop to travel between the two cross hairs varied from 14.2 seconds to 29.0 seconds. Evidently the charge on the oil drop changed. If electricity is indeed composed of small particles, each with an identical charge, it should be possible to estimate that charge from

Table 12-1 *Observations of an Oil Drop*

Time to Fall	Time to Rise	q (coul)	Δq (coul)	n	e (coul)
22.8 sec	29.0 sec	11.5×10^{-19}		7	1.64×10^{-19}
22.0	21.8	13.1×10^{-19}	1.6×10^{-19}	8	1.64×10^{-19}
22.3	17.2		1.7×10^{-9}		
22.4	—	14.8×10^{-19}		9	1.64×10^{-19}
22.0	17.3		1.6×10^{-19}		
22.0	17.3				
22.0	14.2	16.4×10^{-19}	3.3×10^{-19}	10	1.64×10^{-19}
22.7	21.5	13.1×10^{-19}		8	1.64×10^{-19}

the limited data given in this table. It is reasonable to assume that on occasions a single drop may have an excess of one or more electrons, giving it a charge of *ne*, where *n* is the number of excess electrons on the drop and *e* is the charge of each electron. It is also reasonable to assume that the charge on the drop will change by only one electron; that is, in all likelihood electrons will join or leave the drop one at a time. Consequently, the difference between the charges on the drop should give us a good idea of the charge on the electron. The column headed Δq indicates that this charge is close to 1.6×10^{-19} coulomb, and that in the last entry of Table 12-1 (3.3) the charge on the drop changed by two electrons.

The Minimum Electric Charge

If the charge on the electron is 1.6×10^{-19} coulomb, each of the measurements of the charge on the oil drop, 11.5, 13.1, 14.8, 16.4, should be a multiple of 1.6. Thus, for the charge to be measured as 11.5×10^{-19} coulomb, the drop must have had an excess of seven electrons; for the charge to be measured as 13.1×10^{-19} coulomb, the drop must have had an excess of eight electrons, and so on. But $1.6 \times 7 = 11.2$, not 11.5; $1.6 \times 8 = 12.8$, not 13.1; $1.6 \times 9 = 14.4$, not 14.8; $1.6 \times 10 = 16.0$, not 16.4. This discrepancy is explained, however, when we note that each of these numbers is a multiple of 1.64.

Using this technique and thousands of other measurements, Millikan determined that the charge of the electron is 1.640×10^{-19} coulomb and published this value in 1911. Later work, however, revealed that the value he had adopted for the resistance of air to the motion of the drop was incorrect, making the value for the mass of each drop also incorrect. After correcting this value, he determined the charge of the electron to be 1.603×10^{-19} coulomb. The value generally accepted today is 1.60206×10^{-19} coulomb.

Millikan found that in each of the thousands of measurements he made the charge on an oil drop never changed by less than 1.60×10^{-19} coulomb. Consequently, Millikan's oil drop experiment reveals that there is a minimum quantity of electricity and that electricity does indeed consist of the small particles we call electrons. Finally, the experiment proves that each electron does have the same electric charge.

By combining the work of Millikan with that of Thomson, we can determine the mass of the electron.

From Thomson: $\dfrac{e}{m} = 1.76 \times 10^{11}$ coul/kg

From Millikan: $e = 1.60 \times 10^{-19}$ coul

Therefore, the mass of the electron must be

$$m = 9.11 \times 10^{-31} \text{ kg}$$

It is no wonder that the acquisition of one, two, or even ten electrons by an oil drop did not alter the mass of that oil drop appreciably.

Once it was shown that an electric current in a metal wire is a flow of negative electrons rather than a flow of "positive electric fluid," it was evident that "Ben's current" (see p. 153) moved in the wrong direction. Since "Ben's current" passes through the wire from the positive to the negative pole of a battery or generator, it actually represents the flow of positive charge. There are many situations in which positive particles do flow. They flow in the solution of the battery and in the solution during electrolysis; they move through the air when lightning strikes. The sun emits streams of positive and negative particles that flow out into the solar system, passing the Earth and affecting its magnetic field. In a metal wire, however, the positive particles are restrained; only the negative particles (the electrons) can flow.

Despite the error in the direction of Ben's current, the right-hand rule based on Ben's current, predicting the direction of the magnetic field, may still be used. If one has to determine the direction of the magnetic field about a flow of electrons, the left hand may be substituted for the right hand in the rule. If the thumb of the left hand points in the direction of electron flow, the fingers wrapped about the line of flow point in the direction of the magnetic field produced about that line of flow (which may be a wire or even a cathode beam). Since the universe has both positive and negative particles (presumably in equal numbers), and since either positive or negative particles can be made to flow, either hand may be used. The universe is ambidextrous.

RUTHERFORD AND THE ATOMIC NUCLEUS

The studies of chemistry and the kinetic theory of gases adequately demonstrated that atoms are about 10^{-10} meter in diameter and that they do combine to form molecules. But the discovery of the electron challenged the idea makers. How are the atom and the electron associated? It seemed reasonable to assume that the positive portions of the atom neutralize the negative electrons. But how?

J. J. Thomson proposed that perhaps the atom is put together something like plum pudding or raisin bread. The bread represents the positively charged portion of the atom, and the raisins distributed throughout the bread represent the negative electrons. The Thomson model of the atom was being seriously considered when Ernest Rutherford (1871–1937), British physicist and a former student of Thomson, produced a surprise.

The Scattering of Alpha Particles

Rutherford had learned from his studies of radioactivity that at least two kinds of particles are emitted by radioactive substances. He called these *alpha particles* and *beta particles*. The alpha particles proved to be the more massive, and Rutherford learned to use them as probes to investigate the structure of matter.

FIGURE 12-6

Ernest Rutherford (1871–1937). (From *Rutherford and the Nature of the Atom* by E. N. da C. Andrade. Copyright 1964 by Educational Services Incorporated. Reprinted with permission of Doubleday & Company, Inc.)

By placing a bit of radium, a good emitter of alpha particles, in a container with a small hole, a fairly narrow beam of alpha particles can be defined. Rutherford directed this beam onto a piece of gold foil to see whether the beam could penetrate the foil. It did, but after the beam of alpha particles traveled through the foil, it was no longer so well defined (Figure 12-7). The foil made the beam spread out somewhat, as if some of the alpha particles

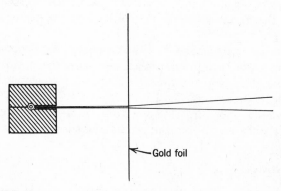

Gold foil

FIGURE 12-7

A lead box with a small hole in it permits only a narrow beam of radiation to escape from a radioactive source inside. The narrow beam can pass right through gold foil, but it spreads out slightly on passing through.

were being deflected or scattered from their original direction by the atoms of metal from which the foil was made. Rutherford realized that this was a good technique for learning something about atoms, for certainly the manner in which alpha particles are scattered should depend on the size and shape of the gold atoms in the foil.

As a result of some work done by one of his assistants, Hans Geiger (1882–1945), a German physicist, Rutherford had decided that the scattering of the alpha particles can be accounted for by assuming that each alpha particle undergoes many small deflections as it passes through the foil. The many small deflections scatter the beam of alpha particles through only a few degrees. This analysis seemed to be consistent with Thomson's "raisin bread" model of the atom; this model can account for deflections of the alpha particles through small angles. But as later narrated by Rutherford in one of his lectures, the picture changed.

One day Geiger came to me and said, "Don't you think that young Marsden, whom I am training in radioactive methods, ought to begin a small research?" Now I had thought that too, so I said, "Why not let him see if any alpha particles can be scattered through a large angle?" I may tell you in confidence that I did not believe that there would be, since we know that the alpha particle was a very fast massive particle, with a great deal of energy, and you could show that if the scattering was due to the accumulated effect of a number of small scatterings the chance of an alpha particle being scattered backwards was very small. Then I remember two or three days later Geiger coming to me in great excitement and saying, "We have been able to get some of the alpha particles coming backwards. . . ." It was quite the most incredible event that has ever happened to me in my life. It was almost as incredible as if you fired a 15-inch shell at a piece of tissue paper and it came back and hit you (5).

All this happened in 1909, and the problem confronted Rutherford for the next two years. Like all of us, he found it difficult to change his preconceived notions. He had been brought up to "look at the atom as a nice little hard fellow, red or grey in color, according to taste." The initial scattering experiments had forced Rutherford to abandon this notion of atoms and adopt the Thomson model. He could imagine alpha particles passing right through "raisin bread" atoms but not through "hard" atoms. The discovery that some of the alpha particles are occasionally scattered through large angles, however, again forced him to change his mental image of the atom.

Large-Angle Scattering and the Atomic Nucleus

Rutherford apparently wrestled with this problem for two years before he finally found the solution, for it was not until 1911 that, as Geiger puts it,

one day Rutherford, obviously in the best of spirit, came into my room and told me that he now knew what the atom looked like and how to explain the large deflections of alpha particles. On the very same day I began an experiment to test the relation expected by Rutherford between the number of scattered particles and the angle of scattering (6).

During those two years Rutherford had come to realize that in order to explain scattering through a large angle, the atom must be composed of a small, very massive center, called the nucleus, with one kind of electric charge, surrounded by a much larger sphere with the opposite charge (Figure 12-8). In an article published in May 1911, Rutherford wrote

It has generally been supposed that the scattering of a pencil of α or β rays in passing through a thin plate of matter is the result of a multitude of small scatterings by the atoms of matter traversed. The observations, however, of Geiger and Marsden on the scattering of α rays indicate that some of the α-particles must suffer a deflection of more than a right angle at a single encounter. They found, for example, that a small fraction of the incident α-particles, about 1 in 20,000, were turned through an average angle of 90° in passing through a layer of gold-foil about 0.00004 cm. thick. . . . A simple calculation based on the theory of probability shows that the chance of an α-particle being deflected through 90° is vanishingly small. In addition . . . the distribution of the α-particles for various angles of large deflection does not follow the probability law to be expected if such large deflections are made up of a large number of small deviations. It seems reasonable to suppose that the deflection through a large angle is due to a single atomic encounter, for the chance of a second encounter of a kind to produce a large deflection must in most cases be exceedingly small. A simple calculation shows that the atom must be a seat of an intense electric field in order to produce such a large deflection at a single encounter. . . .

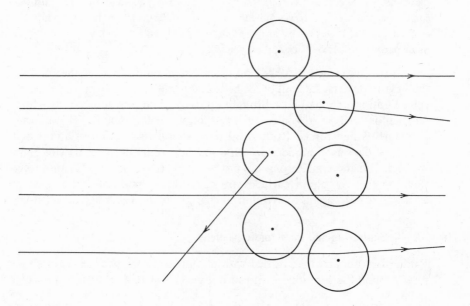

FIGURE 12-8

Rutherford found that the only way he could account for an occasional large-angle scatter of alpha particles by the gold foil was to assume that the gold atoms have small very massive nuclei at their centers.

The theory of Sir J. J. Thomson is based on the assumption that the scattering due to a single atomic encounter is small, and the particular structure assumed for the atom does not admit a very large deflection of an α-particle in traversing a single atom, unless it be supposed that the diameter of the sphere of positive electricity is minute compared with the diameter of the sphere of influence of the atom (7).

And so was born the idea that the atom is composed mostly of space. A very minute and high-density nucleus in which most of the mass of the atom resides, and which has a positive charge, is surrounded by a very much larger and ill-defined volume of negative electricity. This rather nebulous model of the atom, however, only created new questions. How is this sphere of negative electricity structured? What is the structure of the nucleus? The outer structure of the atom is the subject of the remainder of this chapter; the study of the nucleus is the subject of Chapter 13.

BOHR AND THE HYDROGEN ATOM

While this work was being performed, Rutherford had been in charge of the physics laboratory at the University of Manchester (1907–1919). Since he was director of the laboratory and had earned a considerable reputation, students from all over the world came to study under him. It is astounding to realize that at least seven physicists who studied under Rutherford during his professorship at McGill University in Montreal, at Manchester, and at the Cavendish Laboratory of Cambridge University went on and, like Rutherford, received a Nobel prize. One of these students, Niels Bohr (1885–1962) of Denmark, studied with Rutherford in 1912. Upon his return to Denmark, Bohr announced a solution to the atomic structure that combined the Planck-Einstein work on radiation (see Chapter 10) with Rutherford's model of the atom. If both the photon concept of radiation and Rutherford's model of the atom were revolutionary, clearly Bohr's more detailed model of the atom was also revolutionary.

FIGURE 12-9

Niels Bohr (1885–1962). (Courtesy of Friends of the Niels Bohr Library, American Institute of Physics, New York.)

Since we live on a planet that rotates on its axis and revolves about the sun, it was natural to speculate that, because the atom has a small massive nucleus, perhaps the electrons revolve about that nucleus as the planets revolve about the sun. Presumably a force of attraction between the nucleus and the electron supplies the centripetal force necessary to maintain the electron in an elliptical orbit.

The Electron, the Proton, and Centripetal Acceleration

Let us consider the simplest atom of all, the hydrogen atom. It has one electron, and presumably its nucleus has a positive charge equal in magnitude to the charge on the electron. This nucleus, like the electron, is a particle, and Rutherford named it the *proton*. The mass of the proton can be determined by substituting protons for electrons in a modified version of the experiment that permitted Thomson to measure the charge-to-mass ratio of the electron. Knowing the charge-to-mass ratio of the proton and assuming its charge is the same magnitude as that of the electron, the proton's mass can be calculated. The rest mass of the proton, it turns out, is very close to 1836 times that of the electron.

If the electron is to revolve about the nucleus in an orbit, there must be, according to Newton, a centripetal force acting on that electron. Without this centripetal force the electron would travel in a straight line. Since the proton and electron both have mass, the centripetal force might be supplied by gravitational attraction. Or, since the electron and proton are of opposite electric charge, the centripetal force might be supplied by the Coulomb force of electrical attraction.

The radius of the hydrogen atom, and thus presumably the radius of the orbit of the electron, is about 10^{-10} meter. The mass of the electron is 9.1×10^{-31} kilogram, and the mass of the proton is 1.7×10^{-27} kilogram. The force of gravitational attraction must therefore be

$$F = G\frac{m_1 m_2}{d^2}$$

$$F = 6.7 \times 10^{-11}\frac{(9.1 \times 10^{-31})(1.7 \times 10^{-27})}{(10^{-10})^2}$$

$$F = 1.0 \times 10^{-47} \text{ nt}$$

The charge on the proton is $q_1 = +1.6 \times 10^{-19}$ coulomb, the charge on the electron $q_2 = 1.6 \times 10^{-19}$ coulomb. So the Coulomb force must be

$$F = k\frac{q_1 q_2}{d^2}$$

$$F = (9.0 \times 10^9)\frac{(1.6 \times 10^{-19})(1.6 \times 10^{-19})}{(10^{-10})^2}$$

$$F = 2.3 \times 10^{-8} \text{ nt}$$

It may come as a bit of a surprise to learn that the electric force of attraction is about 2×10^{39} times stronger than the gravitational force of attraction.

Furthermore, with the equipment we use today, our measurements certainly are not so accurate that we need worry about the miniscule gravitational force between the two particles composing the hydrogen atom. Therefore, the gravitational force is neglected.

Niels Bohr suggested that the Coulomb force holds the electron in an orbit about the nucleus of the hydrogen atom. But the electron is a particle with an electric charge, and *Maxwell had maintained that if any electric charge is accelerated, it will radiate electromagnetic energy.* An electron revolving about the proton is accelerated by the centripetal force. According to Maxwell, it must radiate energy. Then, by the principle of conservation of energy the electron would lose energy. By losing energy, the electron would have to revolve in a smaller orbit, from which it would continue to radiate energy, causing the electron to seek a still smaller orbit. Eventually it would simply spiral into the nucleus. But obviously this does not happen; the hydrogen atom is a perfectly good and stable atom. Perhaps the electron does not revolve in an orbit.

The Hydrogen Spectrum

There are other data to consider. Bohr was very much aware that under the right conditions each chemical element emits a bright-line spectrum (see p. 271). The spectrum of each element can be unique only if the atoms of each chemical element are themselves unique.

The complexity of the spectra of the various elements disturbed the physicists of the late nineteenth century. They could not imagine that the pre-Rutherford atoms were able to vibrate in such a complex fashion. Newtonian mechanics had no explanation to offer.

The hydrogen spectrum is particularly simple, and so it is not surprising to find that this sequence of bright lines was the first to be expressed mathematically. The sequence of lines runs from one line in the red region of the spectrum to one in the blue, then several in the violet, and finally a great many in the ultraviolet, where the sequence comes to an end (Figure 12-10). From the first line in the red to the last in the ultraviolet, the distance be-

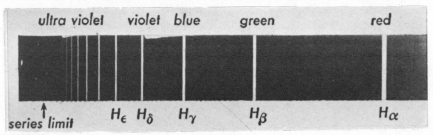

FIGURE 12-10

The Balmer spectrum of hydrogen. The first line in the spectrum appears in the red (far right of photograph), the next line is in the blue-green. Each succeeding line is closer to the next. (Reprinted with permission from *Introduction to Atomic and Nuclear Physics* by Harvey E. White, D. Van Nostrand Company, Inc., 1964.)

tween adjacent lines becomes less and less, until finally the lines are so close they seem virtually to merge.

J. J. Balmer (1825–1898), a Swiss physicist, showed that the wavelengths of this sequence of lines could be matched by a mathematical expression,

$$\lambda = B \frac{n^2}{n^2 - 4}$$

where $B = 3645.6$, a constant value, and n takes on the values of integers beginning with 3, that is, 3, 4, 5, 6, ...

Balmer could give neither rhyme nor reason why his equation worked, but work it did. Spectroscopists were even able to locate lines in the hydrogen spectrum whose wavelengths had not yet been measured. When their wavelengths were then measured, they agreed quite well with those predicted by Balmer (see Table 12-2). The Balmer equation gained tremendous support because it not only accounted for the wavelengths of the known spectral lines of hydrogen but also predicted the wavelengths of then unknown lines in the Balmer series.

Table 12-2   The Balmer Series of Hydrogen Lines

n	Calculated	Observed
3	6562.1	6562.1
4	4860.8	4860.7
5	4340.	4340.1
6	4101.3	4101.2
7	3969.	3968.1
8	3888.	3887.5
⋮	⋮	⋮
16	3702.9	3699

Similar equations were soon found for spectral sequences of elements other than hydrogen, and the race was on. Finally, in 1890 J. R. Rydberg (1854–1919), a Swedish physicist, was able to write an equation for a number of elements with hydrogenlike spectra. That equation, when written specifically for hydrogen, is

$$\frac{1}{\lambda} = R\left(\frac{1}{4} - \frac{1}{n^2}\right)$$

in which R is called Rydberg's constant and n takes on the same values given it by Balmer.

Rydberg's equation was extended for hydrogen and hydrogenlike elements to include other series of spectral lines. For example, in 1908, F. Paschen (1865–1947), a German physicist, discovered a second series of lines in the infrared region of the hydrogen spectrum. Bohr pointed out that if Rydberg's equation is written

$$\frac{1}{\lambda} = R\left(\frac{1}{n_l^2} - \frac{1}{n_u^2}\right)$$

the wavelengths of the Balmer series of spectral lines are obtained by letting $n_1 = 2$ and $n_u = 3, 4, 5, \ldots$ Furthermore, by letting $n_l = 3$ and $n_u = 4, 5, 6, \ldots$, the wavelengths of the Paschen series of spectral lines are also obtained.* Theodore Lyman (1874–1954), an American physicist, discovered a series of lines in the ultra-violet region whose wavelengths are obtained by letting $n_l = 1$ and $n_u = 2, 3, 4, \ldots$ The relationship between the Rydberg equation and the hydrogen atom was clarified by Bohr.

Bohr's Postulates

Although Bohr was convinced that Rutherford's work was correct, he still had to consider the spectroscopic work of Balmer and Rydberg. He also had to find some way around the stipulation of Maxwell's equations that the electron should radiate energy when revolving about the proton. He accomplished this by enunciating three postulates.

Postulate 1. The hydrogen atom consists of an electron revolving about the nucleus with the centripetal force supplied by the electric force of attraction. This system is a stable configuration; the electron does not radiate electromagnetic energy even if it is continually accelerating while it revolves.

No reason is given why the electron does not so radiate. The postulate simply states that it does not.

Postulate 2. The electron is restricted to only certain orbits, that is, only certain discrete orbits are permissible.

If limitations are to be set on orbits, however, the rules governing these limitations must be given. Bohr stipulated that the *angular momentum* of the electron can have only certain values. Angular momentum is akin to linear momentum mv. The angular momentum of an electron moving in a circular orbit is simply its linear momentum mv multiplied by the radius r of the orbit in which it revolves, that is, mvr. Bohr's restriction was that angular momentum can have only values that are a multiple of Planck's constant divided by 2π,

$$mvr = n\left(\frac{h}{2\pi}\right)$$

where h is Planck's constant and n takes on the values $1, 2, 3, 4, \ldots$.

After Planck's initial work on radiation in which he stipulated that light is emitted only in certain discrete amounts, and after Einstein's statement that light is absorbed in bundles in the photoelectric process, Bohr took the third step leading toward the formation of a quantum theory. He suggested that the atoms themselves are quantized (see p. 278f).

---

*The meaning of the subscripts *l* and *u* will be made clear on p. 332.

An atom in which an electron is allowed only certain discrete values of angular momentum is thus at the same time allowed only certain discrete amounts of energy. Because the electron is some distance from the proton, it has a certain amount of potential energy; because it has mass and is revolving about the nucleus, it has kinetic energy. But the atom as a whole is permitted only certain values of the total energy. With the solar system as a model, this means that the electron can reside only in certain discrete orbits, each with a different amount of energy.

Postulate 3. *The hydrogen atom will emit a quantum or photon of light when the electron makes a transition from an orbit of greater energy to one of less energy. Conversely, the electron may absorb energy and so make a transition from an orbit of one energy to one of a greater energy.*

By the principle of conservation of energy, if an electron is to lose energy in making a transition from one orbit to another, that energy must reappear elsewhere. Bohr suggested that the energy is converted into a photon of light, the same photon Planck and Einstein were talking about.

A photon is, so to speak, created by the electron "jumping" from an orbit of greater energy (greater radius) to one of less energy (shorter radius). The amount of energy that the photon carries away should be equal to the difference between the energies of the two orbits in question; according to Planck that energy is

$$E = hf$$

Expressed mathematically, Bohr's relationship becomes

$$E_u - E_l = hf$$

where E_u is the energy of the atom with the electron in an upper (u) orbit, and E_l is the energy with the electron in a lower (l) orbit.

Through a detailed study of the values of the atom's energy with the electron in each of the quantized orbits, Bohr was able to show that the wavelength of the photon emitted by a downward transition of the electron is given by the formidable equation

$$\frac{1}{\lambda} = \frac{2\pi^2 m k^2 e^4}{ch^3} \left(\frac{1}{n_l^2} - \frac{1}{n_u^2} \right)$$

where m is the mass of the electron, k the constant in Coulomb's law, e the charge of the electron, h Planck's constant, and n_u and n_l the numbers of the upper and lower orbits.

We now recall the Rydberg equation

$$\frac{1}{\lambda} = R \left(\frac{1}{n_l^2} - \frac{1}{n_u^2} \right)$$

The similarity of these two equations certainly must have encouraged Bohr to consider his postulates seriously. Furthermore, because he knew the numerical values of m, k, e, and h, he could calculate the numerical value of the factor in front of the parenthesis:

$$\frac{2\pi^2 mk^2 e^4}{ch^3} = 1.10 \times 10^7$$

He compared this number with the numerical value for the Rydberg constant first calculated twenty-three years earlier:

$$R = 1.10 \times 10^7$$

Precisely the same value!

It is this amazing agreement that persuaded Bohr to make such a bold suggestion about the structure of the hydrogen atom. And his boldness paid off. The Rutherford-Bohr model led the way to our present-day concept of the atom.

One of the many startling aspects of Bohr's postulates is that the electron can travel in orbits of only certain sizes. The radii of those orbits must be

$$r = n\left(\frac{h}{2\pi mv}\right) \qquad n = 1, 2, 3, \ldots$$

This means that the electron in a hydrogen atom cannot fall into the nucleus and collide with the proton. The Coulomb force continues to act, but rather than falling into the proton, the electron assumes an orbit about the proton. Bohr did not attempt to explain why.

DE BROGLIE'S PARTICLES AND WAVES

Bohr published his work in 1913; another major step in the development of quantum theory was made in 1923. In that year Louis de Broglie (b. 1892), French physicist and Nobel prizewinner, was a graduate student working on his doctoral dissertation. For this dissertation he studied the relationships or rather the similarities between matter and light. De Broglie was fully aware of Planck's stipulation that light is radiated in particles called photons and of Einstein's successful explanation of the photoelectric effect in which he assumed that light is absorbed as photons. He also realized that Young's double-slit experiment and other interference and diffraction phenomena required some form of wave theory to explain them. Surely the two must be reconciled in some way.

Momentum and Wavelength

The manner in which de Broglie reconciled the wave and the particle theories is rather startling, but then so were the successes of his new theory. Recognizing that mass is one form of energy and that the energy of a photon is related to its frequency, he was able to derive an equation for the momentum of a photon. Using the symbol p to represent momentum, $p = mv$, de Broglie's expression for the momentum of a photon is

$$p = \frac{h}{\lambda}$$

Since a photon of light is supposedly pure energy, to imagine that it also had momentum seemed difficult.

De Broglie's truly bold suggestion, however, was the extension of the dual wave-particle aspect of light to include *all matter*. He suggested that electrons, for example, ought to exhibit a wave nature when viewed under conditions similar to those in which light exhibits a wave nature! If light can exhibit either a wavelength or a momentum, depending on how it is observed, electrons can exhibit either a wavelength or a momentum. The momentum and wavelength are related by substituting mv in place of p in the previous equation:

$$mv = \frac{h}{\lambda}$$

When this equation is solved for the wavelength, we obtain

$$\lambda = \frac{h}{mv}$$

De Broglie suggested that electrons with a velocity v and a mass m have a wavelength given by this equation. But the idea of an electron having a wavelength seems ridiculous! And it was, but de Broglie was willing to suggest it anyway.

A wave, a disturbance in some medium, carries energy. Yet mass, a property we associate with particles, is one form of energy. It may sound foolish to speak of a particle with a wavelength, yet we must think in such terms. For instance, if we follow de Broglie's idea further and apply it to the Bohr atom, we have immediate success.

Standing Waves and the Atom

De Broglie suggested that the electron be considered not a particle revolving about the nucleus of the atom but a standing wave oscillating about the nucleus. Standing waves can easily be set up in a taut string held stationary at both ends (Figure 12-11). The manner in which these standing waves oscillate is limited. In Figure 12-11a the string is vibrating with one loop. This single loop comprises half a wavelength. In Figure 12-11b the string is vibrating with two loops; the ends do not vibrate noticeably, and the center of the string does not take part in the oscillation. The parts that do not oscillate are called *nodes*. With three nodes and two loops, this standing wave consists of one full wavelength. In Figure 12-11c there are three loops, four nodes, and consequently, one and a half wavelengths.

Standing waves in taut strings vibrate only when the length of the string is some multiple of half a wavelength: $\frac{1}{2}\lambda$, $\frac{2}{2}\lambda$, $\frac{3}{2}\lambda$, $\frac{4}{2}\lambda$, ... There is no other possibility. Standing waves are, by their very nature, quantized. A standing

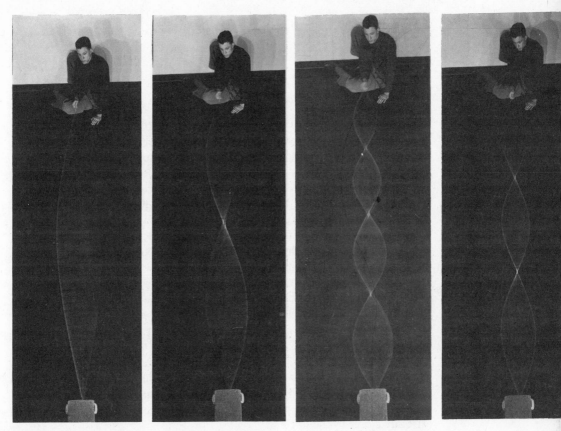

FIGURE 12-11

Standing waves set up in a taut string. (From *Physics,* Physical Science Study Committee, D. C. Heath and Company, 1960.)

wave in a string will not oscillate if the length of the string is 1.38 times the wavelength, 3.72 times the wavelength, or any multiple except those of half a wavelength. Even though standing waves are quantized, they are an integral part of nineteenth-century classical physics. The importance of the concept of quantization was not made evident until the turn of the twentieth century. Standing waves in strings (and in air columns) can indeed be described very well by Newtonian physics.

Standing waves can also be set up in rings as well as in taut strings, and again the waves are quantized. The circumference of the ring, however, must be a multiple of one full wavelength: 1λ, 2λ, 3λ, ... (Figure 12-12). That is,

$$n\lambda = 2\pi r$$

If we now substitute the equivalent wavelength from de Broglie's equation $\lambda = h/mv$, we have

$$\frac{nh}{mv} = 2\pi r$$

or

$$mvr = n\frac{h}{2\pi}$$

This equation indicates that the angular momentum is quantized; it is the same equation Bohr used to quantize the atom (see p. 331). The electrons seem actually to be standing waves about the nucleus. Only those waves will exist whose wavelength can fit around the circumference of the orbit an integral number of times. No other waves will exist. Hence, the electron wave is quantized, just as a vibrating guitar string is quantized.

An electron transition is equivalent to a change in the mode or method of vibration. A downward transition amounts to a decrease in the number of wavelengths that fit around the circumference (Figure 12-12); an upward transition amounts to an increase in the number of wavelengths in the standing wave.

De Broglie's wave-particle theory answers the question why the electron never falls any closer to the nucleus of the atom than the first orbit. A standing wave can exist in a ring only if the circumference of the ring can contain an integral number ($n = 1, 2, 3, \ldots$) of wavelengths of that standing wave. The smallest integral number is one wavelength. A standing wave cannot exist in a ring if the wavelength is longer than the circumference of the ring. The smallest possible ring (orbit) for the electron standing wave is therefore one in which $\lambda = 2\pi r$. No smaller orbit is possible, and the electron cannot exist any closer to the nucleus.

Electron standing waves about a nucleus, however, are actually not as simple as standing waves set up in a ring. To account for all the properties of the atom, the electron standing waves are actually treated as existing on the surface of a sphere or a similar three-dimensional figure. This view of atomic structure accounts for many more features of the atom than does the simple standing wave in a ring.

Energy Levels

Whether we think of the electron as existing in an orbit or as a standing wave, the electron, in fact, has greater or lesser amounts of energy, and these amounts of energy are quantized. Hence we conveniently refer to the various quantized electron states as energy levels and speak of the transitions between these levels. The transition of an electron from one energy level to a level of less energy results in the emission of one photon. If the electron is to make a transition to a level of greater energy, energy must be given to the electron. This energy may be supplied by a collision with a passing electron which will then lose some of its kinetic energy. Or energy may be supplied by a collision with a photon, in which case all the photon's energy must be used. The photon then ceases to exist as a photon; its energy reappears as increased energy in the electron which makes a transition to a level of greater energy.

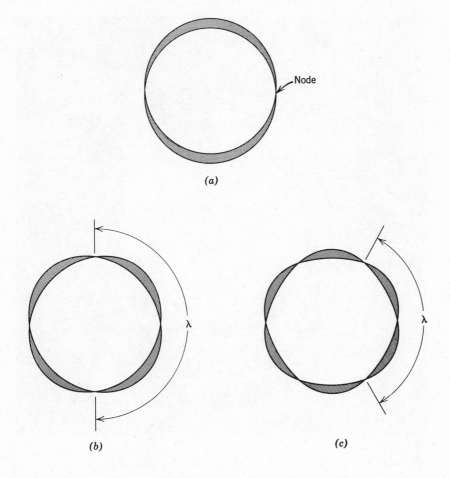

FIGURE 12-12

A ring may vibrate such that its circumference equals: *(a)* one full wavelength of the standing wave, *(b)* two wavelengths, *(c)* three wavelengths.

Observational Support for Electron Waves

In 1927 C. J. Davisson, Nobel prizewinner, and L. H. Germer, both American scientists, showed that electrons reflect from the surface of a single large crystal in a manner similar to light reflecting from a thin film. The atoms in a crystal line up to form what amounts to many layers of thin films, and thus electrons reflecting from the crystal behave as light reflecting from a film of air between two plates of glass (compare Figures 12-13 and 8-22). The measured wavelength of the electrons agrees with that predicted by de Broglie's equation,

$$\lambda = \frac{h}{mv}$$

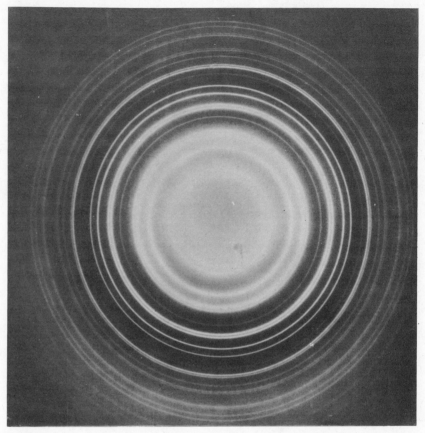

FIGURE 12-13

A photograph of an interference pattern of electrons after passing through a thin film of gold. Compare this with Figure 8-22, a photograph of an interference pattern of light reflecting from a thin film of air. (Reprinted with permission from *Fundamentals of Physics* by Henry Semat, Holt, Rinehart and Winston, 1966; used through the courtesy of Professors Row and Mukherjee.)

Photographs of electrons passing through an apparatus equivalent to that used in Young's double-slit experiment reveal an interference pattern just as light does; compare Figures 8-15 and 12-14. Electrons do indeed exhibit a wave nature; their action can be explained by a wave theory.

Let us calculate the wavelength of an electron after it has fallen through a potential difference of 150 volts. The energy acquired in falling through a potential difference V is converted into kinetic energy (see p. 236):

$$eV = \frac{1}{2}mv^2$$

$$v^2 = \frac{2eV}{m}$$

FIGURE 12-14

Electron interference pattern from an experiment which is equivalent to Young's double-slit experiment. (Photograph through the courtesy of Professor G. Moenstedt, University of Tubingen.)

$$v^2 = \frac{2(1.6 \times 10^{-19})(150)}{9.1 \times 10^{-31}}$$

$$v^2 = 53 \times 10^{12} \text{ m}^2/\text{sec}^2$$

$$v = 7.3 \times 10^6 \text{ m/sec}$$

With this velocity the de Broglie wavelength is

$$\lambda = \frac{h}{mv}$$

$$\lambda = \frac{6.7 \times 10^{-34}}{(9.1 \times 10^{-31})(7.2 \times 10^6)}$$

$$\lambda = 1.0 \times 10^{-10} \text{ m}$$

This is the wavelength of fairly energetic x rays.

We may also calculate the wavelength of an automobile. A small car has a mass of roughly 10^3 kilograms; when it travels with a velocity of 20 meters per second (nearly 50 mph), its de Broglie wavelength is

$$\lambda = \frac{h}{mv}$$

$$\lambda = \frac{6.7 \times 10^{-34}}{(10^3)(20)}$$

$$\lambda = 2.2 \times 10^{-30} \text{ m}$$

The diameter of an atom is 10^{20} times greater than the de Broglie wavelength of this auto. Clearly the wavelength is much too small for us to consider seriously, much less to measure. It should therefore be clear that since Planck's constant is so extremely small, the de Broglie wavelength is meaningful only on the scale of atomic particles.

HEISENBERG'S UNCERTAINTY PRINCIPLE

Although Young and Fresnel showed that certain properties of light can best be explained by a wave theory, Planck and Einstein showed that certain other properties require a particle theory. Similarly, although J. J. Thomson and Millikan showed that electricity is composed of small particles called electrons, de Broglie proved that certain electron behavior can be described by a wave theory. We need a dualistic description, a wave-particle description.

The problem lies in our inability to picture the electron, the photon, the proton, or the atom. We may conceive of these as fuzzy particles or we may picture them as waves, but we cannot be sure which is correct, if indeed either is correct. This fundamental fact was eloquently stated in 1927 by Werner Heisenberg (b. 1901), a German physicist and Nobel prizewinner, in his enunciation of the *uncertainty principle*.

Wavelength versus a Narrow Beam of Waves

The work of Huygens and Fresnel (see p. 214f and p. 219f) can now be interpreted to mean that the wave nature of light limits the size of a beam of light. A beam of light can be defined by a narrow slit placed some distance from the source. The narrower the slit, the smaller the beam of light and the smaller the elongated spot of light produced when the beam strikes a screen. A very narrow beam of light can indeed be defined, but according to Fresnel, if the slit is too narrow, diffraction effects finally prevent a very narrow beam from being defined.

Fresnel showed that the extent of diffraction is directly related to the width of the slit and to the wavelength of light used (see Figure 8-20). The diffraction pattern of a single slit is composed of a central fringe with less intense fringes on either side (see Figure 8-18). The width of the central fringe is an indication of the extent of diffraction.

Wave theory predicts, and observations have confirmed the prediction, that as the slit defining a beam of light becomes narrower and narrower, the beam itself becomes narrower and narrower until the slit width approaches the wavelength of the light in the beam. At a slit width approximately the wavelength of light used, diffraction causes the beam to fan out through nearly 180 degrees and the beam ceases to exist. The slit acts like a point source of light. The same diffraction effects occur with all waves (see Figure

8-11); hence all regions of the electromagnetic spectrum (light waves, radio waves, x rays, and so on) exhibit diffraction. It is therefore impossible to obtain a beam of any type of wave, water wave, sound wave, or light wave, as narrow as the wavelength of the waves in that beam.

Momentum Versus a Narrow Beam of Particles

Fresnel's analysis of *waves* passing through a slit must also apply to *particles* passing through a slit. Let us examine a parallel beam of monochromatic light impinging upon a slit. That is, initially all the photons have the same momentum (same wavelength and direction of travel). The limitations placed on the narrowness of the beam transmitted can be interpreted in terms of particles. Since diffraction does in fact occur, we conclude that some of the particles acquire a transverse momentum while passing through the slit. In other words, they are deflected by this slit. The narrower the slit, the more the beam spreads out and the greater the transverse momentum acquired by some of these photons.

In order to predict where a particle passing through the slit will strike a screen placed beyond that slit, we must know two quantities: (1) the particle's position when passing through the slit, that is, its position in the x-direction, the direction of the slit width (Figure 12-15a); (2) how much transverse momentum Δp_x the particle acquires by passing through the slit. If we know both of these quantities precisely, we can predict precisely where a given particle will strike the screen. We simply add vectorially the initial momentum p_0 and the transverse momentum Δp_x (Figure 12-15b). If the particle passes through the slit at x and acquires a transverse momentum Δp_x, it will arrive at point C on the screen.

We therefore need to find both the particle's position x in the slit, and its transverse momentum Δp_x. How can we do this? Let us first consider its position. We can be sure of the particle's position as it passes through the slit very simply and precisely by narrowing the slit so that its width is just a trifle more than the particle's diameter. We can then be sure of the location of every particle of that size as it passes through the slit. Each particle of that size will have the same or nearly the same position.

The particles we are referring to are photons. What do we mean, therefore, when we say that the width of the slit is "just a trifle more than the diameter of the particles passing through it"? What is the diameter of a photon? Could it possibly be *less* than its de Broglie wavelength $\lambda = h/mc$? No, of course not. In fact, both theory and observation clearly indicate that if a photon has a diameter, it must be equal to its de Broglie wavelength. Therefore, when the slit is "just a trifle" wider than the diameter of a photon, the slit width is almost equal to the wavelength of light, at which point diffraction effects cause the beam to fan far out through nearly 180 degrees.

Can we now answer the question about how much transverse momentum a given photon acquires as it passes through the slit? No, we cannot. Some

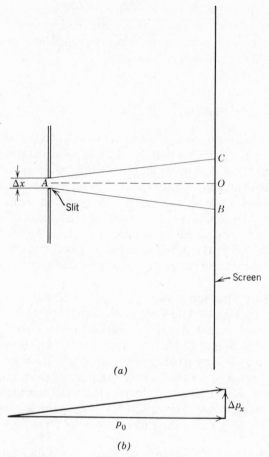

(a)

(b)

FIGURE 12-15

(a) Most of the particles entering a slit whose width is Δx, will be diffracted within the angle BAC, which defines the central fringe of the diffraction pattern (see Figure 8-18). *(b)* By passing through the slit, the particles acquire a transverse momentum Δp_x.

photons pass through undeflected, some are deflected through small angles, and others are deflected through angles of nearly 90 degrees. Before entering the slit, each of these photons has the same momentum, and each has nearly the same position while passing through the slit. Yet we find that the photons actually acquire a whole range of transverse momenta while passing through. We are at a loss to predict how much transverse momentum any one particular photon will acquire. It appears that if we know the position of a photon rather precisely, we have very little knowledge of its momentum.

On the other hand, if we make the slit much wider so that there is very little diffraction, we produce a very definite beam. Therefore, we know that the transverse momentum of nearly all the particles passing through is zero.

Nearly all the particles pass straight through undeflected. We have a very good idea of the transverse momentum, but what can we say about the position of a given particle in the slit? Next to nothing. If we have fairly good knowledge of a photon's momentum, we are left with very little knowledge of its position.

We seem to be caught in a trap. If we know the position of a given particle, we are not able to say what its momentum is. Conversely, if we are able to say what its momentum is, we are at a loss to give its location. It seems that we are not able to know simultaneously both the particle's precise position and its precise momentum. We are therefore at a loss to predict where a given particle will strike the screen after passing through a narrow slit. Can we *guess* where it might hit the screen? Yes, we can, and that is the extent of our knowledge of very small particles.

Before proceeding with our guesswork about the particle's motion after it passes through the slit, let us indicate more clearly the limits of our knowledge of the particle's position and its momentum. These limits are called *Heisenberg's uncertainty principle* and are mathematically expressed as follows:

$$\Delta p_x \, \Delta x \approx h$$

The product of the uncertainty in the particle's momentum in the x-direction, Δp_x, and the uncertainty of the particle's position, Δx, must be approximately equal to h. Therefore, when the uncertainty in the position Δx is very small, the uncertainty in momentum Δp_x is very large, and vice versa. The price we pay to know one of the quantities with great accuracy is great uncertainty about the other quantity.

Experiments prove conclusively that electrons also undergo diffraction effects when passing through narrow slits. How else could we account for the interference pattern of the electron equivalent of Young's double-slit experiment (Figure 12-14)? If the electrons had passed through the two slits without diffracting, two well-defined electron beams would have emerged from the two slits, and two well-defined fringes would have appeared on the photographic plate. But an interference pattern appeared instead. The electrons were diffracted; the beams spread out and interfered with one another. Therefore, the uncertainty principle must also apply to electrons and to protons, to alpha particles, in fact, to all particles.

From knowing a particle's momentum we can calculate its energy. Similarly, knowledge of a particle's position implies a knowledge of what time it occupied that position. Heisenberg's uncertainty principle can therefore be written in terms of energy and time:

$$\Delta E \, \Delta t \approx h$$

If we know the energy of an electron very accurately, we are not sure when the electron had that energy. If we know its momentum very accurately, we are not sure where it was when it had that momentum.

Limitations to Particle-Waves

Another aspect of the uncertainty principle must be pointed out. If a particle's position is uncertain because of its wave nature, that particle's ability to be used as a probe is limited. If billions of particles are used to locate both the position and shape of some unknown chunk of matter, what these particles can reveal depends on their wavelength.

For example, imagine a stick placed in the mud of a shallow pond so that it projects above the surface of the water. We can learn something of the size and location of the stick by sending little waves past it. We simply need to analyze the diffraction pattern of these waves after they have passed the stick (see Figure 12-16). If the stick is big, much bigger than the wavelength of the waves, a great deal can be learned about that stick. But if the stick is small,

FIGURE 12-16

A study of the diffracted waves would reveal the location and something of the structure of the rocks (John Shelton).

not so much can be learned; in fact, if a fine wire replaces the stick, there is no diffraction pattern at all. The waves will not even reveal the existence of the wire, much less its shape.

When used as a probe, waves are capable of revealing information about only the objects that are nearly the same size or larger than the wavelength of the probing wave. Clearly, the de Broglie wavelength of the alpha particles used by Rutherford to discover the nucleus of the gold atom was smaller than the nucleus itself.

Unfortunately, however, if the particle's wavelength is small compared to the size of the object being investigated, the particle's momentum must be large. The de Broglie wavelength $\lambda = h/mv$ decreases as the particle's momentum increases. As the momentum increases, so must the energy. In elucidating the structural details of the atom and its nucleus, particles of very small wavelength must be used. And if the probing particles have a very small wavelength, they correspondingly have a great deal of energy. Consequently, in the process of investigation these energetic little particles disturbed the object being studied. By the time we have learned something of the object's position, it is no longer there; and we are not sure where it went after being disturbed. Another investigation in search of the object will only disturb it again. It is the same old question. How do you study something without disturbing it?

This limitation on using particle-waves as probes for the investigation of nature is very frustrating; it indicates that we will never find any sharp boundaries in the world of atoms. But then perhaps there are no sharp boundaries in the microcosm. Perhaps electrons, protons, and the like do not have fixed and definite radii. But why should they? Didn't de Broglie show that particles can be considered to have a wavelength? The world of atoms is a fuzzy place.

SCHRÖDINGER AND PROBABILITY WAVES

By considering the electron as a wave, the uncertainty principle does indeed make sense. It is really quite difficult to say precisely where a wave is. A wave spreads out over a region; it cannot be confined to a point, that is, where $\Delta x = 0$.

On the other hand, if we consider an electron to be a point center of force in the manner of Boscovich, we find that when we start to locate the electron, we can only say where it is more likely to exist, but not where it "really" exists. Our guesswork comes into play here. The problem is really one of probability. In 1926, the German physicist and Nobel prizewinner E. Schrödinger (1887–1961) developed a very complex mathematical equation, called the wave equation, which describes the wave-particle duality of matter and light. Later in 1926, Max Born (1882–1970), still another German physicist and Nobel prizewinner, showed that Schrödinger's wave equation really describes the probabilistic nature of the particles. The question is not "where is the particle?" but "where is the particle most likely to be?"

One of the solutions to Schrödinger's wave equation states that when an electron is confined to a small region of space, such as the region about a proton, it can exist only with certain particular discrete energies. These discrete amounts of energy correspond to the energy levels developed by Bohr. The electron is more likely to be found at certain distances from the proton, not very likely to be found at others.

The 1920's were a very active time indeed for theoretical physicists. De Broglie, Heisenberg, Schrödinger, Born, and others all focused the attention of the world of science on the probabilistic nature of physics. Of course, not all physicists agreed. Einstein quite emphatically stated that he did not care for Heisenberg's uncertainty principle which seems to limit our means of investigation. According to this principle, it would appear that Nature has restricted man and told him that he shall not pass beyond a certain point in his intellectual quest to understand the physical world in which he lives.

The present picture, however, is not as bleak as all this may imply. Studies have proceeded far beyond the dreams of the 1920's and even those of the 1930's. The nucleus of the atom has been dissected and studied bit by bit. The uncertainty principle has not even been overthrown, but investigations can still be carried out, both in the laboratory and in the study. For every question answered, many new questions continue to appear.

References

1. M. R. Shamos, ed., *Great Experiments in Physics*, Henry Holt and Company, New York (paperback), 1959, p. 219.
2. *Ibid.*, pp. 220 f.
3. *Ibid.*, pp. 221 ff.
4. R. A. Millikan, *Physical Review*, Vol. 32, 1911, pp. 349–397.
5. E. N. da C. Andrade, *Rutherford and the Nature of the Atom*, Doubleday Anchor Books, Garden City, N.Y. (paperback), 1964, p. 111.
6. *Ibid.*, p. 114.
7. *The Collected Papers of Lord Rutherford of Nelson*, Vol. II, Interscience Publishers, New York, 1963, pp. 238 f.

Questions

1. Explain why the motion of an electron traveling at right angles to a uniform electric field is the same as the motion of the stone Galileo's sailor dropped from the mast of a moving ship.

2. Explain how the motion of an electron traveling at right angles to a uniform magnetic field differs from the motion of a stone dropped from the mast of a moving ship.

3. A television tube is a refined cathode ray tube. Do you think that the placement of the picture on the face of the tube will change if the television set (which is emersed in the Earth's magnetic field) is rotated about the vertical axis?

4. Describe the experiments and Rutherford's reasoning leading to the concept that the atom has a small but massive nucleus.

5. What aspect of Maxwell's work with electromagnetic waves troubled Bohr the most in his model of the atom?

6. Describe the process by which an atom (*a*) emits a photon, (*b*) absorbs a photon.

7. What principle limits our ability to predict the motion, that is, the position and momentum, of an electron or other atomic particle?

8. It has been said that the reason one good billiard ball cannot be balanced on top of another one is given by the uncertainty principle. Comment on this statement, but do not waste much time testing it by experiment.

Problems

1. Calculate the centripetal acceleration of an electron traveling at right angles to a uniform electric field of strength $\mathbf{E} = 1.2 \times 10^4$ nt/coul.

2. An electron of J. J. Thomson's experiment enters an electric field of strength $\mathbf{E} = 1.2 \times 10^4$ nt/coul and initially travels perpendicular to that field. Calculate (*a*) the force acting on the electron, (*b*) the acceleration of the electron, and (*c*) the distance the electron will move parallel to the field in 3×10^{-8} sec.

3. An electron travels at right angles to the Earth's magnetic field in a region where that field is $B = 6 \times 10^{-5}$ nt-sec/coul-m. The velocity of the electron is 2×10^7 m/sec. Find (*a*) the magnetic force on that electron, and (*b*) the acceleration of that electron.

4. If the electron in problem 3 is in a television tube, how far will the electron be deflected if it has to travel 0.2 m before reaching the screen?

5. Using the first entry in Table 12-1 in which an oil drop with an excess of seven electrons falls and rises between the plates of Millikan's experiment, calculate the mass of the oil drop. The electric field strength is 5.0×10^5 nt/coul, and the distance the drop moved during those time intervals is 1.0×10^{-2} m.

6. Calculate the wavelength of the spectral line resulting from an electron falling from the energy level $u = 4$ to $l = 2$ of the hydrogen atom.

7. Calculate the energy of photons resulting from the transition in problem 6.

8. Using the work of Balmer, which presupposes that to ionize a hydrogen atom the electron must at least make a transition to the last energy level, $u = \infty$, calculate the minimum amount of energy needed to ionize a hydrogen atom whose electron is in the ground state.

9. A meteorite with a mass of 4.0×10^{-5} kg enters the Earth's atmosphere with a velocity of 3.3×10^4 m/sec. What is the wavelength of that meteorite?

10. An electron falls from rest (or essentially so) through a potential difference of 220 volts. Find (*a*) its kinetic energy in both electron volts and joules, (*b*) its velocity, and (*c*) its wavelength.

11. If the uncertainty in the position is equal to one wavelength, what is the uncertainty in the momentum of the electron in problem 10?

Chapter Thirteen

Radioactivity and the Nucleus

The most active field of research in physics today is the study of elementary particles, supposedly the basic building blocks of the universe. To study them, physicists use these same particles as probes in special equipment designed to discover how the elementary particles react to man's manipulations. Like Galileo, the modern physicist makes Nature perform, and from this performance he learns more of her secrets.

Since the time of Galileo, physics has proceeded from an investigation of the tangible and obvious to a probing of the more intangible and subtle. This aspect of the science intrigues the physicist, for he enjoys the precision of mathematics and the great demands physics places on his imagination and ingenuity. Certainly contemporary theoretical physics is extremely complex and both mathematically and conceptually rigorous. In addition, the research physicist must be able to master the highly complex and technically sophisticated equipment used in modern research. The research physicist, who is in one sense a manipulator of the elementary particles, directs electrons, protons, and other particles to do his probing and computers to do his counting and calculations.

The student untrained in the field, however, is often wary of mathematical precision; he is likely to question the various interpretations of the apparent jumble of numbers pouring out of the computer, not because he feels he can better interpret them, but because the mathematics may alienate him. The nonscience student asks "What is it all for?"

The scientist contemplates the workings of nature because he is basically curious. He does not necessarily wish that his work be immediately or directly applicable to the practical world of the house, the television set, and the automobile. He knows the society that maintains an active curiosity about both man and nature will benefit in the long run. Newton's study of the moon did not lead directly to a specific practical invention, but application of the fundamental principles elucidated in his "impractical" investigation has certainly benefited mankind.

Although the methods and equipment of contemporary research are complex, we can arrive at an understanding of them by first studying earlier investigating techniques, techniques upon which present research is based. With this background the layman can then follow the reports of contemporary research that he encounters in newspapers and magazines.

RUTHERFORD AND THE ALPHA PARTICLE

The work of J. J. Thomson and Millikan was the first of the deeper searchings into individual elementary particles. These two men convinced the world that electricity exists only in certain discrete packets, each with a specific and measurable rest mass and electric charge. Then Rutherford and Bohr proved that atoms are composed of a small massive nucleus surrounded by electrons. The structure and actions of the atomic nucleus were also investigated by Rutherford.

351

OLDSCHMIDT   PLANCK   RUBENS   LINDEMANN   HASENOHRL
NERNST   BRILLOUIN   SOMMERFELD   DE BROGLIE   HOSTELET
                SOLVAY   KNUDSEN   HERZEN JEANS   RUTHERFORD
                    LORENTZ   WARBURG   WIEN   EINSTEIN   LANGEVI
                              PERRIN   Mme CURIE   POINCARÉ   KAMERLINGH ONNES

FIGURE 13-1

The Solvay conference in 1911, called to discuss the relationship between specific heats and the quantum theory, brought together many of the great men in physics and chemistry of that time. The de Broglie appearing in this picture is Maurice, one of the scientific secretaries for the conference. His younger brother, Louis—19 years old at the time—read the discussion as the manuscripts were being prepared for publication. Rutherford returned from the conference to Manchester and there passed on his enthusiastic reaction to Bohr, then working in his laboratories. (Courtesy of Solvay et Cie.)

After Becquerel had discovered radioactivity, a number of people, notably Becquerel himself and the French physicists Marie Curie (1867–1934) and her husband Pierre Curie (1859–1906), unraveled the mystery of the rays emanating from radioactive substances; they discovered new chemical elements. The Curies, through ingenuity and industry, were able to separate from the natural ore of uranium a new chemical element called radium. As a strong emitter of radioactive rays, radium then became one of the standard sources of radioactive material for other investigators.

Alpha and Beta Particles

Only three years after Becquerel discovered radioactivity in 1896, young Rutherford, in his first year as professor of physics at McGill University in Montreal, determined that the radioactive rays, unlike the x rays discovered by Roentgen (see pp. 249ff), were complex. Rutherford wrote

The remarkable radiation emitted by uranium and its compounds has been studied by its discoverer, Becquerel, and the results of his investigations on the nature and properties of the radiation have been given in a series of papers in the *Comptes Rendus*. He showed that the radiation, continuously emitted from uranium compounds, has the power of passing through considerable thicknesses of metals and other opaque substances; it has the power of acting on a photographic plate and of discharging positive and negative electrification to an equal degree. The gas through which the radiation passes is made a temporary conductor of electricity and preserves its power of discharging electrification for a short time after the source of radiation has been removed. . . .

It is the object of the present paper to investigate in more detail the nature of uranium radiation and the electrical conduction produced. . . . Most of the results obtained have been interpreted on the ionization theory of gases which was introduced to explain the electrical conduction produced by Röntgen radiation [x rays] . . . (1).

Rutherford used the conductivity of air to investigate radioactivity by determining the reaction to radioactivity of the air between two electrically charged, parallel metal plates (a parallel plate *capacitor*; see Figure 13-2). He placed a battery and an electrometer across the plates, and when the air between the plates was made conducting, an electric current was set up in the circuit which consisted of the two plates, the conducting air, the electrometer, and the battery. The current was measured by the electrometer (Figure 13-3). Rutherford also relied on the fact that the effect of radioactivity is diminished if the radiation is first directed through metal sheets.

In order to test the complexity of the radiation, an electrical method was employed. The general arrangement is shown in Fig. [13-3].

The metallic uranium or compound of uranium to be employed was powdered and spread uniformly over the center of a horizontal zinc plate *A*, 20 cm square. A zinc plate *B*, 20 cm square, was fixed parallel to *A* and 4 cm from it. Both plates were insulated. *A* was connected to one pole of a battery of 50 volts, the other pole of which was [connected] to earth; *B* was connected to one pair of quadrants of an electrometer, the other pair of which was connected to earth.

Under the influence of the uranium radiation there was . . . a leak [an electrical current] between the two plates *A* and *B*. The rate of movement of the electrometer needle, when the motion was steady, was taken as a measure of the current through the gas.

Successive layers of thin metal foil were then placed over the uranium compound and the rate of leak determined for each additional sheet. . . .

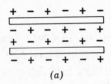

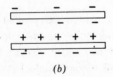

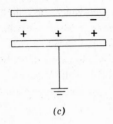

(c)

FIGURE 13-2

(a) Two metal plates separated by an air space form a capacitor. (b) If an excess of one charge— in this instance negative—is placed on the top plate, the electrons in the bottom plate will be repelled to opposite sides. (c) If the bottom plate is attached to ground those electrons will escape to ground, leaving the bottom plate with a net positive charge and the upper plate with a net negative charge.

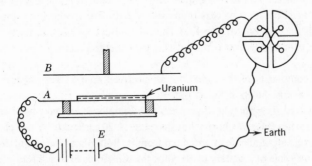

FIGURE 13-3

One of the early ionization chambers built by Rutherford. Radiation from a radioactive substance placed between the two plates caused a brief electric current to pass from one plate to the other. The electrometer in the upper right measured that current.

354

Rutherford found (see Table 13-1) that

the rate of leak [the electric current] diminishes in a geometrical progression with the thickness of metal . . . [and consequently] we see . . . that the intensity of the radiation falls off in a geometrical progression, i.e. according to an ordinary absorption law. This shows that the part of the radiation considered is approximately homogeneous.

Table 13-1   Absorption of Radiation (Metal leaf 0.00008 cm thick)

Number of Layers	Rate of Leak (Electric Current)	Ratio of Current for Successive Layers
0	91	0.85
1	77	0.78
2	60	0.82
3	49	0.86
4	42	0.79
5	33	0.75
6	24.7	0.79
8	15.4	0.77
10	9.1	0.86
13	5.8	

With increase of the number of layers the absorption commences to diminish. This is shown more clearly by using uranium oxide with layers of thin aluminum leaf [see Table 13-2].

It will be observed that for the first three layers of aluminum foil, the intensity of radiation falls off according to the ordinary absorption law, and that, after the fourth thickness, the intensity of the radiation is only slightly diminished by adding another eight layers.

Table 13-2   Identification of Complex Radiation (Aluminum leaf 0.0005 cm thick)

Number of Layers	Rate of Leak (Electric Current)	Ratio of Current for Successive Layers
0	182	0.42
1	77	0.43
2	33	0.44
3	14.6	0.65
4	9.4	
12	7	

The aluminum foil in this case was about 0.0005 cm. thick, so that after the passage of the radiation through 0.002 cm. of aluminum the intensity of the radiation is reduced to about one-twentieth of its value. The addition of a thickness of 0.001 cm. of aluminum has only a small effect in cutting down the rate of leak. The intensity is, however, again reduced to about half of its value after passage through an additional thickness of 0.05 cm., which corresponds to one hundred sheets of aluminum foil.

These experiments show that the uranium radiation is complex, and that there are present at least two distinct types of radiation—one that is very readily absorbed, which will be termed for convenience the α radiation, and the other of more penetrative character, which will be termed the β radiation (2).

The two separate rays, the alpha and the beta rays, were found by Rutherford because their penetrabilities of matter were different (Figure 13-4). This observation seems simple enough, yet its consequences are great. It was soon learned that the beta-ray beam can be deflected by a magnetic field and also by an electric field. The deflection is in the same direction as that of the cathode ray beam studied by J. J. Thomson. In fact, in 1900 Becquerel showed that the beta-ray beam has the same charge-to-mass ratio as the "electrons" only recently discussed by Thomson. It became clear that the beta ray is, in fact, a stream of electrons.

A French physicist, P. Villard (1860–1934), also working in 1900, identified a third type of radiation, called gamma rays, which cannot be deflected by either a magnetic field or an electric field. These gamma rays have been assigned the position at the short-wavelength end of the electromagnetic spectrum. They proved to be the most penetrating of the three kinds of rays.

Identification of Alpha Particles

Only with some difficulty was the nature of the alpha ray learned, however, Rutherford wrote in 1903 that

The determination of the mass of the α body, taken in conjunction with the experiments on the production of helium by . . . [radioactive compounds], supports the view that the α particle is in reality helium. In addition, the remarkable experiment of Sir William and Lady Huggins in which they found that the spectrum of the phosphorescent light of radium consisted of bright lines, some of which within the limits of error were coincident with the lines of helium in the ultra-violet, strongly supports such a view (3).

By 1908, after much experimentation, Rutherford could write

The experimental evidence collected during the last few years has strongly supported the view that the α-particle is a charged helium atom, but it has been found exceedingly difficult to give a decisive proof of the relation. In recent papers, Rutherford and Geiger have supplied still further evidence of the correctness of this point of view. The number of α-particles from one gram of radium have been counted, and the charge carried by each determined. The values of several radioactive quantities, calculated on the assumption that the α-particle is a helium atom carrying two unit charges, have been shown to be in good agreement with the experimental numbers. . . .

FIGURE 13-4

The tracks of alpha particles of two distinct energies are seen in this photograph from one of the early Wilson cloud chambers.. The more energetic particles are the more penetrating. (Reprinted with permission from *Radiations from Radioactive Substances* by Rutherford, Chadwick, and Ellis, Cambridge at the University Press, 1951.)

The methods of attack on this problem have been largely indirect, involving considerations of the charge carried by the helium atom and the value of e/m of the α-particle. . . .

We have recently made experiments to test whether helium appears in a vessel into which the α-particles have been fired, the active matter itself being enclosed in a vessel sufficiently thin to allow the α-particles to escape, but impervious to the passage of helium [by diffusion processes] or other radioactive products . . . (4).

These experiments thus show conclusively that the helium could not have diffused through the glass walls [of the vessel], but must have been derived from the α-particles which were fired through them. In other words, the experiments give a decisive proof that the α-particles after losing its charge is an atom of helium (5).

Had Rutherford written only a few years later, after his own work and that of Niels Bohr on the structure of the atom had been completed, he might not have said "after losing its charge . . ." but "after picking up two electrons." The alpha particle is the nucleus of the helium atom. Rutherford's work became very closely associated with the alpha particle since, as we have seen, he used it to great advantage in his discovery of the nucleus of the atom.

TECHNIQUES OF OBSERVATION

When using alpha particles as projectiles with which to probe the secrets of the atom, Rutherford needed many different techniques for observing these alpha particles. His measurements of penetrability through matter and of the ability of radiation to discharge a capacitor have been mentioned. Alpha particles were shown to carry a positive electric charge by their deflection; when traveling in an electric or a magnetic field they are deflected in the opposite direction from that of electrons.

Photography

The beams that Rutherford deflected were generally observed by the darkening effect alpha particles (and beta particles) have on a photographic emulsion. For convenience of observation and accuracy of measurement, a narrow beam of alpha particles is used, producing a line on the photographic emulsion instead of a spot. The beam is made narrow by placing the radioactive source on a wire and then using only the radiation that passes through two successive narrow slits, each parallel to the wire.

Scintillations

Other materials besides photographic emulsions indicate the location of alpha particles. Roentgen discovered that certain chemicals, such as zinc sulfide, emit light when radiated by x rays (see p. 249). Zinc sulfide will also emit light when a beam of alpha particles (or beta particles) strikes it. As each particle hits a surface coated with a layer of this material, there is a brief flash of light. Each flash of light is called a *scintillation*.

Scintillations are easy to see, but counting them manually is tedious; better methods of counting have been devised. Today the flashes of light are "seen" by photoelectric cells, cells based on the photoelectric effect. The electric currents produced by the photoelectric cells are registered by a recording galvanometer. Not only are the scintillations counted automatically, but a permanent record is made and may be referred to at any later date for further study or verification.

The Wilson Cloud Chamber

Another of the very productive workers in Rutherford's laboratory, C. T. R. Wilson (1869–1959), Scottish physicist and Nobel prizewinner, discovered a dramatic way of observing both alpha and beta particles. While investigating the possibilities of producing fog and clouds in the laboratory, he found that when moist, cool air is suddenly cooled even further, tiny particles of water condense if centers of "impurities" upon which the water may collect are provided.

The moist air can be cooled suddenly by expanding it in a chamber above a piston. When the piston is suddenly lowered, the air expands and cools, and tiny clouds form. In such a device, known as the *Wilson cloud chamber,* the tiny particles upon which the water condenses may be simple dust particles, or as Wilson discovered, they may be charged particles. He found, for instance, that water particles will condense on molecules ionized by x rays.

In 1911, however, Wilson noticed that alpha and beta particles leave a track in the cloud chamber if they happen to pass through the chamber as it is expanding (Figure 13-4). As either of these high-speed particles pass through the air, they ionize molecules by collision. The stream of ionized molecules provides numerous particles upon which the water can condense. This streak of tiny water particles is not unlike the vapor trail left by a high-flying jet plane.

The Geiger Tube

The trail of ionized gas molecules led Rutherford and two other colleagues, Hans Geiger and Ernest Marsden (see p. 325), to invent still another means of detecting the passage of either alpha or beta particles or, in fact, even of gamma rays. The final instrument, called a Geiger tube, consists of a fine wire running down the axis of a metal cylinder which is enclosed in a glass tube. An appropriate amount of air has been evacuated from the glass tube. An electrical potential difference of about 1000 volts is placed across the wire and the cylinder, with some device for detecting an electric current through

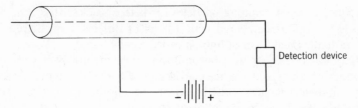

FIGURE 13-5

A schematic diagram of a Geiger tube. The wire carries a negative charge, the cylinder a positive charge.

the circuit (Figure 13-5). As long as the gas in the tube is nonconducting, there is no current. A high-energy particle passing through the tube ionizes some of the molecules of the gas, however, and the electrons freed by this ionization are attracted to the positive wire along the center of the tube. Since the electrons are in a strong electric field, they accelerate and very quickly acquire a velocity great enough to ionize still more gas molecules. More freed electrons cause further ionization, producing an avalanche of electrons. These electrons strike the central wire and set up a current in the circuit. This current in turn is detected by a device that may be as simple as an earphone; a click is heard when an avalanche of electrons occurs. Or the detection device may be a counting mechanism; hence the name Geiger counter.

ALPHA PARTICLES DISRUPT THE NUCLEUS

The Geiger counter, the Wilson cloud chamber, and the instruments developed from them have become invaluable tools of the physicist. The simpler method of scintillation counting, however, informed Rutherford that he had disrupted and altered the nucleus of a nitrogen atom.

Rutherford, while investigating the effect of directing a beam of alpha particles through a gas containing such elements as nitrogen, hydrogen, oxygen, and so on, picked up spurious scintillations, ones he could not readily explain. He found that these spurious scintillations occurred when he bombarded nitrogen gas with alpha particles. The objects causing the scintillations were shown to have a positive charge by their reaction in a magnetic field, but they traveled much farther through air than alpha particles. After intensive study of this phenomenon, Rutherford stated in 1919 that

it is difficult to avoid the conclusion that the long-range atoms arising from collision of α-particles with nitrogen are not nitrogen atoms but probably atoms of hydrogen, or atoms of mass 2. If this be the case, we must conclude that the nitrogen atom is disintegrated under the intense forces developed in a close collision with a swift α-particle, and that the hydrogen atom which is liberated formed a constituent part of the nitrogen nucleus. . . .

Taking into account the great energy of motion of the α-particle expelled from radium C, the close collision of such an α-particle with a light atom seems to be the most likely agency to promote the disruption of the latter; for the forces on the nuclei arising from such collisions appear to be greater than can be produced by any other agency at present available. Considering the enormous intensity of forces brought into play, it is not so much a matter of surprise that the nitrogen atom should suffer disintegration as that of the α-particle itself escapes disruption into its constituents. The results as a whole suggest that, if α-particles—or similar projectiles—of still greater energy were available for experiment, we might expect to break down the nucleus structure of many of the lighter atoms (6).

This was the first announcement of the "splitting" of an atomic nucleus. Further investigations proved Rutherford's predictions correct, and projectiles with very high energy are now used to explore the subatomic world.

CHADWICK AND THE NEUTRON

In 1919, just before discovering that the nitrogen nucleus can be disrupted by bombarding alpha particles, Rutherford assumed the directorship of the Cavendish Laboratories in Cambridge, a famous and productive laboratory to this very day. He was also invited (for the second time) to give the well-known Bakerian Lecture to the Royal Society. In this 1920 lecture, Rutherford stated quite clearly that it ought to be possible for a particle with the same (or nearly the same) mass as the proton but with no electric charge to exist. That particle was discovered in the Cavendish laboratories by James Chadwick (b. 1891), English physicist and Nobel prizewinner, in 1932.

Chadwick had studied under Rutherford in Manchester and joined him as soon as Rutherford went to Cambridge. Proceeding with Rutherford's technique of bombarding atoms with alpha particles, Chadwick used a radioactive source called radium F (or polonium) which emits only alpha particles with no interference from beta particles.

The polonium was coated on a metal disk in front of which was placed a disk of beryllium (Figure 13-6). The results of the bombardment, observed with an ionization chamber (similar to a Geiger tube), proved rather startling. Whatever the form of radiation produced by this bombardment of beryllium atoms with alpha particles, it was extremely penetrating. It could be detected after traveling through an inch of lead.

Suspecting that the radiation might be composed either of the neutral particles predicted by Rutherford or of what he himself called high-energy quantum radiation, that is, gamma-ray photons, Chadwick devised an experiment in which the powerful principles of conservation of energy and momentum would be brought to bear. He placed a piece of paraffin between the disk of beryllium and the ionization chamber and noticed that the counting rate of the chamber increased. Apparently, the mysterious particles

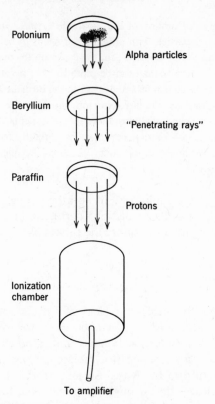

Polonium

Alpha particles

Beryllium

"Penetrating rays"

Paraffin

Protons

Ionization
chamber

To amplifier

FIGURE 13-6

Alpha particles from the polonium bombard the beryllium resulting in the ejection of neutrons ("penetrating rays"), which in turn knock protons from the paraffin. These protons are easily observed by the ionization chamber.

coming from the beryllium ejected particles from the paraffin. These secondary particles from the paraffin were presumably more easily detected by the ionization chamber.

The object of the test devised by Chadwick was to calculate the total mass and kinetic energy as well as the momentum of the original alpha particles, and the total mass, the kinetic energy, and the momentum of the secondary particles from the paraffin. From these calculations he hoped to determine what passed from the beryllium to the paraffin; the particle must have a mass, kinetic energy, and momentum that permit the total energy and momentum for the entire reaction to be conserved.

His supposition that these particles were high-energy gamma-ray photons led him to conclude that

it is evident that we must either relinquish the application of the conservation of energy and momentum in these collisions or adopt another hypothesis about the nature of the radiation. If we suppose that the radiation is not a quantum radiation, but consists of particles of mass very nearly equal to that of the proton, all the difficulties connected with the collisions disappear, both with regard to their frequency and to the energy transfer to different masses. In order to explain the great penetrating power of the radiation we must further assume that the particle has no net charge. We may suppose it to . . . [be] the "neutron" discussed by Rutherford in his Bakerian Lecture in 1920.

When such neutrons pass through matter they suffer occasionally close collisions with the atomic nuclei and so give rise to the recoil atoms which are observed. Since the mass of the neutron is equal to that of the proton, the recoil atoms produced when the neutrons pass through matter containing hydrogen [in the paraffin] will have all velocities up to a maximum which is the same as the maximum velocity of the neutrons. The experiments showed that the maximum velocity of the protons ejected from paraffin was about 3.3×10^9 cm/sec. This is therefore the maximum velocity of the neutrons emitted from beryllium bombarded by $\propto$-particles of polonium. From this we can now calculate the maximum energy which can be given by a colliding neutron to other atoms, and we find that the results are in fair agreement with the energies observed in the experiments (7).

Chadwick's decision to rely on the principles of conservation of energy and of momentum permitted him to discover a third elementary particle.

RADIOACTIVITY AND THE NUCLEUS

During all these years Rutherford had been working on other projects, one of which was also very important to the future development of physics.

Many investigators had helped determine that a number of different chemical elements are involved in the radioactive process, for example, uranium, radium, thorium, and others, and that these can be separated out and identified. But the relationship between these various chemical elements and radioactivity was obscure until 1902 when Rutherford and his colleague F. Soddy explained it. Referring to their experimental observations, they wrote

The foregoing experimental results may be briefly summarized. The major part of the radioactivity of thorium—ordinarily about 54 percent—is due to a non-thorium type of matter, ThX, possessing distinct chemical properties, which is temporarily radio-active, its activity falling to half value in about four days. The constant radioactivity of thorium is maintained by the production of this material [the thorium X] at a constant rate. Both the rate of production of the new material and the rate of decay of its activity appear to be independent of the physical and chemical condition of the system . . . [Figure 13-7].

Turning from the experimental results to their theoretical interpretation, it is necessary to first consider the generally accepted view of the nature of radioactivity. It is well established that this property is the function of the atom and not of the molecule. Uranium and thorium, to take the most definite cases, possess the property in whatever molecular condition they occur, and the former also in the elementary state. So far as the radioactivity of different compounds of different density and states of division can

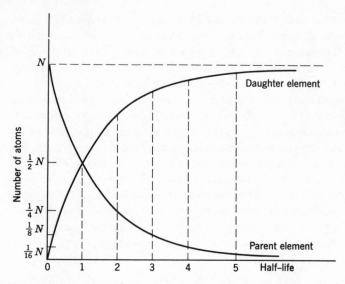

FIGURE 13-7

As one radioactive substance, the "parent element," decays into another, the "daughter element," the total number of atoms N remains the same.

be compared together, the intensity of the radiation appears to depend only on the quantity of active element present. . . .

All the most prominent workers in this subject are agreed in considering radioactivity an atomic phenomenon. M. and Mme. Curie, the pioneers in the chemistry of the subject, have recently put forward their views. They state that this idea underlies their whole work from the beginning and created their methods of research. M. Becquerel, the original discoverer of the property of uranium, in his announcement of the recovery of the activity of the same element after the active constituent had been removed by chemical treatment, points out the significance of the fact that uranium is giving out cathode rays. These, according to the hypothesis of Sir William Crookes and Professor J. J. Thomson, are *material* particles of mass one thousandth of the hydrogen atom [more accurately $1/1836$].

Since, therefore, radioactivity is at once an atomic phenomenon and accompanied by chemical changes in which new types of matter are produced, these changes must be occurring within the atom, and the radioactive elements must be undergoing spontaneous transformation. The results that have so far been obtained, which indicate that the velocity of this reaction is unaffected by the conditions, make it clear that the changes in question are different in character from any that have been before dealt with in chemistry. It is apparent that we are dealing with phenomenon outside the sphere of known atomic forces. Radioactivity may therefore be considered as a manifestation of subatomic chemical change (8).

This final statement demonstrates the insight and intuition of a great man. Rutherford was already speaking of subatomic processes in 1902, fully nine years before he unexpectedly proved the existence of the atomic nucleus, subsequently identified as the very place where these subatomic processes occur.

Speaking of the various chemical elements produced, Rutherford wrote in 1903 that

these results find their explanation if it is supposed that the α-particles projected form integral portions of the atom of the radioactive element. Thus ThX is thorium minus one or more projected α-particles. The emanation [later shown to be still another chemical element, a gas now called either radon or thorium emanation] similarly is ThX less a further α-particle, and so on. The nonseparable activity is due to the atoms of the original radio-element disintegrating at a constant rate. The whole of the processes take place unaltered in velocity, agitation, and chemical combination. This is to be expected of a subatomic change in which one system only is involved at each change. On this view the spontaneous heat-emission of solid radium-salts, discovered by Curie, is explained by the internal bombardment by the α-particles shot off and absorbed in the mass of the substance (9).

It had become clear that there is a radioactive series. For example, uranium decays into another element by emitting an alpha particle, the second element decays still further by radioactive processes into a third element, and so on.

By 1914 Rutherford wrote:

Soddy has pointed out that the recent generalizations of the relation between the chemical properties of the elements and the radiation can be interpreted by supposing that the atom loses two positive charges by the expulsion of an α-particle, and one negative by the expulsion of a high speed electron [beta particle]. From a consideration of the series of products of the three main radioactive branches of uranium, thorium, and actinium, it follows that some of the radioactive elements may be arranged so that the nucleus charge decreases by one unit as we pass from one element to another (10).

Evidence that the electric charge on the nucleus distinguishes radioactive elements one from the other was accumulating. If this was true of the radioactive elements, all of which are very massive, it must also be true of the lighter elements. Rutherford referred to work done by H. G. J. Moseley (1887–1915), one of the brilliant students who worked with Rutherford at the University of Manchester. Moseley obtained spectra of x rays emitted by various substances (Figure 13-8) and found that

by examination of the wavelength of the characteristic x-rays emitted by twelve elements varying in atomic weight between calcium (40) and zinc (65.4) he [Moseley] has shown that the variation and wavelength can be simply explained by supposing that the charge on the nucleus increases from element to element by exactly one unit. This holds true for cobalt and nickel, although it has long been known that they occupy an anomalous relative position in the periodic classification of the elements according to atomic weight . . . [see Appendix D] (8).

It is clear on the nucleus theory that the physical and chemical properties of the ordinary elements are for the most part dependent entirely on the charge of the nucleus,

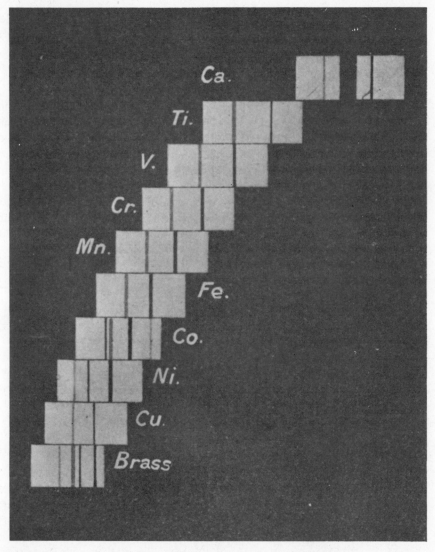

FIGURE 13-8

The x-ray spectral lines of each of these elements, copper, nickel, cobalt, etc., form an obvious sequence when arranged according to wavelength. Moseley concluded correctly that the nuclei of these elements have successively one more positive charge. (H. G. J. Moseley, *Philosophical Magazine*, Vol. 26, Series 6, Plate XXIII, 1913)

for the latter determines the number and distribution of the external electrons on which the chemical and physical properties must mainly depend (10).

The energy levels of the heavier elements have been grouped into what are often called "shells." The names for the shells, *K, L, M,* and so on, derive from the designations Moseley gave the various series of x-ray spectra, *K*, *L*, *M*, and so on, when he discovered them. An x-ray photon is emitted by one of these heavier elements when an electron makes a downward transition from one shell to another. Many such photons produce an x-ray spectral line, and different transitions produce an x-ray spectrum.

THE NUCLEUS

The relationship between nuclear charge and the distinctness of each chemical element became clear when it was discovered that each element has a different nuclear charge. This means that the various neutral atoms have different numbers of electrons in the electron shells surrounding the nucleus. Thus neutral carbon with a charge of 6^+ on the nucleus must have six electrons with a net charge of 6^- in the surrounding electron shells. The chemical properties of the elements depend on the number and location of the electrons in the electron shells. With this realization, the periodic table of the chemical elements was better understood (see Appendix D).

The Nucleons

To account for the positive charge on the nucleus changing by one from one element to the next in the periodic table, the number of protons in the nucleus is assumed to change by one from one element to the next. Thus the hydrogen nucleus has one proton, the helium nucleus two, the lithium nucleus three, beryllium four, and so on, to as many elements as the table lists. The electric charge on the nucleus is called the *atomic number*.

To account for the mass of the nucleus changing in a manner different from that of the charge, the number of neutrons in the nucleus is also assumed to change, but in a less systematic manner. For example, the hydrogen nucleus is composed of one proton only, but there is also another hydrogen nucleus that has one proton and one neutron. Because the two forms of the hydrogen nucleus each have only one proton, each needs to attract only one electron to become electrically neutral. Therefore, these two forms of hydrogen have the same chemical properties, occupy the same place in the periodic table, and are called *isotopes* (Greek: *isos*, equal; *topos*, place).

The hydrogen with both a proton and a neutron in its nucleus is called *deuterium*. A third isotope of hydrogen, with one proton (of course) and two neutrons in its nucleus, is called *tritium* and is radioactive.

Because deuterium has both a proton and a neutron in its nucleus, its mass is greater than that of regular hydrogen, but it still has the same charge. Since its charge is the same and the mass is different, its charge-to-mass ratio is different from that of regular hydrogen. The proton and the neutron, because of their role in the nucleus, are known collectively as *nucleons*.

Atomic Mass Unit

In order to measure the mass of such minute particles, a new unit of mass has been chosen. Clearly the kilogram is too large to be convenient. By agreement the mass of the carbon atom which has six protons and six neutrons in its nucleus constitutes the standard. The unit of mass, the *atomic mass unit* (amu) is one twelfth the mass of the carbon atom.

The mass, in grams, of the carbon atom with twelve nucleons in its nucleus, called carbon-12, can be determined by performing experiments similar to Thomson's determination of the charge-to-mass ratio of the electron. One electron can be removed by bombarding the carbon atoms with free electrons of just the right amount of energy. The carbon atoms with one electron missing carry a net electric charge and are called *ions*. These carbon ions can then be accelerated through a potential difference and their mass determined by sending them through electric and magnetic fields of known field strength and following their motions. The mass of the carbon-12 atom has been determined to be 1.99×10^{-23} gram.

The mass of one carbon atom is also 12 amu, therefore

$$12 \text{ amu} = 1.99 \times 10^{-23} \text{ g}$$

and

$$1 \text{ amu} = 1.66 \times 10^{-24} \text{ g}$$

or in kilograms

$$1 \text{ amu} = 1.66 \times 10^{-27} \text{ kg}$$

The proton and the neutron have masses of very nearly one amu.

The proton mass is, strangely enough, not exactly one. It is 1.007 276 amu. The mass of the electron is 0.000 549 amu. Hence the mass of the hydrogen atom is 1.007 825 amu. The mass of the deuterium atom is 2.014 103 amu, that of tritium 3.016 049 amu, and that of beryllium 9.012 186 amu.

To simplify numbers, another term, the *mass number*, which is the number of protons and neutrons that together comprise the nucleus, has come into general use. The mass number designates the number of nucleons in a given nucleus.

The pertinent data concerning nuclear charge and mass are conveniently written in symbolic form. Hydrogen with two nucleons is written as $^{2}_{1}\text{H}$; the subscript is the charge on the nucleus, the superscript the mass number.

Radioactive Series

The concept that the atomic nucleus is composed of both protons and neutrons finally explained the radioactive series that had been unfolded by Rutherford and others. For example, the uranium series begins with a uranium atom that has a mass number of 238 and an atomic mass of 92. This isotope is written as $^{238}_{92}\text{U}$, and it decays radioactively by emitting an alpha particle. The reaction may be written

$$^{238}_{92}\text{U} \rightarrow {}^{4}_{2}\text{He} + {}^{234}_{90}\text{Th}$$

Half of the uranium-238 will decay in 4.5×10^9 years. The period of time in which one half the radioactive substance will decay is called the *half-life*. Each radioactive substance has a distinct half-life which, as Rutherford pointed out, is constant and unique to that isotope.

Electric charge is shown to be conserved in this reaction by the fact that the charge on the uranium nucleus 92 equals the charge on both the helium and thorium nucleus, $2 + 90$. The mass numbers, $238 = 4 + 234$, show that the number of particles is conserved.

But thorium-234 is also radioactive, and it decays by giving off a beta particle. Its half-life is 24.1 days. The reaction is written as

$$^{234}_{90}\text{Th} \rightarrow {}^{0}_{-1}\beta + {}^{234}_{91}\text{Pa}$$

By losing one negative charge from the nucleus, the positive charge increases by one, $90 = -1 + 91$. The nucleus changes from thorium into protactinium. The electron has a zero mass number.

But protactinium-234 is also radioactive and decays by giving off a beta particle. Its half-life is 1.18 minutes or 6.7 hours, depending on the energy of the beta particle. This reaction is written

$$^{234}_{91}\text{Pa} \rightarrow {}^{0}_{-1}\beta + {}^{234}_{92}\text{U}$$

And so we are back to uranium, but a different isotope. Uranium-234, which decays by emitting an alpha particle, has a half-life of 2.5×10^5 years.

$$^{234}_{92}\text{U} \rightarrow {}^{4}_{2}\text{He} + {}^{230}_{90}\text{Th}$$

Thorium-230, which also decays by emitting an alpha particle, has a half-life of 8.0×10^4 years.

$$^{230}_{90}\text{Th} \rightarrow {}^{4}_{2}\text{He} + {}^{226}_{88}\text{Ra}$$

And thus radium is formed. But radium-226 is also radioactive; it decays into another radioactive substance, which in turn decays into still another and another, until finally lead-206 ($^{206}_{82}\text{Pb}$) is reached. This nucleus is stable and does not decay; it is the end product. The entire radioactive sequence is shown in Figure 13-9 and Table 13-3.

Other radioactive sequences begin with other naturally found radioactive substances, thorium-232 and actinium-235. These series end in lead-208 and lead-207, respectively. In each series lead is the stable element, the end product. But why should one nucleus be unstable and decay by emitting an alpha particle, another nucleus be unstable and decay by emitting a beta particle, and still another nucleus be stable?

Nuclear Binding Energy

For a long time the unanswerable question was why is any nucleus stable at all? How can a helium nucleus, for instance, be stable if it contains two

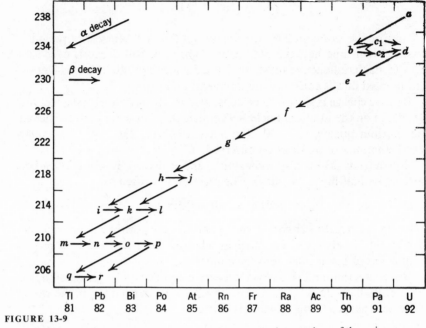

FIGURE 13-9

The uranium 238 series. The *a*, *b*, *c*, . . . refer to successive members of the series.

Table 13-3   The Uranium Series

	Nuclide	Type of Disintegration	Half-life ^{a}	Principal Particle Energy (Mev)
(*a*)	$^{238}_{92}$U	α	4.5×10^9 y	α, 4.2; γ, 0.05
(*b*)	$^{234}_{90}$Th	β	24.1 d	β, 0.2; γ, 0.03–0.09
(c_1)	$^{234}_{91}$Pa	β	1.18 m	β, 2.3; γ, 1.0
(c_2)		β	6.7 h	β, 0.5; γ, 0.10–0.57
(*d*)	$^{234}_{92}$U	α	2.5×10^5 y	α, 4.8; γ, 0.02–0.12
(*e*)	$^{230}_{90}$Th	α	8.0×10^4 y	α, 4.7; γ, 0.07–0.18
(*f*)	$^{226}_{88}$Ra	α	1620 y	α, 4.8; γ, 0.19–0.66
(*g*)	$^{222}_{86}$Rn	α	3.82 d	α, 5.5; γ, 0.51
(*h*)	$^{218}_{84}$Po	α, β	3.05 m	α, 6.0
(*i*)	$^{214}_{82}$Pb	β	26.8 m	β, 0.65; γ, 0.05–0.35
(*j*)	$^{218}_{85}$At	α	1.5–2.0 s	α, 6.6
(*k*)	$^{214}_{83}$Bi	α, β	19.7 m	α, 5.4; β, 1.65; γ, 0.61–2.43
(*l*)	$^{214}_{84}$Po	α	1.65×10^{-4} s	α, 7.7
(*m*)	$^{210}_{81}$Tl	α	1.32 m	β, 1.8
(*n*)	$^{210}_{82}$Pb	β	22 y	β, 0.03; γ, 0.05
(*o*)	$^{210}_{83}$Bi	α, β	5.0 d	α, 4.8; β, 1.17
(*p*)	$^{210}_{84}$Po	α	138.3 d	α, 5.3; γ, 0.89
(*q*)	$^{206}_{81}$Tl	β	4.2 m	β, 1.51
(*r*)	$^{206}_{82}$Pb	stable		

^{a}Half-lives are designated in years (y), days (d), hours (h), minutes (m), and seconds (s).

protons? Why does the Coulomb force of repulsion not make the nucleus fly apart? Some other force must bind the nucleus together.

The results of Rutherford's work with the scattering of alpha particles indicate that, whatever the nuclear force holding the nucleus together may be, it must indeed be a short-range force, for it makes itself felt over only about 10^{-15} meter. Since the entire atom has a diameter of about 10^{-10} meter, the nucleus has a diameter only 1/100,000 that of the atom. The Coulomb force has a much longer range, but since the protons do remain bound in a nucleus, the Coulomb force must not be as strong as the nuclear binding force at very short distances.

We can gain some idea of the strength of this nuclear binding force if we examine the constituents of the nucleus more closely. The helium nucleus is composed of two protons and two neutrons. A proton has a mass of 1.007 267 amu, a neutron a mass of 1.008 665 amu. Hence two protons and two neutrons taken separately have a total mass of 4.031 882 amu. But an alpha particle, the nucleus of the helium atom, is 4.001 506 amu.

If we were to build a helium nucleus from the basic particles, we would apparently lose some mass (that is, rest mass). But the conservation principles make us look suspiciously at such an outcome. Yet there is every evidence that helium is in fact being made from hydrogen in the very hot central regions of the stars. How can this be? Special relativity makes it clear that the mass which appears to be lost has simply been converted into another form of energy:

$$E = \Delta mc^2$$

Apparently during the process of forming a helium nucleus, mass is converted into other forms of energy, such as electromagnetic energy (as photons) and kinetic energy of the nuclei and other particles involved. To break the nucleus into its four constituent particles, an amount of energy equal to that produced by the conversion of mass in the formation process must be supplied in one form or another. That is, work must be done on the four nucleons, two protons and two neutrons, to separate them.

But by what units of measure are we to designate this amount of energy? If we determine the loss of mass in kilograms, the amount of energy will be determined in joules, but the numbers will be so small as to be cumbersome. Consequently, more convenient units of energy have been devised.

Because bombarding particles must be used to break up nuclei, it has become convenient to establish a unit of energy in terms of the energy of the bombarding particle. The electron is a good standard. The amount of kinetic energy gained by an electron falling through an electrical potential difference of one volt is the standard and is called the *electron volt* (ev).

The electron volt can be related to the mks system of units in a very simple way. The kinetic energy gained by an electric charge falling through a potential difference of V is given by (see p. 236)

$$\Delta E = QV$$

Since the ΔE is the increase in kinetic energy E_k, and the charge $Q = e$, we can write the equation

$$E_k = eV$$

The charge e is 1.6×10^{-19} coulomb, and V is one volt, so the kinetic energy gained must be

$$E_k = (1.6 \times 10^{-19})(1) \text{ joule}$$

$$E_k = 1.6 \times 10^{-19} \text{ joule}$$

Hence one electron volt equals 1.6×10^{19} joule.

The amount of mass lost in any particular reaction can be expressed as energy. The amount of energy equivalent to 1 amu (1.66×10^{-27} kg) is given by the relationship

$$E = mc^2$$

$$E = (1.66 \times 10^{-27})(3.0 \times 10^8)^2$$

where the velocity of light

$$c = 3.0 \times 10^8 \text{ m/sec}$$

Therefore

$$E = 1.5 \times 10^{-11} \text{ joule}$$

This amount of energy may not seem like much, but it is the amount released by the conversion of only 1 amu of matter into other forms of energy. How much is converted from a reasonable amount of hydrogen gas? The mass of one atom of hydrogen is 1.66×10^{-24} gram, so one gram of hydrogen atoms must contain 6.0×10^{23} atoms.* Hydrogen gas, however, is usually composed of hydrogen molecules, each of which consists of two hydrogen atoms. Therefore, 2 grams of hydrogen gas must contain 6.0×10^{23} hydrogen molecules. Consequently, the amount of energy released by the complete conversion of 2 grams of hydrogen gas into energy would be

$$E = (1.5 \times 10^{-11})(2)(6.0 \times 10^{23})$$

$$E = 18 \times 10^{12} \text{ joules}$$

And this is indeed a prodigious amount of energy.

The energy equivalent of 1 amu can also be expressed in electron volts. One atomic mass unit is equivalent to 1.5×10^{-11} joule and one electron volt equals 1.6×10^{-19} joule, so 1 amu expressed as electron volts must be

$$1 \text{ amu} = (1.5 \times 10^{-11} \text{ joule}) \frac{1 \text{ ev}}{1.6 \times 10^{-19} \text{ joule}}$$

$$1 \text{ amu} = 0.9 \times 10^8 \text{ ev}$$

More precisely,

$$1 \text{ amu} = 931 \times 10^6 \text{ ev}$$

* This number 6.0×10^{23} is called Avogadro's number.

The conversion of 1 amu yields an amount of energy equivalent to letting one electron fall through a potential difference of 931 million volts! One million electron volts is called an Mev, and one billion electron volts used to be called a Bev. The term billion has different meanings, however. In England and most other European countries it means 10^{12}, that is, a million million. In the United States and in France a billion is equal to a thousand million. To avoid confusion, the term *giga* has been adopted to stand for 10^9 or what we in this country call a billion. Consequently, 10^9 electron volts is now called a Gev, not a Bev.

To return to our example of the four nucleons of the helium nucleus, taken separately they have a mass of 4.031 882 amu; the four combined into the helium nucleus have a mass of 4.001 506 amu. Hence the loss of mass is 0.030 376 amu. This loss of mass is released as energy, and the amount can be determined by the following proportion (rounding off 0.030 376 to 0.0304 and recalling that 1 amu = 931 Mev):

$$\frac{E}{0.0304 \text{ amu}} = \frac{931 \text{ Mev}}{1 \text{ amu}}$$

Solving for E we find that

$$E = (9.31 \times 10^2)(3.04 \times 10^{-2}) \text{ Mev}$$
$$E = 28.3 \text{ Mev}$$

An electron would therefore have to fall through a potential difference of 28.3 million volts before it acquires enough energy to shatter the nucleus of the helium atom into its four constituent parts.

On the average 28.3/4 or 7.1 Mev *per nucleon* would be required to shatter the helium nucleus. But much more than 7.1 Mev would be required to remove just one proton or neutron. Removing one neutron would produce an isotope of helium, $^3_2$He. The total mass of the helium-3 atom (electrons included) is 3.016 030 amu, the mass of the freed neutron 1.008 665 amu. The sum of these two is 4.024 695 amu. The mass of the helium-4 atom (electrons included) is 4.002 604 amu. Consequently, there is a difference of mass between the helium-4 atom and the two resulting pieces—the helium-3 atom and the neutron—of 0.022 091 amu. (The electron masses cancel out by taking the difference.)

The amount of energy equivalent to 0.022 091 amu is

$$2.21 \times 10^{-2} \text{ amu} \times 9.31 \times 10^2 \text{ Mev/amu} = 21 \text{ Mev}$$

It would therefore require a minimum of 21 Mev of energy to remove a single neutron from the nucleus of the helium atom. Similar calculations will show that 21 Mev are required to remove a proton as well. But if either the proton or neutron is to have any kinetic energy once it is knocked out of the nucleus, more than 21 Mev of energy must be supplied.

Now it can be shown why, to the surprise of Rutherford, the alpha particle remained intact after bombarding the nitrogen atom. It was the nitrogen atom that lost the proton.

By losing a proton, nitrogen-14 is converted into carbon-13, $^{13}_{6}C$, a stable isotope of carbon. The mass of the nitrogen atom is 14.003 074 amu, the mass of the carbon-13 atom 13.003 354 amu; the mass of the hydrogen atom (proton removed plus one electron) is 1.007 825 amu. The total mass of the resultant pieces is 13.003 354 + 1.007 825 = 14.001 179 amu. The difference in mass is therefore 0.008 105 amu. Multiplying this by 931 Mev/amu gives us the amount of energy required to dislodge only one proton from the nitrogen nucleus. That amount of energy is 7.5 Mev, roughly one-third the amount of energy necessary to remove a proton from an alpha particle. In fact, the alpha particle is one of the most stable configurations of the nuclear structures. It is no wonder then that an entire alpha particle is emitted during the decay of many radioactive nuclei.

Each radioactive isotope that decays with the emission of an alpha particle does so with a predictable amount of energy. For example, the alpha particle emitted by the decay of uranium-238 has an amount of kinetic energy equal to 4.18 Mev. There is also a gamma ray with energy of 0.05 Mev. The alpha particle emitted by the decay of the radium-226 isotope has an energy of 4.78 Mev.

Often an alpha decay is accompanied by the emission of a gamma-ray photon. The nucleus, as well as the atom itself, appears to have energy levels. If after a radioactive decay the nucleus is left in an energy level above the lowest level, it will drop to that lowest energy level, and in the process of this nuclear transition one photon of energy will be emitted. The difference in the energy levels of the nucleus is so great that the photon emitted has more energy than the photons emitted by most electron transitions. The term gamma rays has come to designate photons emitted as the result of nuclear transitions.

From its discovery to the present day, the alpha particle has played an extremely important role in the physicist's investigation of the atomic world. Rutherford learned how to use the energetic particle to great advantage in studying the atom, even before he knew what it was. The techniques developed by Rutherford form the foundation of atomic and particle studies today.

A beam of particles can reveal details of material through which they are projected by their scattering or their absorption in that material. A beam of particles can also alter atoms in the material through which it travels. As particles of greater and greater energy are used, finer and finer details are revealed, for as the momentum of the particle increases, its wavelength decreases. Similarly, with particles of greater energy more profound changes are wrought in the material through which the beam of particles is projected.

References

1. *The Collected Papers of Lord Rutherford of Nelson,* Vol. I, 1962, p. 169.
2. *Ibid.,* Vol. I, pp. 174 f.
3. *Ibid.,* Vol. I, p. 610.
4. *Ibid.,* Vol. II, p. 163.
5. *Ibid.,* Vol. II, p. 166.
6. *Ibid.,* Vol. II, pp. 589 ff.
7. J. Chadwick, "The Existence of a Neutron," in *Foundations of Nuclear Physics*, Robert T. Beyer, ed., Dover Publications, New York, 1949.
8. Rutherford, *op. cit.*, Vol. I, pp. 492 f.
9. *Ibid.,* Vol. I, p. 615.
10. *Ibid.,* Vol. II, p. 430.

Questions

1. Describe the observations by which Rutherford was able to demonstrate that at least two distinct types of radiation are emitted by uranium.

2. Indicate clearly the similarities and the differences between the instrument Rutherford used to distinguish the two types of radiation from uranium (see Figure 13-3) and the Geiger tube (Figure 13-5).

3. How did Rutherford ultimately identify the alpha particle as the nucleus of a helium atom?

4. What observation led Rutherford to believe that his alpha particles had disrupted a nitrogen nucleus?

5. Describe the observations and logic that enabled Chadwick to announce he had discovered the neutron.

6. Describe the steps that led to the realization that the charge on the nuclei of adjacent atoms as listed in the periodic table changes by one.

7. What is meant by a radioactive series?

8. The Earth is very close to 4.5×10^9 years old. Why is it, then, that naturally formed radioactive elements are still found in rocks?

Problems

1. Complete the following equations of radioactive decay:

(a) $^{235}_{92}\text{U} \to \alpha +$

(b) $^{232}_{91}\text{Pa} \to \beta +$

(c) $^{249}_{97}\text{Bk} \to \quad + \ ^{249}_{98}\text{Cf}$

(d) $\quad \to \alpha + \ ^{206}_{82}\text{Pb}$

(e) $\quad \to \beta + \ ^3_2\text{He}$

(f) $^{224}_{90}\text{Th} \to \quad + \ ^{220}_{88}\text{Ra}$

2. Calculate the binding energy per nucleon of the hydrogen-3 nucleus.

3. How much energy is required to remove one proton from the lithium-6 nucleus? Compare this with the amount of energy required to ionize a hydrogen atom (similar to problems 6 and 7, Chapter 12, but let $l = 1$ and $u = \infty$).

4. If 10^{23} atoms of helium were converted entirely into energy, how much energy would be released? (Give the answer in either electron volts and joules.)

5. Using classical physics, calculate the change in energy (in either joules and electron volts) if each of the following particles were to fall through a potential difference of 1.8×10^5 volts: (*a*) an electron, (*b*) a proton, (*c*) an alpha particle.

6. Referring to problem 5, calculate the velocities of each particle and estimate whether relativistic mechanics should have been used instead of classical mechanics.

7. Complete the following nuclear reactions that result from bombardment:

(*a*) $^4_2\text{He} + ^{11}_5\text{B} \rightarrow ^{14}_7\text{N} +$ (*d*) $^4_2\text{He} + ^{14}_7\text{N} \rightarrow ^{17}_8\text{O} +$

(*b*) $^1_1\text{H} + ^7_3\text{Li} \rightarrow ^4_2\text{He} +$ (*e*) $^4_2\text{He} + ^{27}_{13}\text{Al} \rightarrow ^1_1\text{H} +$

(*c*) $^1_1\text{H} + ^{19}_9\text{F} \rightarrow ^{16}_8\text{O} +$ (*f*) $^2_1\text{H} + ^{16}_8\text{O} \rightarrow ^4_2\text{He} +$

Chapter Fourteen

The Nucleus and Particles

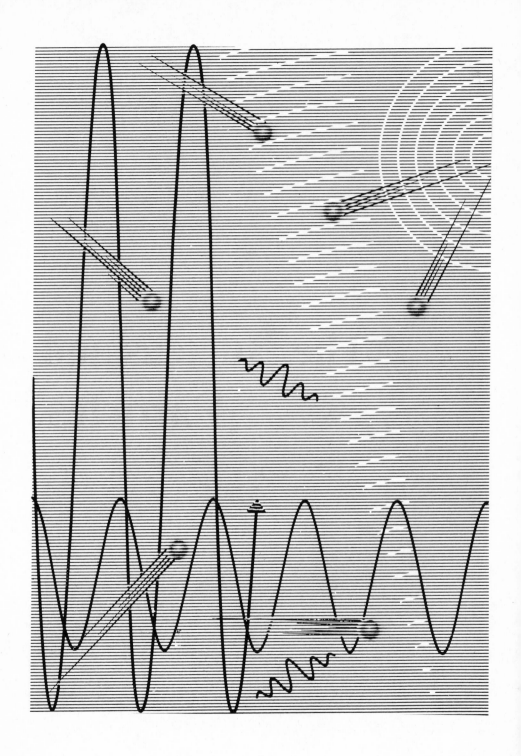

The atomic theory enabled the nineteenth-century scientists to develop both the principle of conservation of energy and the kinetic theory of gases and to explain chemical reactions. These scientists discovered that atoms combine in many and complex ways to form molecules. They were sure that the atom was the smallest part of matter and that it was indestructible. In fact, the name *atom* had been selected from the writings of the ancient Greeks to denote an *indivisible* particle. (*Atomos* is the Greek word for indivisible.)

In the opening years of the twentieth century, however, the atom itself became the subject of a continuing and detailed study. The discovery was made that, in fact, the atom is not indestructible; its electrons can be stripped off, its nucleus bombarded. Some nuclei break up spontaneously; others can be broken apart with bombarding particles of greater and greater energy. Will these new studies of the atom permit us to develop new theories and principles to describe the submicroscopic world?

PAULI, FERMI, AND THE NEUTRINO

Beta Decay

Studies of the process of alpha decay reconfirmed the conservation principles and extended their application to the subatomic world of the nucleus. During any alpha decay the energy, charge, and momentum are all conserved and, in addition, for any particular nucleus the energy of the alpha particle emitted is always the same. For example, the alpha particle emitted by a uranium-238 nucleus always has an energy of 4.18 Mev, and that emitted by a polonium-212 nucleus always has an energy of 8.78 Mev. During any alpha decay momentum is also conserved. Beta decay, however, appears to be quite a different process. Beta particles emitted by the same isotope have an unpredictable amount of energy (within a certain range) and, more important, the total amount of energy after decay appears to be *less* than the amount of energy before decay.

The beta particles emitted during any beta decay have a range of energy from a very small value to an upper limit. Those emitted by bismuth-210, for example, have a maximum energy of 1.16 Mev, but more particles are emitted with an energy of 0.15 Mev than with any other energy (Figure 14-1). Let us examine this apparent contradiction more closely.

Bismuth-210 has an atomic mass of 209.984 110 amu. By the process of beta decay it gains one positive charge, and the nucleus becomes polonium-210 with an atomic mass of 209.982 866 amu.

$$^{210}_{83}\text{Bi} \rightarrow {}^{0}_{1}\beta + {}^{210}_{84}\text{Po}$$

The difference in mass is 0.001 244 amu, equivalent to 1.16 Mev. The energy conservation principle thus predicts that the beta particle should have 1.16 Mev of energy; but this is the upper limit of the observed range of beta-

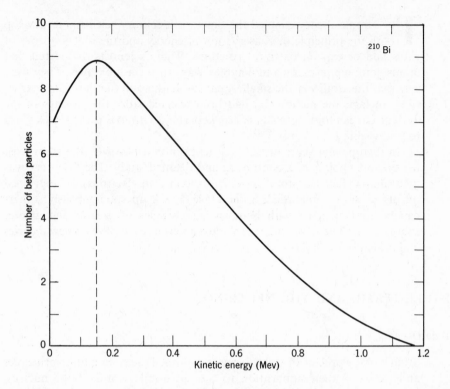

FIGURE 14-1

The number of observed beta particles of each energy released during decay of ²¹⁰Bi varies according to the curve shown. More particles have 0.15 Mev of energy than any other energy; no particle has more than 1.16 Mev.

particle energies for bismuth-210. If the nucleus loses 1.16 Mev during this process, how can a beta particle be emitted with less energy than 1.16 Mev? Where is the missing energy? Moreover, bismuth-210 is not the only trouble-maker, for other beta emitters seemingly defy the principle of conservation of energy. The average energy of the beta particles emitted in the beta decay of many different isotopes is roughly one-third to one-fourth the amount predicted by the energy conservation principle.

In the late 1920's it seemed that to accept the observations made of beta decay, the principle of conservation of energy would have to be altered in some way. Niels Bohr himself suggested that perhaps energy is not conserved on a subatomic scale.

In the 1820's, as we know, it had similarly appeared that Newton's law of gravity might have to be altered because of the unexplainable motion of the planet Uranus (see p. 260f). Then Leverrier dispelled doubts by accepting Newton's law of gravity and predicting the existence, location, and size of an unknown planet. The planet Neptune was found where Leverrier had hoped it would be, and it was about the size he had estimated.

FIGURE 14-2

Enrico Fermi (1901–1954). (Courtesy of Los Alamos Scientific Laboratory, University of California, Los Alamos, New Mexico.)

Wolfgang Pauli (1900–1958), an Austrian physicist and Nobel prize-winner, proposed in 1931 that the principle of conservation of energy could be saved if during the process of beta decay another *as yet undetected* particle was also emitted. Enrico Fermi (1901–1954), an Italian-American physicist and another Nobel prizewinner, named the particle the *neutrino,* for he, along with Pauli, predicted that when found the particle would be electrically neutral and essentially without mass. The two physicists expected that the neutrino, like the photon, would have zero rest mass. In essence, their theory predicted that during beta decay the neutron spontaneously decays into a proton, an electron, and a neutrino.

The Neutrino Discovered

In the August 1956 edition of *Scientific American* the discovery of the neutrino was announced in an article titled "Little Neutral One." It reads, in part,

A long and exciting adventure in physics has come to a triumphant end. The neutrino has been found. Frederick Reines and Clyde L. Cowan, Jr., of the Los Alamos Scientific Laboratory trapped the ghostly particle in an underground chamber near the Savannah River atomic pile. . . .

The existence of the particle was suggested by Wolfgang Pauli and Enrico Fermi to account for the puzzling phenomenon of beta-decay—the spontaneous conversion of a neutron into a proton and an electron. In this process a certain amount of mass is lost, which should turn up as energy of the product particles. However, the proton and electron almost never have enough energy to balance the account. Pauli and Fermi proposed that the missing energy is carried away by an uncharged, virtually weightless particle which Fermi named neutrino—"little neutral one."

A consequence of the theory is that the hypothetical particle should interact hardly at all with other forms of matter. The average neutrino would pass through 50 light-years* of solid lead before reacting with another particle. Thus it is intrinsically almost undetectable. Reines and Cowan undertook to detect it by recording its extremely rare interactions. Their experiment, whose successful outcome was reported at a meeting of the American Physical Society in New Haven last month, depends on the reverse of beta-decay: a proton captures a neutrino and is converted into a neutron and a positron. Both of the products are easily detectable. . . .

Reines and Cowan had run their experiment for 1,371 hours at the time of their report. In a number of variations of the experimental setup they obtained average counts of from one-half to three neutrinos per hour. The ratio of these counts to un-wanted background from cosmic rays and other radiation . . . was about 10 to one (1).

* One light-year is the distance light travels in one year, 6×10^{12} miles.

Reliance on the principle of conservation of energy paid large dividends. As it turned out, the discovery of the neutrino saved other conservation principles as well, such as the principle of conservation of momentum. As so often happens in science, theory is challenged by observations; if the observations are reliable, if the theory is to be trusted, other as yet unde-tected factors must be involved. In a search for such factors, the neutrino was found, and it now plays a major role in the study of nuclear and particle physics.

ANDERSON AND THE POSITRON

Another new particle named the *positron* was discovered in 1932 by C. D. Anderson (b. 1905), an American physicist and Nobel prizewinner.

Cloud Chamber Tracks and Electric Charges

When Anderson and his colleague R. A. Millikan (see p. 318) began their study, they intended to investigate the secondary particles of cosmic radia-tion. Very high energy "rays" or particles are indeed produced somewhere in the universe, and some of these do strike the Earth and its atmosphere. The exact nature of cosmic rays has been the subject of a great deal of study and speculation. One of the most interesting aspects of cosmic radiation is that it is extremely energetic and offers a field of research that cannot be duplicated in the physical laboratory. Whenever a cosmic ray particle strikes an atomic nucleus on the Earth, a great shower of secondary particles is produced (Figure 14-3). Among this debris the positron was first detected.

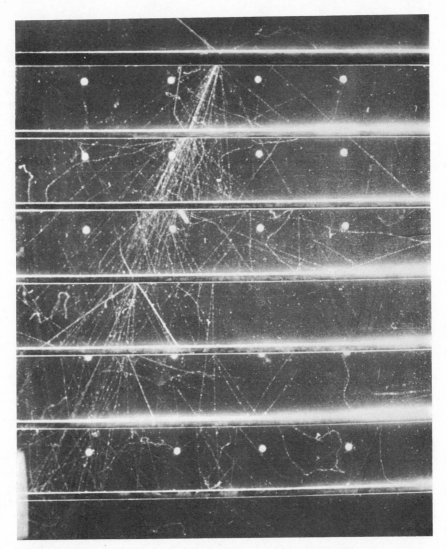

FIGURE 14-3

A shower of different particles results from a cosmic ray particle colliding with the nucleus of an atom in the top lead plate. (Courtesy of Professor William Fretter, University of California, Berkeley.)

Secondary particles that carry an electric charge can be studied by the tracks they leave in a Wilson cloud chamber. The more energetic the particle, the greater the extent of ionization, and therefore the wider the track.

If the cloud chamber is placed in a strong magnetic field, any charged particle traveling at right angles to that field will move in a curved path. The curvature of the path permits the observer to determine the energy of the particle if its charge is known. But the nature of the charge (positive or negative) cannot be determined from the curvature unless the direction in which the particle is traveling is known. It is impossible to distinguish between a positively charged particle traveling at right angles to a magnetic field and a negatively charged particle traveling in exactly the opposite direction in that same field. Both particles curve in the same direction (Figure 14-4a), for each reacts oppositely in a magnetic field.

Anderson contrived a very clever device to determine the direction of the particles he observed in the cloud chamber. He realized that the force on a charged particle moving in a magnetic field depends on that particle's velocity (see p. 317),

$$F = qvB$$

where F is the magnetic force, q the particle's charge, v its velocity, and B the magnetic field strength. The magnetic force, however, acts as a centripetal force on the particle, F_c

$$F_c = \frac{mv^2}{r}$$

Since the magnetic force supplies the centripetal force, we can set these two equal to each other,

$$qvB = \frac{mv^2}{r}$$

and further simplify this equation by solving it for r:

$$r = \frac{mv}{qB}$$

With the equation in this form, it is clear that the radius of curvature r of the path of a given charged particle is directly proportional to the particle's velocity and inversely proportional to the magnetic field B. For a constant magnetic field, then, the radius of curvature of the path of a particle is proportional to its velocity; if the velocity is cut in half, the radius is also cut in half.

Anderson realized that he could change the particle's velocity by letting it pass through a thin lead sheet. Because the particle loses some kinetic energy during its passage through this sheet, its velocity decreases, and therefore the radius of curvature of its path also decreases. With a decrease in radius, the particle travels in a more sharply curving path. When such a change in curvature is observed, the direction of travel of the particle is revealed. The particle photographed in Figure 14-4b is traveling from the bottom to the top. By knowing the direction of the magnetic field, the charge on the particle is easily determined.

The Positron Discovered

Out of a group of some 1300 photographs of cosmic ray events, Anderson found one that enabled him to make the following announcement:

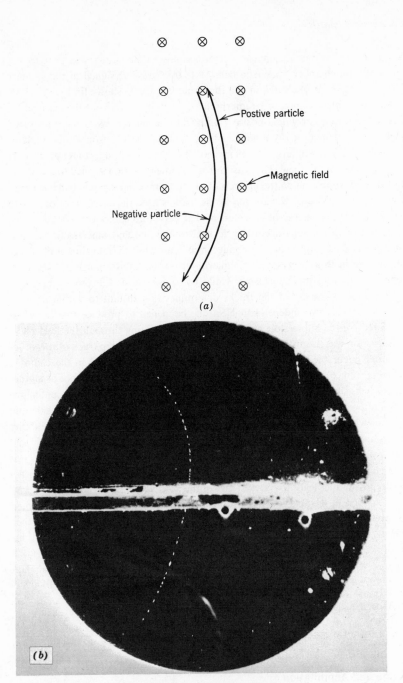

FIGURE 14-4

(a) The path of a negative particle and a positive particle traveling in opposite directions in the same magnetic field (indicated by the x's) will curve in the same direction. (b) The existence of the positron was first identified with this photograph. The positron entered the cloud chamber from the bottom with an energy of 63 Mev. By passing through the lead plate, its energy was reduced to 23 Mev and the radius of curvature of its path decreased. By knowing the direction of the magnetic field, the charge is then fixed as positive. (Courtesy of Professor C. D. Anderson, California Institute of Technology.)

On August 2, 1932, during the course of photographing cosmic-ray tracks produced in a vertical Wilson chamber (magnetic field of 15,000 gauss) designed in the summer of 1930 by Professor R. A. Millikan and the writer, the tracks shown [in Figure 14-4b] were obtained, which seemed to be interpretable only on the basis of the existence in this case of a particle carrying a positive charge but having a mass of the same order of magnitude as that normally possessed by a free negative electron. Later study of the photograph by a whole group of men of the Norman Bridge Laboratory only tended to strengthen this view. The reason that this interpretation seemed to inevitable is that the track appearing on the upper half of the figure cannot possibly have a mass as large as that of a proton for as soon as the mass is fixed the energy is at once fixed by the curvature. The energy of a proton of that curvature comes out 300,000 volts, but a proton of that energy according to well established and universally accepted determinations[1] has a total range of about 5 mm in air while that portion of the range actually visible in this case exceeds 5 cm without a noticeable change in curvature. The only escape from this conclusion would be to assume that at exactly the same instant (and the sharpness of the tracks determines that instant to within about a fiftieth of a second) two independent electrons happened to produce two tracks so placed as to give the impression of a single particle shooting through the lead plate. This assumption was dismissed on a probability basis, since a sharp track of this order of curvature under the experimental conditions prevailing occurred in the chamber only once in some 500 exposures, and since there was practically no chance at all that two such tracks should line up in this way. We also discarded as completely untenable the assumption of an electron of 20 million volts entering the lead on one side and coming out with an energy of 60 million volts on the other side. A fourth possibility is that a photon, entering the lead from above, knocked out of the nucleus of a lead atom two particles, one of which shot upward and the other downward. But in this case the upward moving one would be a positive particle of small mass so that either of the two possibilities leads to the existence of the positive electron.

In the course of the next few weeks other photographs were obtained which could be interpreted logically only on the positive-electron basis, and a brief report was then published[2] with due reserve in interpretation in view of the importance and striking nature of the announcement (2).

[1] Rutherford, Chadwick, and Ellis, *Radiations from Radioactive Substances,* p. 294. Assuming $R \propto v^3$ and using data there given the range of a 300,000 volt proton in air S.T.P. [standard temperature and pressure] is about 5 mm.
[2] C. D. Anderson, *Science* 76, 238 (1932).

And so another elementary particle was discovered.

Pair Production and Annihilation

The life history of a positron is interesting, but so also is it a definite confirmation of the principle of special relativity. The positron is born along with an electron when a gamma-ray photon travels close to the nucleus of a heavy atom. If that gamma-ray photon has enough energy, it may spontaneously change into two particles, an electron and a positron (Figure 14-5).

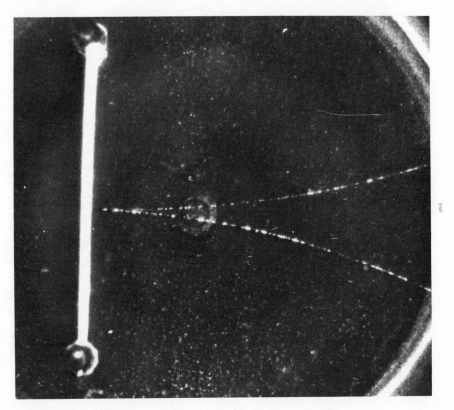

FIGURE 14-5

"Pair production." A gamma ray photon entering the field from the left is transformed into an electron and a positron in the metal plate. In a magnetic field the paths of the two high-speed particles curve in opposite directions. (Courtesy of the Lawrence Radiation Laboratory, University of California, Berkeley.)

During this conversion of energy from one form, a photon, into another form, two particles, all the conservation principles are adhered to. An electrically neutral photon is converted into a positron and an electron, each with equal but opposite charges; electric charge is therefore conserved. The de Broglie momentum of the original photon is conserved (the heavy nucleus is required to absorb some of the momentum), and so of course is the energy.

From the energy-mass relationship of special relativity, $E = mc^2$, the energy equivalent to the mass of one electron is 0.51 Mev. Since the positron has the same mass as the electron, the original gamma-ray photon must have at least 2×0.51 Mev or 1.02 Mev, or this pair of particles will not be created for want of energy. And since the two particles invariably carry away some kinetic energy, the photon will need to have even more energy than 1.02 Mev.

The process of converting one gamma-ray photon into two particles is called *pair production*. But the positron does not last long in this world of ours; there are too many electrons about. As soon as the positron slows down enough, it joins with an electron and the two mutually annihilate each other! The result of this *mass annihilation* is two gamma-ray photons. To conserve momentum, two such photons must be produced.

If, for example, the electron and positron are traveling with exactly the same speed but in diametrically opposite directions, the sum of their momenta will be zero. (Momentum, it will be recalled, is a vector quantity.) If these two particles collide, there being no momentum to start with, there must be zero momentum afterward. Therefore, one photon alone cannot be created, for that one photon would carry away momentum. Two photons must be created, each with equal momentum, and the two must travel in opposite directions.

This process obviously confirms the equivalence of energy and mass. One photon with zero rest mass is converted into two particles, each with a nonzero rest mass. One of these particles, the positron (with nonzero rest mass) collides with an electron, and the two are converted into two gamma-ray photons with zero rest mass. The confirmation is dramatic; the questions raised are many!

What happens to the electric charge? What is electric charge? What happens to rest mass? What is rest mass? These questions and many more are being asked and some are being answered in the current research on elementary particles.

YUKAWA AND THE PION

It is not surprising that research into the intricacies of the atom raised new questions; it was a new adventure. For example, what holds the nucleus together? By Newton's law of gravity any two nucleons in the nucleus should exert an attractive force of about 2×10^{-36} newton on each other; but by Coulomb's law of electric forces, any two protons in the nucleus should exert a repulsive force of about 2 newtons! The repulsive force is about 10^{36} times greater than the attractive force. Why don't the protons fly out of the nucleus?

Could accepted theories somehow be extended, as they were with the prediction and discovery of both Neptune and the neutrino, to explain how the nucleus stays together? Or would an entirely new theory be required, as with special relativity and quantum mechanics which were developed in response to specific observations? If these two forces of attraction and repulsion are not capable of holding the nucleus together, could an unknown force be operating?

Particles and Forces

Fermi had suggested that since photons do carry momentum, and thus can exert a force, they may be the means by which electric and magnetic forces are exerted. Two electrons as they approach one another supposedly exchange photons, which causes them to repel each other. By way of analogy, two boys who are each on a skate board and are playing catch with a heavy ball may be said to repel one another.

In 1935 Hideki Yukawa (b. 1907), Japanese physicist and Nobel prizewinner, suggested that perhaps the forces binding the nucleus together could also be accounted for by an exchange of particles. But which particles? It was clear to him that the photon was incapable of exerting enough force. After all, in this line of reasoning the photons are assumed to carry the Coulomb forces that cause the protons in the nucleus to repel one another. So photons cannot both bind the nucleus together and cause the repulsion of protons. Another more massive particle, yet one not as massive as the proton itself, is needed.

The Uncertainty Principle and Energy Conservation

Suppose, Yukawa suggested, that the proton itself is a complex particle, one composed of a core with a "particle cloud" about it. Given enough energy, the cloud might reveal a particle that could leave the region about the core. Again, by way of analogy, if an atom is given enough energy, an electron may leave the region of the nucleus and the atom be left ionized. But the nucleus must be given energy before it will reveal the existence of the supposed particle; the spontaneous formation and escape of such a particle from the cloud about the core of the proton (leaving the proton unaltered) would constitute a violation of the energy conservation principle. Yukawa therefore proposed that these particles only begin to form and only nearly escape, but that they never succeed for want of energy. Because they only "seem" to exist in the proton cloud, they were dubbed virtual particles.

The particle that Yukawa proposed has since been named the *pion* (pi meson). Its supposed "appearance" about the nucleus has suggested the analogy of a man inside a large balloon, a man who is punching and kicking to get out. As each fist and foot hits the balloon, the surface is pushed way out; the balloon does not break, however, although it may appear to do so momentarily. To break the balloon would violate the principle of conservation of energy.

But, realized Yukawa, the amount of energy is in fact uncertain! Heisenberg's uncertainty principle maintains that the amount of energy cannot be known with great precision during any very small interval of time. Mathematically (see p. 343),

$$\Delta E \, \Delta t \approx h$$

If the amount of energy is really uncertain, perhaps the principle of conservation of energy can be violated to that extent, but no more!

To just what extent the energy is uncertain can be estimated from the mass of the nucleon and by assuming a mass for the pion. If the Coulomb force of proton repulsion must be overcome, and if the force of gravity is too weak to be of any use, Yukawa assumed that the mass of the pion must be between 200 and 300 times that of an electron. If this amount of mass is created from a proton, an amount of energy $E = mc^2$ will have to be supplied. This amount of energy is roughly 140 Mev. Hence, the conservation of energy principle must be violated to this extent.

The time interval during which the principle may be violated can be calculated from the uncertainty principle. The value of h is about 4.1×10^{-21} Mev-sec, so the interval of time is given by

$$\Delta t \approx \frac{h}{E}$$

$$\Delta t \approx \frac{4.1 \times 10^{-21} \text{ Mev-sec}}{1.4 \times 10^2 \text{ Mev}}$$

$$\Delta t \approx 2.9 \times 10^{-23} \text{ sec}$$

Certainly this is not a very long interval of time, and perhaps Dame Nature does not object to having the principle of energy conservation violated for such a brief instant. At any rate, if these pions are traveling at nearly the speed of light—which is a reasonable assumption—we can calculate how far they will travel during this short interval of time:

$$s = vt$$

$$s = (3.0 \times 10^8 \text{ m/sec})(2.9 \times 10^{-23} \text{ sec})$$

$$s = 8.7 \times 10^{-15} \text{ m}$$

$$s = 10^{-14} \text{ m}$$

Presumably this distance is roughly the diameter of the "pion cloud" about the nucleon; it is also about the same as the diameter of the atomic nucleus!

According to Yukawa, then, the nucleon is a busy sort of particle; it has a core with a pion cloud. From this cloud pions start to form and escape, doing so momentarily, then dissolving back and becoming totally lost in the cloud once more. Each pion lives for less than 10^{-23} second, and it lives only as a virtual particle.

Pion Exchange and the Binding Force

Yukawa made additional predictions about the behavior of the nucleon and its pions. If two or more nucleons are close together, as they must be inside the nucleus, a pion trying to escape from one nucleon may venture far enough out to become trapped by a neighboring nucleon. It may find itself transferred from one nucleon to the other. Hence, a *positive pion* may escape from a proton and be captured by a neutron. What was originally a proton becomes a neutron; the particle originally a neutron captures a positive pion and becomes a proton. Yukawa suggested that the nucleons in the nucleus are held together by just this transference of pions from one nucleon to the other.

The short range over which nuclear forces are effective seemed puzzling at first, but Yukawa of course proposed an explanation. Since the pions cannot venture very far from the nucleon during their brief lifetime as a virtual particle, they are not able to interact with nucleons more distant than roughly 10^{-14} meter. The nuclear binding force is therefore not very effective beyond that distance.

The Pion Observed

Since 1935, when Yukawa made all these predictions, the pion has been found; in fact three pions, a positive, a negative, and a neutral pion, have been discovered. Their masses are respectively 273.2, 273.2, and 264.2 times the rest mass of an electron. The charged pions were discovered in 1947; the neutral pion was found in 1950.

In 1947 C. F. Powell and his group in England found the first pion, like the positron, in the debris of cosmic radiation. One year later the pion was detected by the track it left in a Wilson cloud chamber at the Lawrence Radiation Laboratory at the University of California in Berkeley. By accelerating protons to very high velocities, and firing them at a quantity of protons at rest, the relativistic kinetic energy of the high-velocity protons is more than enough to supply the 140 Mev required to form pions should a high-velocity proton collide with a proton originally at rest. In fact, the amount of energy available is enough both to form the pion and to give it kinetic energy. If a positive pion is created from a stationary proton, a neutron is left. The bombarding proton goes its way. The reaction may be written

$$p^+ + p^+ \longrightarrow p^+ + n^0 + \pi^+$$

If there seems to be too much mass on the right-hand side of this equation, we must remember that one of the protons on the left-hand side has enough kinetic energy to balance the equation. Perhaps we should write the equation

$$E_k + p^+ + p^+ \longrightarrow p^+ + n^0 + \pi^+$$

and let E_k, the kinetic energy of the bombarding proton, balance the equation. In fact, this reaction is a striking confirmation that mass is one form of energy. The relativistic kinetic energy of the bombarding proton is converted into a pion with rest mass.

If the bombarding proton has even more kinetic energy, more than one pion can be formed from the pion cloud. For example, a positive and negative pion may be formed, and it is even possible to create a positive, a negative, and a neutral pion:

$$p^+ + p^+ \longrightarrow p^+ + p^+ + \pi^+ + \pi^- + \pi^0$$

This last reaction is revealed in the striking photograph taken at the Lawrence Radiation Laboratory (Figure 14-6). Within the framework of the principles of conservation, that is, as long as energy is conserved, any number of pions can be created in this process by ever more energetic bombarding protons.

Pion Cloud Observed

Experiments using electrons as scattering particles (as Rutherford used alpha particles) have been conducted by Robert Hofstadter (b. 1915), American physicist and Nobel prizewinner. His research group at Stanford University bombarded protons with electrons, and from the pattern of electron scatter produced they concluded that protons are indeed complex. Protons appear to have a core of some kind or other, and they do seem to have a cloud of pions about that core. The diameter of the pion cloud is estimated to be about 10^{-14} meter, about what was expected from Yukawa's work.

There seems little question that pions exist, that protons and neutrons have complex natures, and that pions play a major role in supplying the binding force necessary to hold the nucleus together.

Other observations may give us more clues about the nucleus. For example, the number of nuclides (nuclei considered as particles) having an odd and even number of nucleons, as listed in Table 14-1, indicate that certain nuclides are favored over others.

Table 14-1

Number Protons	Number Neutrons	Known Stable Nuclides
odd	odd	4
odd	even	50
even	odd	55
even	even	165

Data indicate very clearly that stable nuclides with an even number of constituent nucleons, such as carbon with six protons and six neutrons, are highly favored over those with an odd number of nucleons. If these nucleons do indeed "play catch" with pions, we would expect the nuclides with an even number of nucleons to be more stable. Apparently there are no effective backstops against which one nucleon can play pion catch all by itself.

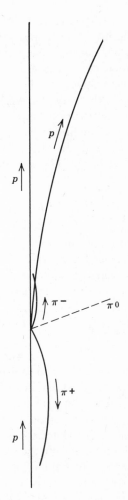

FIGURE 14-6

A very energetic proton entering the photograph from the bottom collides with a relatively stationary proton in the chamber. The result of this collision is that both protons proceed upward (see sketch accompanying the photograph) with a slight curvature to the right. But a positive pion leaves the scene of the collision heading downward and curving toward the left. A negative pion proceeds upward with a curvature to the left. A neutral pion was also created, but its existence is made evident only by calculating the amount of the energies involved in the visible tracks. The little pigtail curlicues are electrons losing energy. (Courtesy of Lawrence Radiation Laboratory, University of California, Berkeley.)

Another line of research gives us additional information about the structure of the nucleus. It appears that nuclides with certain numbers of nucleons are favored. For example, the nucleus, like the electron structure surrounding it, appears to be composed of shells. The electron structure has shells that are successively filled with two, eight, eighteen, and thirty-two electrons; hence the periodicity of chemical properties and the resulting periodic table. The first shell holds no more than two electrons, the second no more than eight, and so on. Similarly, nuclear shells appear to be filled at $N = Z = 20$, 50, and 82, and $N = 126$, where N equals the number of neutrons, Z the number of protons. There also seems to be a minor shell filled at $N = Z = 28$.

PARTICLE PHYSICS

Accelerators produce particles with energies in the order of many gigaelectron volts (see Figure 14-7). Techniques devised to observe the reactions of these energetic particles are ingenious. For example, the cloud chamber has to some extent been supplanted by the bubble chamber.

The Bubble Chamber

It will be recalled that in a cloud chamber the charged particles produce liquid droplets in a supercooled vapor. D. H. Glaser (b. 1926), an American physicist and Nobel prizewinner, therefore reasoned that a charged particle should hasten bubble formation in a superheated liquid. High pressure can prevent a liquid from boiling. If the pressure is suddenly released, however, the liquid is superheated for a moment, since it is hot enough to boil but has not yet done so. During the moment it is superheated, charged particles will supply centers for bubble formation.

Ions have a net electric charge, and they are formed when a high-energy charged particle, such as a proton traveling at speeds close to that of light, collides with an atom. After the collision the proton, possessing a little less kinetic energy, goes its way, and the atom, with one or two electrons stripped off, becomes a positively charged ion. If the high-energy proton should pass through the bubble chamber an instant before the pressure is released, it will leave a trail of free electrons and positive ions. As the pressure is released, bubbles form on each trail of ions.

The bubble chamber has several advantages over the cloud chamber. The path in a liquid has more atoms along each unit length than does a path in a vapor. Therefore, more ions are formed along each unit length. Since the high-energy proton loses some energy during each ion formation, it loses its kinetic energy faster in a bubble chamber than in a cloud chamber. The proton's trail is therefore shorter, with more ions produced per unit length and probably more details revealed. Furthermore, if liquid hydrogen is used in a bubble chamber, there is an ample supply of proton targets, more than with hydrogen gas in a cloud chamber.

FIGURE 14-7

The Bevatron at the Lawrence Radiation Laboratory, University of California, Berkeley, can accelerate protons up to 6.3 Gev. The protons are injected into the Bevatron with an energy of 19 Mev from the linear accelerator in the right foreground. (Courtesy of Lawrence Radiation Laboratory, University of California, Berkeley.)

A collision of elementary particles occurs strictly by chance. No one proton, for example, is ever aimed at another particular proton. The beam from a particle accelerator sprays billions of protons at the target—a shotgun with its few buckshot cannot do nearly as well. Of the billions of protons in the beam, surely some will strike a proton in the target. So a bubble chamber filled with liquid hydrogen makes a better target than hydrogen vapor in a cloud chamber, even if that hydrogen vapor is under pressure.

In 1953 Glaser built his first bubble chamber, and since that time bubble chambers have become a vital tool in the investigation of the elusive particles studied by physicists.

Particles and Conservation Principles

With the help of these machines a bewildering array of elementary particles has been discovered. The electron and the proton, the neutron and the neutrino, the pion and the positron—these are only a beginning. It turns out that the pion is unstable, lasting on the average for only about 2.5×10^{-8} second before it decays into other particles. The positive pion π^+ decays into a particle called a positive *muon* μ^+ and a neutrino v. Actually, the muon was discovered before the pion and was confused with that particle for a time. In equation form,

$$\pi^+ \to \mu^+ + v$$

The negative pion decays into a negative muon μ^- and a neutrino:

$$\pi^- \to \mu^- + v$$

The neutral pion decays into two gamma-ray photons:

$$\pi^0 \to \gamma + \gamma$$

The muon has a rest mass 206 times that of an electron. But it too is unstable; the negative muon decays into an electron and two neutrinos, after an average lifetime of 2.2×10^{-6} second.

Another particle, which as yet exists only in theory, is associated with gravitational force. Current particle theory suggests that for every force there is a particle. The pion supplies the nuclear binding force, the photon the electromagnetic force, and the postulated *graviton* the gravitational force.

And there are other particles, the kaons, the lambda particles, the sigma particles, and the xi particles. To arrange these particles into a meaningful pattern and to understand their inter-relationships is a tremendous task. The formulation of a theory that will encompass all these many diverse particles is the aim of men working in the forefront of one of the most advanced fields of modern physics. Certain characteristics do permit these particles to be studied in a fairly orderly manner.

One of the most important properties is mass, for it is the particle's rest mass that indicates the amount of energy necessary to create that particle, or the amount of energy given up if that particle decays or is converted into another form of energy. Of all the particles, only the photon, the neutrino, and the graviton have zero rest mass. Each of these particles may therefore travel at only one velocity, the velocity of light. No matter how slight a force is applied to any one of them, because they have zero resistance to that force, a velocity less than that of light is impossible. The mass of the other particles varies over a considerable range.

Consulting Table 14-2, we see that the masses of the particles have permitted them to be grouped. The most massive particles are called the *baryons*, those with the next less mass the *mesons*, then the family of smaller *muons*, the *electron* family, and finally the photon and the graviton. During any reaction between particles, mass-energy is conserved.

Table 14-2 *The Known Elementary Particles*

Family Name	Particle Name	Symbol	Mass	Spin	Electric Charge	Antiparticle	Number of Distinct Particles	Average Lifetime (seconds)	Typical Mode of Decay
	photon	γ (gamma ray)	0	1	neutral	same particle	1	infinite	—
	graviton	—	0	2	neutral	same particle	1	infinite	—
Electron family	electron's neutrino	ν_e	0	$\tfrac12$	neutral	$\bar\nu_e$	2	infinite	—
	electron	e^-	1	$\tfrac12$	negative	e^+ (positron)	2	infinite	—
Muon family	muon's neutrino	ν_μ	0(?)	$\tfrac12$	neutral	$\bar\nu_\mu$	2	infinite	—
	muon	μ^-	206.77	$\tfrac12$	negative	μ^+	2	2.212×10^{-6}	$\mu^- \rightarrow e^- + \bar\nu_e + \nu_\mu$
Mesons	pion	π^+	273.2	0	positive	π^- $\Big\}$ same as the particles	3	2.55×10^{-8}	$\pi^+ \rightarrow \mu^+ + \nu_\mu$
		π^-	273.2	0	negative	π^+		2.55×10^{-8}	$\pi^- \rightarrow \mu^- + \bar\nu_\mu$
		π^0	264.2	0	neutral	π^0		1.9×10^{-16}	$\pi^0 \rightarrow \gamma + \gamma$
	kaon	K^+	966.6	0	positive	$\overline{K^+}$ (negative)	4	1.22×10^{-8}	$K^+ \rightarrow \pi^+ + \pi^0$
		K^0	974	0	neutral	$\overline{K^0}$		1.00×10^{-10} and[a] 6×10^{-8}	$K^0 \rightarrow \pi^+ + \pi^-$
Baryons	nucleon	p (proton)	1836.12	$\tfrac12$	positive	$\bar p$ (negative)	4	infinite	
		n (neutron)	1838.65	$\tfrac12$	neutral	$\bar n$		1013	$n \rightarrow p + e^- + \bar\nu_e$
	lambda	Λ^0	2182.8	$\tfrac12$	neutral	$\overline{\Lambda^0}$	2	2.51×10^{-10}	$\Lambda^0 \rightarrow p + \pi^-$
	sigma	Σ^+	2327.7	$\tfrac12$	positive	$\overline{\Sigma^+}$ (negative)	6	8.1×10^{-11}	$\Sigma^+ \rightarrow n + \pi^+$
		Σ^-	2340.5	$\tfrac12$	negative	$\overline{\Sigma^-}$ (positive)		1.6×10^{-10}	$\Sigma^- \rightarrow n + \pi^-$
		Σ^0	2332	$\tfrac12$	neutral	$\overline{\Sigma^0}$		about 10^{-20}	$\Sigma^0 \rightarrow \Lambda^0 + \gamma$
	xi	Ξ^-	2580	$\tfrac12$	negative	$\overline{\Xi^-}$ (positive)	4	1.3×10^{-10}	$\Xi^- \rightarrow \Lambda^0 + \pi^-$
		Ξ^0	2570	$\tfrac12$	neutral	$\overline{\Xi^0}$		about 10^{-10}	$\Xi^0 \rightarrow \Lambda^0 + \pi^0$

33

[a] The K^0 meson has two different lifetimes. All other particles have only one.
[b] Reprinted with permission from Kenneth W. Ford, *The World of the Elementary Particles*, Blaisdell Publishing Company, New York, 1963.

Another conservation principle, the conservation of "family number," becomes evident with this grouping of particles into families. For example, particles in the electron family are conserved. During any reaction, the number of particles belonging to the electron family must be the same before and after the reaction occurs. Nor can the members of the muon family and of the baryon family change in number.

Another characteristic of the elementary particles is called *spin,* the name given to the property recognized as intrinsic angular momentum. But the particle spin is strictly limited. A given particle has a particular spin and only that spin; its spin does not change. If the photon is arbitrarily assigned a spin of 1, all other particles have a spin of either 1/2 or 0, except for the postulated graviton which should have a spin of 2. During any reaction between particles spin is conserved.

Electric charge is a very distinct property of the elementary particles. Only three charges are observed: unit positive charge, unit negative charge, and zero charge. Each particle that has a negative charge has the same identical negative charge; the positive charges of particles are also identical. Some physicists consider it possible but improbable that a particle will be found with a charge less than that of an electron or positron, but no one is holding his breath waiting.

All the known particles can be divided into two equal groups, *particles* and *antiparticles.* In 1928 P. A. M. Dirac (b. 1902), an English physicist, proposed the existence of pairs of particles, all the properties of one particle of the pair identical with those of the other except for one property. Dirac suggested that these particles could be created by pair production and that they would annihilate each other on collision. He actually predicted the existence of the positive electron, the positron, before Anderson discovered it.

It is now clear that each particle has its own antiparticle. For example, in 1955, with particles accelerated to nearly 2 Gev (2×10^9 electron volts), proton-antiproton pairs were created in the Lawrence Radiation Laboratory. The antiproton has the same mass and spin as the proton and an equal but opposite charge. Its electric charge is identical with that of the electron. The antiparticle for the negative muon is the positive muon. Neutrinos and antineutrinos have spins of the same magnitude but opposite direction.

Theory predicts and observations confirm that particles and their antiparticles will mutually annihilate one another and produce two photons if they come close. Consequently, since the world in which we live is composed of matter, "antimatter"—that is, matter composed of antiparticles—cannot exist for very long in our world. But there seems to be no reason why other regions of our universe cannot be composed entirely of antimatter. An antihydrogen atom, for example, is composed of a negative proton for a nucleus and a positron in the outer shells.

The neutrino has continued to pay a very interesting role in the studies of particle physics. In 1962 a research group at Columbia University found what is now called the muon's neutrino (a member of the muon family), for which there is an antiparticle. But there is also an electron's neutrino (a member of the electron family), the one postulated by Pauli and Fermi. It too has an antiparticle. Since the neutrino is so extremely penetrating, and since it forms in the process of neutron decay, we suppose that a good deal of the energy of the hotter stars escapes in the form of neutrinos.

SPACE-TIME CONTINUUM

The prediction, from theory, of the existence of the particle later found and named the pion was based on the concept of very small intervals of time and very small intervals of space. The question may arise, if atomic energies and nuclear energies are quantized when atomic particles are localized in a very small volume of space, whether space itself is quantized? Is time quantized? What is the relationship between space and time?

Space and time are so inextricably bound together that one without the other is meaningless. To describe an event in the universe, we must know not only its location on some chosen frame of reference but also the time when that event occurred. To describe any motion, we must consider both space and time. The transformation equations of special relativity, for example, the Lorentz-FitzGerald contraction, are descriptive of both time and space. With galaxies so far away that the light they emit travels through space for thousands of millions of years before reaching us, there seems little doubt that we must consider space and time together to describe the universe.

Einstein considered both space and time to be continuous. There are not, according to the principles of relativity, any *smallest* portions of space and time, no "atoms" of space and time with gaps between them. Although space and time can be subdivided into indefinitely small segments, each is still continuous with itself and they are continuous with each other. Einstein referred to the union of space and time as the *space-time continuum*.

World Diagrams

The unity of space and time can be demonstrated graphically. Let us consider a particle that moves along a straight line. Its location along the line must be referred to some arbitrarily chosen point that we generally call the origin. To describe the actions of the particle, we need to say *when* it was at each location along the line, which we can do very nicely by plotting its position against time. If we call the position line the x-axis and draw the time or t-axis at right angles to the x-axis, we form a space-time coordinate system (Figure 14-8a). The particle whose actions are depicted in the figure is moving with a constant velocity to the right; its x-coordinate is increasing by equal increments during equal intervals of time. A particle may travel in the positive or the negative direction of the x-axis, but only in the positive direction of the time axis.

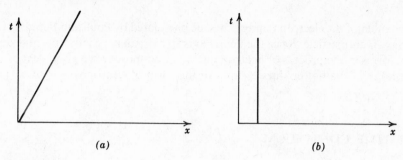

(a)   (b)

FIGURE 14-8

(a) The particle whose position and time are plotted in this world diagram is moving with a uniform velocity in the positive *x* direction. *(b)* The particle whose position and time are plotted in this world diagram is at rest relative to the *x*-axis.

Position-time diagrams of this nature were devised by Hermann Min-kowski (1864–1909) in 1908. He called them *world diagrams*. The line representing the position of any object is a *world line*, the point defining the location (in both space and time) of an event is a *world point*. In the coordinate system of the world diagram the world line for a particle at rest is a straight line parallel to the time axis. Its *x*-coordinate does not change (Figure 14-8*b*).

If the motion of a particle is uniform, its world line is straight; the slope of a world line indicates speed. Particle *A* (world line *AA'* of Figure 14-9) is traveling faster than particle *B*. During the time interval between t_1 and t_2 particle *A* traveled farther (x_a) than did particle *B* (x_b). It would appear that particle *A* will overtake particle *B*, for their world lines are about to

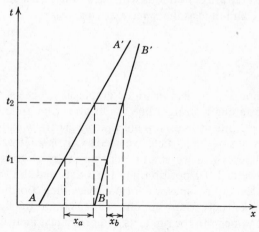

FIGURE 14-9

Particle *A* is traveling faster relative to the *x*-axis than is particle *B*. They will presumably collide.

intersect. They will soon attempt to occupy the same position at the same time.

A particle whose velocity changes cannot be represented on a world diagram by a straight line. If the velocity increases, its world line must become ever more nearly parallel to the x-axis, for it is covering a greater and greater distance in each succeeding time interval (particle D of Figure 14-10). If the speed of a particle is decreasing, the slope of its world line becomes ever more nearly parallel to the time axis (particle E of Figure 14-10).

The locations in the space-time continuum of particles that move only along the x-axis are fairly simple to represent graphically, but very few particles restrict their motion in this manner for our convenience. A complete description must usually place particles in three dimensions of space and one dimension of time, that is, in the four dimensions of the space-time continuum. Although we can easily draw a two-dimensional space-time continuum, it is impossible to draw or even imagine a four-dimensional space-time continuum. The four-dimensional continuum, however, can be described and studied with mathematics.

FEYNMAN DIAGRAMS

Photon Exchange

Interactions among the various particles prove to be the most important aspect of their behavior. Richard Feynman (b. 1918), an American physicist and Nobel prizewinner, suggested in 1949 that to understand these interactions better we can apply the world diagrams invented by Minkowski. For instance, the path of an atom through the space-time continuum can be simplified by considering only the x, t-continuum (Figure 14-11a). Should that atom emit, that is, create by a downward transition and emit,

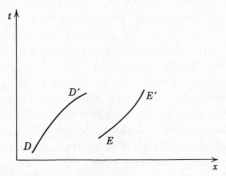

FIGURE 14-10

Particle D is accelerating positively; particle E is accelerating negatively.

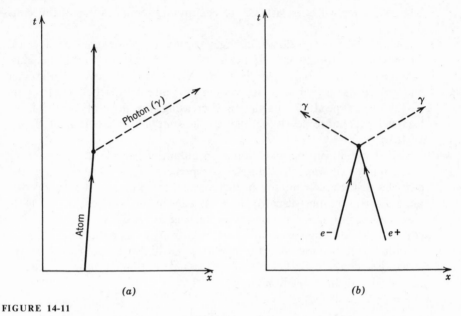

FIGURE 14-11

(a) An atom releasing a photon recoils since momentum is conserved. *(b)* An electron and positron meet, and convert into two gamma ray photons. Energy and momentum are conserved.

a photon, the photon will travel away with the velocity of light. Since the photon's speed is the limiting speed, the angle made by the world line of the photon and the x-axis is *the* limiting angle; no other world line can make such a small angle with the x-axis.

By emitting the photon, however, the atom itself recoils, since momentum is conserved. The entire reaction resembles a Y or fork in a road, the photon in this instance proceeding in the positive x-direction and the atom in the negative x-direction.

The mass annihilation of antiparticles, for example, the collision of an electron and positron (Figure 14-11*b*), can be represented in x, t-space. The two antiparticles meet, and two photons are created in their stead.

The reaction of two electrons approaching and interacting with each other can also be pictured on these Feynman diagrams (Figure 14-12). As the electrons approach one another, one of them emits a photon at A. Again, because a photon is created, the road forks. The electron goes off in one direction, the photon in the other, only to be absorbed by the second electron at B. At juncture B the road, instead of forking, merges with another. The photon brings some momentum with it which it imparts to the electron, and the electron in turn changes its momentum. Hence the two electrons depart in different directions. Presumably many photons are exchanged in such an encounter.

This collision is certainly different from the one described by Boscovich

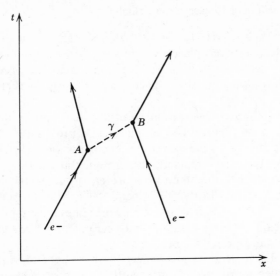

FIGURE 14-12

An encounter between two electrons may result in an exchange of photons. This exchange causes the two particles to repel each other, since momentum is conserved.

(see p. 129f), who based his argument on the principle of continuity. The classical electrodynamic theory of Maxwell applies the principle of continuity in describing the encounter between two electrons; the electrons travel along smooth paths, not a series of straight lines joined together at discontinuous junctions.

Modern quantum and particle theory, however, maintains that many properties of nature appear in discrete bundles. For example, sudden changes in momentum, as when an electron creates or absorbs a photon, are explained in terms of these bundles or particles. The new theory has, for the first time, explained the action-at-a-distance forces of Gilbert and Newton. Gilbert, it will be recalled (see p. 139f), accounted for attraction by proposing that effluvia, a substance emitted by the charged particle, brought the bits of chaff to it. Gilbert's effluvia has now been replaced by photons, and a more substantial and elaborate theory.

The conservation principles can be checked on these Feynman diagrams by making what is in essence a tally or inventory. By moving a straightedge in the positive time direction while keeping it nearly parallel to the x-direction, we pass one of the junctures in the world line of a particle. Before the juncture, energy, momentum, charge, spin, and all other properties that are conserved have certain values; the conservation principles stipulate that these values will not have changed after the juncture, no matter what particles are created or annihilated.

Newtonian Dictates Versus Freedom of Conservation Principles

The conservation principles, however, do not dictate which particles should be created or annihilated; they merely place constraints or rather limitations on the values of certain observed quantities. Any reaction can occur as long as the values of the quantities regulated by the conservation principles are not altered.

As an example, let us illustrate the formation of a virtual pion in the pion cloud about a proton. In Figure 14-13 we represent a proton at rest, with its world line perpendicular to the x-axis. At juncture A we find that a neutron and a positive pion have replaced the proton. But a check of conservation principles indicates that there is more energy immediately after juncture A than before it. Clearly the principle of conservation of energy has been violated, and the neutron and pion world lines must rejoin to form a proton once again. Before juncture A and after juncture B energy is conserved, although momentarily it is not.

The Feynman diagrams indicate in greater detail what is going on in nature than has any earlier approach. All forces appear as an interchange of particles; the world line of any particle is a series of straight lines joined by junctures indicating where particles were either created or annihilated. The only regulating factors are the conservation laws. Newtonian mechanics stipulates what must happen if such and such a force is applied to such and

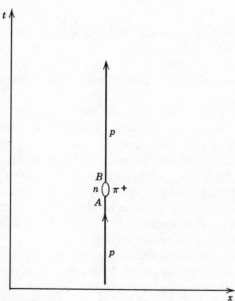

FIGURE 14-13

For only a brief instant of time, a proton may convert into a neutron and positive pion. During the time before A and after B, energy is conserved; but from the time just after A till before B energy is *not* conserved. The time interval from A to B is 10^{-24} sec.

such a mass. It clearly describes the gross effects; nothing else is possible. The conservation principles describe what cannot happen; everything else is possible. The new physics is more open, with a few very important rules which must be obeyed.

Each and every conservation principle, however, was learned through experience. In the late seventeenth century it was observed that the product of the mass times the velocity (momentum) is conserved in any collision. Franklin suggested that electric charge is conserved. Joule and others in the middle of the nineteenth century found that energy is a concept common to everything in the universe, and that in every interaction energy is conserved. The conservation principles stemming from the classification of elementary particles are also empirical. Nothing in current theory dictates that each conservation principle must be part of the "ultimate description."

Indeed, in 1956 Tsung Dao Lee (b. 1926) and Chen Ning Yang (b. 1922), both from China but now living in America and both Nobel prizewinners, predicted and proved that one of the conservation principles, the conservation of parity, is indeed violated under certain conditions. In essence, parity refers to the "mirror image" aspect of nature. It appears that some particles are not mirror images and do have either right-handedness or left-handedness; thus, during some interactions parity is not conserved.

We emphasize again that the present statements of the conservation principles need not be the final or definitive statements. There is a possibility that they may be altered by future observations. We can assume, however, that most physicists would be greatly surprised were the main conservation principles shown to be inaccurate statements of some aspects of Nature's doings.

THE EBB AND FLOW OF IDEAS

Aristotle's physics linked observation with philosophical logic and inappropriate axioms. Newtonian physics combined observation with the rigor and exactness of mathematics and opened up the entire universe as a testing ground for the scientific theories developed by man. Einstein's principle of special relativity expanded Newtonian physics to speeds approaching that of light and showed the equivalence of mass and energy. The principles of quantum mechanics have indicated a direction for the study and understanding of the atomic and subatomic world.

So many questions are asked when one is answered. The process would be frustrating were it not so fascinating. When months and months of research and years and years of thought have culminated in the formulation of a new theory, a new description of Nature's actions, scientists take pride in their achievement. But invariably new theories give rise to new problems, and the solution of one problem creates many more.

New problems promote progress, however. The new observations of Tycho Brahe led Kepler to formulate a better explanation of planetary motion. This explanation and Galileo's work led Newton to propound his laws upon which so much of the work of physics has been based. Then continued and more refined observations indicated that Newtonian mechanics is not enough. The atom is complex and is best described by quantum mechanics. The concepts of space, time, mass, and motion are explained better by the principles of relativity than by Newtonian mechanics.

The experimentalist continues to make observations, he continues to search for data upon which the theoretician can build better theories. Every new theory must be subjected to the close scrutiny of other theoreticians and of the experimentalists who test predictions made by the theory. Theories compete in the open market of ideas; discussions and articles criticize new theories. The most acceptable theory wins out; it does the best job of describing Nature's actions.

For every theory that becomes accepted by the scientific world, however, many fall by the wayside. But each theory that falls makes its contribution toward a better understanding of the universe. Finally, no theory could even be proposed were it not for that vast store of data and observational facts—as cold as they may seem to some. There would be nothing for a theory to describe if it were not for the observations, the data; the loom must be strung before the cloth can be woven.

References

1. *Scientific American,* August 1956, pp. 48 f.
2. C. D. Anderson, "The Positive Electron," *Physical Review,* Vol. 43, March 15, 1933, p. 491.

Questions

1. Describe the steps that led to the discovery of the neutrino.

2. How is it possible to determine both the charge and energy of a particle that leaves a track in a bubble chamber?

3. Compare the concept that all forces result from an exchange of particles with the ideas of collision presented by Boscovich (see pp. 129ff).

4. Compare the concept that electric forces result from an exchange of photons with Gilbert's theory that effluvia accounts for electrical attraction (see pp. 139ff).

5. What part did the uncertainty principle play in the development of the theory of the pion?

6. Why is the bubble chamber so much better for many experiments than the cloud chamber?

7. Indicate how the principle of conservation of momentum can replace the concept of force.

8. During the mass annihilation of an electron and a positron, why must two photons be created?

Problems

1. A 3.30-Mev photon passes close to a heavy nucleus and is converted into an electron and positron. If the kinetic energies of the electron and positron are equal, calculate (*a*) the frequency and the wavelength of the initial photon, (*b*) the kinetic energy of the positron, (*c*) the wavelength of the two equal photons resulting from the mass annihilation should that positron meet an electron at rest. Compare this wavelength with that of the initial photon.

2. A charged particle traveling at essentially the velocity of light makes a track in a bubble chamber of 2 cm after being created and before decaying. For how long a period of time did this particle exist?

3. Using the principles of conservation of energy, charge, and family number, and referring to Table 14-2, indicate which of the following reactions are possible. If a reaction is not permitted, indicate which of the conservation principles is violated.

(*a*) $\pi^- \rightarrow \mu^+ + \nu_\mu$ (*d*) $n \rightarrow \mu^+ + e^- + \gamma$

(*b*) $\mu^- \rightarrow e^- + \nu_e + \nu_\mu$ (*e*) $e^+ + e^- \rightarrow \mu^+ + \pi^-$

(*c*) $e^- \rightarrow \nu_e + \gamma$

4. Draw Feynman diagrams to illustrate the following reactions: (*a*) a proton-proton collision in which no additional particles are created, (*b*) a proton-proton collision in which two pions are created, (*c*) mass annihilation of an electron and a positron, (*d*) a pion decaying into a muon and a neutrino.

Appendix A   Scientific Notation

$$3.4 \times 10^9 = 3{,}400{,}000{,}000$$
$$8.21 \times 10^6 = 8{,}210{,}000$$
$$10^4 = 10{,}000$$
$$4.3 \times 10^3 = 4{,}300$$
$$10^1 = 10$$
$$10^0 = 1$$
$$10^{-1} = 1/10 = 0.1$$
$$5.8 \times 10^{-2} = 0.058$$
$$7.52 \times 10^{-3} = 0.00752$$
$$10^{-4} = 1/10{,}000 = 0.0001$$
$$1.92 \times 10^{-6} = 0.00000192$$

Appendix B   Examples of the Decimal System

	Length
10^{-12}	picometer
10^{-9}	nanometer
10^{-6}	micrometer
10^{-3}	millimeter
10^{-2}	centimeter
10^{-1}	decimeter
10^{0}	meter
10^{3}	kilometer

Appendix C   Constants and Values

Acceleration due to gravity g	32 ft/sec^2
	9.8 m/sec^2 (10 m/sec^2)
Universal gravitational constant G	6.67×10^{-11} nt-m^2/kg^2
Coulomb's constant k	9.0×10^9 nt-m^2/coul^2
Specific heat of steel	0.11 cal/gm°C
Specific heat of aluminum	0.22 cal/gm°C
Atmospheric pressure	14.7 lb/in^2
	1.02×10^5 nt/m^2
Speed of light	3.0×10^8 m/sec
Mass of electron	9.1×10^{-31} kg
Charge of electron	1.6×10^{-19} coul
Mass of neutron	1.67×10^{-27} kg
Mass of proton	1.67×10^{-27} kg
Mass of alpha particle	6.68×10^{-27} kg
Planck's constant h	6.63×10^{-34} joule-sec
One electron volt	1.6×10^{-19} joule
Wien's displacement constant A	2.9×10^{-3} m-°K
Stefan-Boltzmann constant σ	5.7×10^{-8} joule/°K m^2 sec
1 atomic mass unit	1.66×10^{-27} kg
Avogadro's number	6.02×10^{23}

Appendix D Periodic Tables

1 H 1.008																	2 He 4.00
3 Li 6.94	4 Be 9.01											5 B 10.8	6 C 12.0	7 N 14.0	8 O 16.0	9 F 19.0	10 Ne 20.2
11 Na 23.0	12 Mg 24.3											13 Al 27.0	14 Si 28.1	15 P 31.0	16 S 32.1	17 Cl 35.5	18 Ar 39.9
19 K 39.1	20 Ca 40.1	21 Sc 45.0	22 Ti 47.9	23 V 50.9	24 Cr 52.0	25 Mn 54.9	26 Fe 55.8	27 Co 58.9	28 Ni 58.7	29 Cu 63.5	30 Zn 65.4	31 Ga 69.7	32 Ge 72.6	33 As 74.9	34 Se 79.0	35 Br 79.9	36 Kr 83.8
37 Rb 85.5	38 Sr 87.6	39 Y 88.9	40 Zr 91.2	41 Nb 92.9	42 Mo 95.9	43 Tc (99)	44 Ru 101.1	45 Rh 102.9	46 Pd 106.4	47 Ag 107.9	48 Cd 112.4	49 In 114.8	50 Sn 118.7	51 Sb 121.8	52 Te 127.6	53 I 126.9	54 Xe 131.3
55 Cs 132.9	56 Ba 137.3	57 see below 57.71	72 Hf 178.5	73 Ta 180.9	74 W 183.9	75 Re 186.2	76 Os 190.2	77 Ir 192.2	78 Pt 195.1	79 Au 197.0	80 Hg 200.6	81 Tl 204.4	82 Pb 207.2	83 Bi 209.0	84 Po 210	85 At (210)	86 Rn (222)
87 Fr (223)	88 Ra (226)	89 see below															

57 La 138.9	58 Ce 140.1	59 Pr 140.9	60 Nd 144.2	61 Pm (147)	62 Sm 150.4	63 Eu 152.0	64 Gd 157.3	65 Tb 158.9	66 Dy 162.5	67 Ho 164.9	68 Er 167.3	69 Tm 168.9	70 Yb 173.0	71 Lu 175.0
89 Ac (227)	90 Th 232.0	91 Pa (231)	92 U 238.0	93 Np (237)	94 Pu (242)	95 Am (243)	96 Cm (247)	97 Bk (245)	98 Cf (251)	99 Es (254)	100 Fm (253)	101 Md (256)	102	103

The most stable known isotopes are shown in parentheses.

Appendix E   Mass, in amu, of Some of the Lighter Elements

Element (or particle)	Symbol	Mass (amu)
electron	$_{1}^{0}e$	5.49×10^{-4}
proton	$_{1}^{1}p$	1.007 276
neutron	$_{0}^{1}n$	1.008 665
hydrogen	$_{1}^{1}H$	1.007 825
deuterium	$_{1}^{2}H$	2.014 102
tritium	$_{1}^{3}H$	3.016 050
helium-3	$_{2}^{3}He$	3.016 030
helium-4	$_{2}^{4}He$	4.002 603
lithium-6	$_{3}^{6}Li$	6.015 125
lithium-7	$_{3}^{7}Li$	7.016 004
beryllium-7	$_{4}^{9}Be$	9.012 186
boron-10	$_{5}^{10}B$	10.012 939
carbon-12	$_{6}^{12}C$	12.000 000

Index